Preface

Wood polymer composites have generated great interest among material scientists due to their eco-friendly nature and good thermo-mechanical performance. Wood polymer composite is an emerging area which possibly may replace traditional polymer composites in many applications. Polymer composites with synthetic fibers are used for automotive, aerospace, construction and building, etc. However, these materials will not degrade, and also, most of the synthetic fibers used are not recyclable. Therefore, the need of replacing synthetic materials with biodegradable wood is highly appreciable. So this book summarized many recent developments in the field of wood polymer composites. The advances in wood polymer composites and replacement of traditional materials with wood polymer composites in the field of materials and polymer science to achieve sustainable practice are described to the readers. This book also offers a good knowledge of wood polymer composites to the readers with numerous methods, illustrations and results for graduate students, researchers and industrialists. Academics, researchers, scientists, engineers, industrialists and students in the field of wood polymer composites will benefit from this book.

Sanjay Mavinkere Rangappa
Jyotishkumar Parameswaranpillai
Mohit Hemanth Kumar
Suchart Siengchin

Editors are honored to dedicate this book to their parents

Editors
Sanjay Mavinkere Rangappa
Mechanical and Process Engineering
King Mongkut's University of Technology
Bangkok, Krung Thep, Thailand

Mohit Hemanth Kumar
Mechanical and Process Engineering
King Mongkut's University of Technology
Bangkok, Thailand

Jyotishkumar Parameswaranpillai
Center of Innovation in Design
and Engineering
King Mongkut's University of Technology
Bangkok, Thailand

Suchart Siengchin
Department of Mechanical and Process
Engineering
King Mongkut's University of Technology
Bangsue, Bangkok, Krung Thep, Thailand

ISSN 2662-1819 ISSN 2662-1827 (electronic)
Composites Science and Technology
ISBN 978-981-16-1605-1 ISBN 978-981-16-1606-8 (eBook)
https://doi.org/10.1007/978-981-16-1606-8

This Springer imprint is published by the registered company Springer Nature Singapore Pte Ltd.
The registered company address is: 152 Beach Road, #21-01/04 Gateway East, Singapore 189721, Singapore

Sanjay Mavinkere Rangappa ·
Jyotishkumar Parameswaranpillai ·
Mohit Hemanth Kumar ·
Suchart Siengchin
Editors

Wood Polymer Composites

Recent Advancements and Applications

Springer

Composites Science and Technology (CST) book series publishes the latest developments in the field of composite science and technology. It aims to publish cutting edge research monographs (both edited and authored volumes) comprehensively covering topics shown below:

- Composites from agricultural biomass/natural fibres include conventional composites-Plywood/MDF/Fiberboard
- Fabrication of Composites/conventional composites from biomass and natural fibers
- Utilization of biomass in polymer composites
- Wood, and Wood based materials
- Chemistry and biology of Composites and Biocomposites
- Modelling of damage of Composites and Biocomposites
- Failure Analysis of Composites and Biocomposites
- Structural Health Monitoring of Composites and Biocomposites
- Durability of Composites and Biocomposites
- Biodegradability of Composites and Biocomposites
- Thermal properties of Composites and Biocomposites
- Flammability of Composites and Biocomposites
- Tribology of Composites and Biocomposites
- Bionanocomposites and Nanocomposites
- Applications of Composites, and Biocomposites

To submit a proposal for a research monograph or have further inquries, please contact springer editor, Ramesh Premnath (ramesh.premnath@springer.com).

More information about this series at http://www.springer.com/series/16333

Composites Science and Technology

Series Editor

Mohammad Jawaid, Lab of Biocomposite Technology, Universiti Putra Malaysia, INTROP, Serdang, Malaysia

Contents

Introduction to Wood Polymer Composites 1
R. Ruban, H. Mohit, P. Ramesh, V. Arul Mozhi Selvan, and G. H. Kumar

Manufacturing of Wood Polymer Composites 21
Neeraj Dubey, Rajesh Purohit, and R. S. Rana

Wood Treatments and Interfacial Bonding in Wood-Plastic Composites 43
Matheus de Prá Andrade and Matheus Poletto

Property Improvements of Wood and Wood-Plastic Composites 67
Md. Minhaz-Ul Haque

Morphological Characterization of the Wood Polymer Composites 93
M. Chandrasekar, K. Senthilkumar, T. Senthil Muthu Kumar, R. Sabarish, and Suchart Siengchin

Mechanical Properties of Wood Polymer Composites 113
Rupam Gogoi and Gaurav Manik

Structure of Wood Fiber and Factors Affecting Mechanical Properties of Wood Polymer Composites 137
Neeraj Dubey, Rajesh Purohit, and H. Mohit

Thermal Properties and Flammability of Wood Plastic Composites 161
Elammaran Jayamani and Vannethasrriy Balakrishnan

Tribology of Wood Polymer Composites 179
Karthikeyan Subramanian, Senthilkumar Krishnasamy, Chandrasekar Muthukumar, Suchart Siengchin, Kalusuraman Gnaniar, and Arunprasath Kanagaraj

Water Absorption and Swelling Behaviour of Wood Plastic Composites 195
Sabarish Radoor, Jasila Karayil, Jyothi Mannekote Shivanna, and Suchart Siengchin

Long Term Natural Weathering of PP Based WPCs: The Effect of TiO_2 on Selected Color, Physical, Mechanical, Morphological and Chemical Properties 213
Fatih Mengeloğlu and Vedat Çavuş

Life–Cycle Assessment (LCA) of Plastic–Wood Composites 233
Rupesh Kumar Malviya, Rajesh Purohit, and Rahul Kumar Singh

Environmental Impact of Wood Based Biocomposite Using Life Cycle Assessment Methodology 255
P. Ramesh, H. Mohit, and V. Arul Mozhi Selvan

Editors and Contributors

About the Editors

Dr. Sanjay Mavinkere Rangappa is currently working as Research Scientist at King Mongkut's University of Technology North Bangkok, Bangkok, Thailand. He has received the B.E. (Mechanical Engineering) from Visvesvaraya Technological University, Belagavi, India, in the year 2010, M.Tech (Computational Analysis in Mechanical Sciences) from VTU Extension Centre, GEC, Hassan, in the year 2013, Ph.D (Faculty of Mechanical Engineering Science) from Visvesvaraya Technological University, Belagavi, India, in the year 2017 and Post Doctorate from King Mongkut's University of Technology North Bangkok, Thailand, in the year 2019. He is a Life Member of Indian Society for Technical Education (ISTE) and Associate Member of Institute of Engineers (India). He is a reviewer for more than 75 international journals (for Elsevier, Springer, Sage, Taylor & Francis, Wiley), book proposals and international conferences. In addition, he has published more than 150 articles in high-quality international peer-reviewed journals, 4 editorial corner, 30+ book chapters, one book, 15 books as Editor and also presented research papers at national/international conferences. His current research areas include natural fiber composites, polymer composites and advanced material technology. He is a recipient of DAAD Academic exchange-PPP Programme (Project-related Personnel Exchange) between Thailand and Germany to Institute of Composite Materials, University of Kaiserslautern, Germany. He has received a Top Peer Reviewer 2019 award, Global Peer Review Awards, Powered by Publons, Web of Science Group. He has been recognized by Stanford University's list of the world's Top 2% of the Most-Cited Scientists in Single Year Citation Impact 2019 and ranked 97th among 187 THAI Scientist in the list.

Dr. Jyotishkumar Parameswaranpillai is currently working as Research Professor at Center of Innovation in Design and Engineering for Manufacturing, King Mongkut's University of Technology North Bangkok. He received his Ph.D.

in Polymer Science and Technology (Chemistry) from Mahatma Gandhi University. He has research experience in various international laboratories such as Leibniz Institute of Polymer Research Dresden (IPF) Germany, Catholic University of Leuven, Belgium, and University of Potsdam, Germany. He has published more than 100 papers in high-quality international peer-reviewed journals on polymer nanocomposites, polymer blends and alloys, and biopolymer, and has edited five books. He received numerous awards and recognitions including prestigious Kerala State Award for the Best Young Scientist 2016, INSPIRE Faculty Award 2011 and Best Researcher Award 2019 from King Mongkut's University of Technology North Bangkok.

Dr. Mohit Hemath Kumar completed his B.E. (Mechanical Engineering) from Anna University, Chennai, Tamil Nadu, India, in the year 2012, and M.E. (Thermal Engineering with specialization in Refrigeration and Air Conditioning Engineering) from College of Engineering Guindy Campus, Anna University, Chennai, in the year 2014. He completed his Ph.D. from Department of Mechanical Engineering, National Institute of Technology Tiruchirappalli in the year 2019 and he has done Post-Doctoral Researcher in King Mongkut's University of Technology North Bangkok, Thailand. He is a life member of Institution of Engineers (India), Powder Metallurgy Association of India (India), Society for Failure Analysis (India) and Science and Technology Research Association (Singapore). In addition, he has published more than 10 articles in high-quality international peer-reviewed journals, 5+ book chapters, and also presented research papers at national/international conferences. Till now he has filed more than 20 patents and 15 designs, and granted two patents and seven industrial designs from Indian Intellectual Property Rights. His current research areas include natural fiber and inorganic filler-based polymer nanocomposites, metal matrix composites and waste to innovative industrial products design.

Prof. Dr.-Ing. habil. Suchart Siengchin is **President** of King Mongkut's University of Technology North Bangkok. Prof. Dr.-Ing. habil. Suchart Siengchin received his Dipl.-Ing. in Mechanical Engineering from University of Applied Sciences Giessen/Friedberg, Hessen, Germany, in 1999, M.Sc. in Polymer Technology from University of Applied Sciences Aalen, Baden-Wuerttemberg, Germany, in 2002, M.Sc. in Material Science at the Erlangen-Nürnberg University, Bayern, Germany, in 2004, Doctor of Philosophy in Engineering (Dr.-Ing.) from Institute for Composite Materials, University of Kaiserslautern, Rheinland-Pfalz, Germany, in 2008 and Postdoctoral Research from Kaiserslautern University and School of Materials Engineering, Purdue University, USA. In 2016, he received the habilitation at the Chemnitz University in Sachen, Germany. He worked as Lecturer for Production and Material Engineering Department at The Sirindhorn International Thai-German Graduate School of Engineering (TGGS), KMUTNB. He has been full Professor at KMUTNB and became the President of KMUTNB.

He won the Outstanding Researcher Award in 2010, 2012 and 2013 at KMUTNB. His research interests are in polymer processing and composite material. He is Editor-in-Chief: KMUTNB International Journal of Applied Science and Technology and the author of more than 250 peer-reviewed journal articles. He has participated with presentations in more than 39 International and National Conferences with respect to materials science and engineering topics. He has recognized and ranked among the world's top 2% scientists listed by prestigious Stanford University.

List of Contributors

V. Arul Mozhi Selvan Department of Mechanical Engineering, National Institute of Technology, Tiruchirappalli, Tiruchirappalli, India

Vannethasrriy Balakrishnan Faculty of Engineering, Computing and Science, Swinburne University of Technology, Sarawak Campus, Kuching, Sarawak, Malaysia

Matheus de Prá Andrade Postgraduate Program in Engineering of Processes and Technologies (PGEPROTEC), University of Caxias do Sul (UCS), Caxias do Sul, Brazil

Neeraj Dubey Department of Mechanical Engineering, LNCT, Bhopal, M.P., India

Kalusuraman Gnaniar Department of Agricultural Engineering, Kalasalingam Academy of Research and Education, Krishnankoil, Tamilnadu, India

Rupam Gogoi Department of Polymer and Process Engineering, Indian Institute of Technology Roorkee, Saharanpur, UP, India

Elammaran Jayamani Faculty of Engineering, Computing and Science, Swinburne University of Technology, Sarawak Campus, Kuching, Sarawak, Malaysia

Arunprasath Kanagaraj Department of Mechanical Engineering, Kalasalingam Academy of Research and Education, Krishnankoil, Tamilnadu, India

Jasila Karayil Department of Chemistry, Government Women's Polytechnic College, Calicut, Kerala, India

Senthilkumar Krishnasamy Center of Innovation in Design and Engineering for Manufacturing (CoI-DEM), King Mongkut's University of Technology North Bangkok, Bangsue, Bangkok, Thailand

G. H. Kumar Center for Nanoscience, Composite Research Center, Chennai, India

Rupesh Kumar Malviya Department of Mechanical Engineering, Rabindranath Tagore University, Bhopal, MP, India

Gaurav Manik Department of Polymer and Process Engineering, Indian Institute of Technology Roorkee, Saharanpur, UP, India

Fatih Mengeloğlu Faculty of Forestry, Department of Forest Industry Engineering, Kahramanmaras Sutcu Imam University, Kahramanmaras, Turkey; Institute of Natural and Applied Sciences, Department of Materials Science and Engineering, Kahramanmaras Sutcu Imam University, Kahramanmaras, Turkey

Md. Minhaz-Ul Haque Applied Chemistry and Chemical Engineering, Isamic University, Kushtia, Bangladesh

H. Mohit Centre for Nanoscience, Composite Research Center, Chennai, Tamil Nadu, India

Chandrasekar Muthukumar School of Aeronautical Sciences, Hindustan Institute of Technology and Science, Chennai, Tamil Nadu, India

Matheus Poletto Postgraduate Program in Engineering of Processes and Technologies (PGEPROTEC), University of Caxias do Sul (UCS), Caxias do Sul, Brazil

Rajesh Purohit Department of Mechanical Engineering, MANIT, Bhopal, Madhya Pradesh, India

Sabarish Radoor Department of Mechanical and Process Engineering, The Sirindhorn International Thai-German Graduate School of Engineering (TGGS), King Mongkut's University of Technology North Bangkok, Bangkok, Thailand

P. Ramesh Department of Production Engineering, National Institute of Technology, Tiruchirappalli, India

R. S. Rana Mechanical Engineering Department, MANIT, Bhopal, Madhya Pradesh, India

R. Ruban Department of Mechanical Engineering, National Institute of Technology, Tiruchirappalli, Tiruchirappalli, India

T. Senthil Muthu Kumar Department of Mechanical Engineering, Kalasalingam Academy of Research and Education, Krishnankoil, Tamilnadu, India

Jyothi Mannekote Shivanna Center for Nano and Material Sciences, Jain University, Bangalore, Karnataka, India

Suchart Siengchin Department of Mechanical and Process Engineering, The Sirindhorn International Thai-German Graduate School of Engineering (TGGS), King Mongkut's University of Technology North Bangkok, Bangkok, Thailand

Suchart Siengchin Department of Materials and Production Engineering, The Sirindhorn, International Thai-German Graduate School of Engineering (TGGS), King Mongkut's University of Technology North Bangkok, Bangsue, Bangkok, Thailand

Rahul Kumar Singh Department of Mechanical Engineering, Rabindranath Tagore University, Bhopal, MP, India

Karthikeyan Subramanian Department of Automobile Engineering, Kalasalingam Academy of Research and Education, Krishnankoil, Tamilnadu, India

Vedat Çavuş Forest Faculty, Forest Industry Engineering, Izmir Katip Celebi University, Izmir, Turkey

Introduction to Wood Polymer Composites

R. Ruban, H. Mohit, P. Ramesh, V. Arul Mozhi Selvan, and G. H. Kumar

Abstract In the recent year, the needs for environmentally-friendly, low-price, and recyclable with higher strength laminates have been improved significantly. In this regard, the wood fillers based polymer laminates have fascinated the investigators owed to their lower price and eco-friendliness as they are formulated from waste wood fillers and better physical and mechanical characteristics. These laminates were fabricated by reinforcing the waste wood fillers in the polymeric resin using different manufacturing techniques like compression molding, hand layup, injection molding, three-dimensional printing, and extrusion. A considerable quantity of research investigation has been examined on the characterization and testing of wood-based polymer composites for different applications. This chapter presents the thermal, water absorption, and mechanical characteristics of thermoplastics, biopolymers, and thermosets have been discussed. The present chapter's conclusion offers a better understanding of the wood laminates, which will inspire the investigators for further research investigations and advancements of new wood laminates for the developed applications.

Keywords Chemical surface modification · Mechanical properties · Thermal properties · Water absorption · Wood polymer composite

R. Ruban (✉) · V. Arul Mozhi Selvan
Department of Mechanical Engineering, National Institute of Technology, Tiruchirappalli, Tiruchirappalli 620015, India

H. Mohit · G. H. Kumar
Center for Nanoscience, Composite Research Center, Chennai 600053, India

P. Ramesh
Department of Production Engineering, National Institute of Technology, Tiruchirappalli, Tiruchirappalli 620015, India

S. Mavinkere Rangappa et al. (eds.), *Wood Polymer Composites*, Composites Science and Technology, https://doi.org/10.1007/978-981-16-1606-8_1

1 Introduction

The attention to fillers reinforced in plastics has been determined for elastic characteristics enhancement (Liang 2012). The developing awareness in the utilization of plant fibers (Dong 2018; Mohit and Selvan 2018) recognizes the olive wood fillers of main economic and ecological interests. Certainly, among the crops from fruit, olive is mainly generated around the Mediterranean regions. During the process of oil extraction, a considerable quantity of solid and liquid waste offers possible issues of pollution to the environment. Besides, this type of wood used for the fabrication of handicraft components develops a tremendous amount of waste consisting of lignin and cellulose as the major constituents (Naghmouchi et al. 2015; Bouhamed et al. 2020).

Moreover, while comparing with carbon and glass fabrics, different polymers have been examined as potential materials for plant fibers because of their lower price, higher biodegradability, and lower density (Amar et al. 2011). Still, different classical plastics are consistently influencing the region of polymers for wood basic polymer laminates. Certainly, the recent investigation was performed in the scheme of bio-based materials. On the other hand, the investigated wood-based polymer laminate depends on the polymer resin; in another part, wood fibril particles are comprised as a capacity for improving the strength of the obtained laminate. This type of wood-based polymer composite material boosts the query about the properties and control techniques either during the fabrication or even after a specific period. Besides, such a wood-based polymer laminate is forecasted to exhibit the viscoelastic characteristics observed from the mechanical testing process and can be compared with the non-destructive examination. The non-destructive examination may be feedback to these queries gratitude to the right to select specimen structure, information availability, and the capability to collect a higher range of precision, low price, and testing speed. In principle, the utilization of non-destructive testing techniques for the mechanical properties of materials is diverse improving (Bouhamed et al. 2020). These techniques are substitutes for the traditional destructive mechanical properties evaluation.

From the non-destructive methods, the ultrasound technique can be applied to determine the examined specimen's elastic characteristics. Lefebvre et al. (2018) applied the Fabry–Perot resonance method to inspect the bulk and shear modulus of the viscoelastic plastic; El-Sabbagh et al. (2013) investigated the influence of plant fibers concentration on the velocity of ultrasound in a laminate material fabricated from polymeric resin. Castellano et al. (2014) has carried out the bio-based laminate mechanical testing from ultrasound measurements. The examination of transversal and longitudinal wave velocity is a useful measurement technique to isotropic elastic composites (Simonetti et al. 2005), primarily for differentiating the mechanical properties and then for evaluating the elastic modulus essential for the explanation of the mechanical characteristics (Scalerandi et al. 2012). The mechanism of this technique depends on the period for the measurement of flight. This examination can also be carried out in the spectral region using Fourier transform infrared spectra spectroscopy. It is well established that the longitudinal velocity of sound is a formulation of the composites' characteristics, which corresponds to the quality of adhesion and fiber concentration (El-Sabbagh et al. 2013).

Solid wood is a premium flexible material and has been broadly applied in different applications, including the following fields as sporting goods, furniture, and civil construction. It has greater strength to weight proportion, lower acoustic, thermal conductivities, and renewable sources as an essential characteristic. Furthermore, the qualified woods (particularly from forests) have more absorbed because of improper resource and utilization planning. For this reason, there is a high contribution to individual fast-breeding plants like pines. Generally, fast-breeding plants create a considerable proportion of young wood when cultivated in tropical regions, which presents an excellent problem for developed growing schemes for maintaining the quality of wood. Principally, a forest can be set in the method that a decreased gap between trees is used in the initial years of the tree life, which reduces the production of new types of cells from the cambial and thus relived the output of young wood. Palmero et al. (2013), the Pinus Elliott wood initiates to create mature timber at the twentieth year of span if cultivated in dense forest regions. Similar scientists have investigated old trees (35 years) with a diameter of 25 cm and explored that the central part of the trunk with a diameter of 15 cm made up of young wood. Under the ultrastructural level, while comparing the mature wood, the young wood has small cell walls, higher grain angle, lower tangential cells, a smaller ratio of latewood, and higher microfibril angle (Mcdonald and Hubert 2002). Thus, most of the biological, mechanical, photochemical, and hygroscopic properties of the young wood are lesser than the mature wood (Li et al. 2011). Insemination utilizing different polymers preliminary to in situ polymerization is the method to enhance many unfavorable characteristics in the wood. In this step, permeable and lightweight solid woods are inseminated with polymers of higher reactivity, lower viscosity, and a lower molecular weight that must be efficient of inter or intracellular voids in wood or chemically bond themselves to different lignin and polysaccharides available in the wood cell wall (Islam et al. 2014). Presently, various patented components were constructed from this principle and are abundant on the market, such as Ligeia, Vecowood, Indurite, epoxy, styrene, furfuryl alcohol, and methyl methacrylate are some already applied polymers in published literature (Acosta et al. 2020).

Among the plant fibers based polymer laminates, wood flour has been broadly applied to fabricate wood polymer laminates. LLDPE (Linear low-density polyethylene) is a standard selection to fabricate wood-based polymer composites (Government et al. 2019) because of its outstanding flexibility. Generally, polymers are hydrophobic, which leads to poor interfacial bonding with the hydrophilic wood flour, which restricts their mechanical characteristics (Zhou et al. 2015). The interfacial adhesion between the polymer and wood flour can be improved to utilize some crosslinking agents and chemical treatment of matrix/fibers (Arora et al. 2012). Currently, radiation is used to enhance the polymer matrix and fiber interfaces (Güven et al. 2016). Gamma radiation can influence the polymer design of wood fillers and matrix, creating active regions that may share to enhance interaction between polymer and wood (Niang et al. 2017). Gamma irradiation of polymer laminates became more general as no heat was generated during the process (Kismet 2017). This outcome in the crosslinking and degradation of polymers concerning the utilized dose (Ndiaye et al. 2014).

Wood is involved; renewable, durable material consists of various biomacromolecules and biomolecules (Guo et al. 2015). As a biodegradable material, it can be applied from the prehistoric period for a wide range of applications containing tools, transportation, fortifications, funerary objects, dwelling, and weapons (Moise et al. 2019). The necessity of cultural heritage conservation has been controlled and recognized internationally. Preservation of wood components is one of the practices within the rehabilitation of a heritage component. It is assigned to the preservation of both archaeological and historical features of the wood. The necessary regulations are mainly targeted at better conserving the individuality of the component. Many of these techniques are dependent on radiation methodologies like gamma biocide modifications for cultural heritage, polymerization under gamma irradiation, consideration of wooden artifacts from radiation curing polymers. One technique of preserving wooden products is radiative insemination with plastic (Schneider 2016). Although the principle can be freely prepared from general woodworking processes, it can also be fabricated from the typical polymer using injection molding and extrusion and produced from additive production methods via fused deposition modeling (FDM) (Ray and Rathore 2015). They are generally applied as an alternative for ordinary wood, which is involved in fencing, decking, and flooring, and their utilization is beneficial in humid working conditions or ever in close contact with fluid; the hydrophobic resin separates and shields the hydrophilic wood fillers, and hence leads to improves durability and needs lower maintenance interferences up to certain level (Chen et al. 2017). Remarkably are also their utilization in automotive and acoustics (Sawpan 2016). The significant benefits of wood as a filler reinforcement are decrement in cost and an enhancement in the finally developed laminate's environmental features. The erstwhile is utterly evident, as wood fillers are generally low-price components, usually impending from industrial or agricultural waste. The recent investigation is also fair since a significant proportion, sometimes as large as 70 wt% of a non-biodegradable matter of fossil region, is being replaced by an eco-friendly component (Kootsookos and Mouritz 2004). A higher environmental benefit can be collected if the applied resin is also from biodegradable polymers obtained from bioresources (Wang et al. 2018; Quan et al. 2018).

Considering the mechanical characteristics of the laminate, strength, and stiffness are adequate only if expensive coupling agents are reinforced during the formulation of the material: these additives are essential for enhancing the compatibility between the polymer and wood flour, which would not contribute any significant chemical correlations (Prusty et al. 2017). These objectives produce an efficient load movement interface between the covering resin and natural fillers. However, since wood plastic composites are generally brittle, their correlation usually consists of strengthening components like thermoplastic elastomers, ethylene propylene diene rubber, and styrene-butadiene rubber (Zhang et al. 2012). To boost as much potential from the previous benefits, an approximately higher wood flour concentration, more than 50 wt%, is more desirable. In the same manner, a higher quantity of reinforcement improve the viscosity of the resin melt noticeably, hence producing flow and thus making the process relatively complicated. This issue is again engaged by comprising useful additives such as lubricant, which may advertise wall slip and thus reduce the fabrication's energy needs. Note, certainly which the general method of reducing melt viscosity by rising temperature is not possible because of thermal degradation. Available lubricant applied for the improvement are ethylene or esters, thermoplastic silicones, and stearates.

For perfectly designing the preparing equipment, a discreet feature of the molten wood plastic composite is essential. This is not entirely genuine, mainly due to the non-newtonian behavior of these fluids. Even though a considerable quantity of research articles, an extensive review is still absent in the literature, and this chapter is targeted by considering specific aspects. In the upcoming sections, the concept behind the physical, mechanical, and thermal characterization techniques applied before showing the outputs of the scientific articles were outlined. In the final section, the conclusions were drawn and recommended a few discussions on the applications.

2 Fabrication Methods of Wood Plastic Composites

The proper manufacturing technique for wood-based polymer composites is a challenging practice based on the polymer type, the product's geometry, surrounding circumstances, the weight percentage of wood fillers in the matrix, and overall price. The type of polymeric resin, such as thermoplastic and thermosetting, is the highest affecting factor for the chosen manufacturing technique. The traditional manufacturing techniques like injection, extrusion, hot press, compression molding, and wet layup are generally utilized production techniques applied by various investigators as explained in the published article (Chaharmahali et al. 2010; Komal et al. 2020). In recent decades, three-dimensional printing is being used to produce wood polymer-based plastic composites as advanced manufacturing technology. In the current scheme, rapid prototype methodologies have been applied in construction, medical, and manufacturing fields.

Furthermore, three-dimensional printing has also been applied for generating the polymer-based laminates. In this technique, the primary three-dimensional model of reliable components is created in the modeling software, and the produced three-dimensional model was separated into the compilation of individual layers as per the printer's capacity. The solid component is engraved as layer by layer deposition of the composite laminate. This method is efficient for the rapid prototyping of components in producing and examining the product's design without the materials loss and expensive tool (Schwarzkopf and Burnard 2016). This method's primary benefit is straight generating the components in one process using the information from CAD. Selective laser sintering (SLS), fused deposition modeling (FDM), stereolithography, three-dimensional inkjet printing, and three-dimensional plotting are general printing methods for polymer-based fabrication laminates. Other methods are under the advancement region so far (Wang et al. 2017).

Further SLS and FDM 3D printing methods are applied for the development of wood polymer composites. The FDM three-dimensional printing method is a composite extrusion method that involves the produced filaments of wood polymer composites. The filaments are passed into the heating chamber, where the wood polymer composites were melted into semi-liquid form and deposited layer by layer from the nozzle in the regulated condition on the platform's surface. The FDM method's restriction is based on the polymer type applied at the melting temperature below the decomposition temperature of wood fillers applied in the wood polymer composites. However, this technique is restricted to the thermoplastic polymer composites (Schwarzkopf and Burnard 2016; Wang et al. 2017; Martikka et al. 2018; Kain et al. 2020). The SLS three-dimensional printing method utilized the powder that softens at various temperatures and is blended from laser radiation. The SLS method for producing wood polymer composites is under the phase of advancement (Schwarzkopf and Burnard 2016). All the three-dimensional printing methods of the progress of wood polymer composites developed prominently in recent decades. Because of the restrictions such as relatively higher processing price, limitations of materials, higher energy consumption, harmful emission, and slower processing, three-dimensional printing methodology is restricted in the area of wood polymer composites so far. Still, it will be sustainable for producing all kinds of wood polymer composites in the upcoming future. To date, a very restricted survey corresponded to the examination and advancement of three-dimensional printed wood polymer composites were abundant. The FDM method of three-dimensional printing has been applied to promote wood polymer composites components precisely in a unique way. Martikka et al. (2018) estimated three-dimensional printed wood fillers' mechanical characteristics based on PLA laminate. The impact and tensile strength of developed laminates were 53 and 58% reduction than the pure PLA. The mechanical characteristics of three-dimensional printed wood polymer composite components are also based on the pattern of printing, the diameter of the nozzle, the thickness of the layer, build a style of laminate, and speed of feed (Kariz et al. 2018).

3 Mechanical Characteristics of Wood Polymer Composites

The mechanical characteristics of wood polymer composites based on the mechanical and physical features of the wood fillers, characteristics of polymeric resin, particles geometry, content of fillers, wettability, compatibility and interfacial adhesion. The stress concentration, lower filler weight proportion, and good interfacial adhesion outcomes in the higher aspect ratio, good wettability of fillers with polymer, higher tensile strength and weight proportion of wood fillers improved the tensile modulus of laminates. The fiber pull out, higher energy absorption and interfacial adhesion enhanced the impact strength of wood polymer composites (Chand and Fahim 2008). The upcoming sections explain the flexural, tensile, hardness, and impact characteristics of different kinds of wood polymer composites as observed from the recent literature.

3.1 Thermoplastic Wood Polymer Laminates

Thermoplastic based polymers are generally softening at a higher range of temperature and can be formed into any shape or structure, and amorphous or semi-crystalline. The semi-crystalline structured polymers show the variation in crystallinity, and amorphous polymers contain molecular networks that are randomly oriented. The amorphous and crystalline polymers act variously under higher temperatures. Amorphous resin does not interpret the genuine melting but softens to the viscous flow (Wolcott and Englund 1999). These resins have a broad compatibility with the traditional fabrics like carbon and glass, need lower processing energy, ease of manufacturing, softer density, and economical (Espert et al. 2004). Universally, polypropylene and polyethylene were applied to advance thermoplastic polymer laminates that can be utilized for the external components because of better protection from moisture and its availability as waste (Chaharmahali et al. 2010). It has been well-established that few thermoplastic resins were applied for the advancement of wood polymer composites due to the preferred resin should have the softening temperature lower than the thermal decomposition of wood fillers (may be lower than 200 C) (Poletto et al. 2012). Thermoplastic resin-like high-density polyethylene, low-density polyethylene, polyethylene, polypropylene, polyvinyl chloride, and polystyrene have better compatibility with wood (Wolcott and Englund 1999). Polypropylene resin has been generally applied because of lower service temperature, higher mechanical characteristics, and lower cost. The studies of mechanical features for these laminates are performed for different weight concentration of wood fillers, structure of the chemical, size and aspect ratio of wood fillers, and laminate manufacturing method. Nourbakhsh et al. (2010) examined and compared the mechanical characteristics of particles of polypropylene laminates containing different aspect ratios and size of wood fillers for 30% weight proportion of general wood addition. They achieved that mechanical characteristics of laminate improve with the increment of wood particles aspect ratio and size. Chaharmahali et al. (2010) investigated the mechanical characteristics of reused particle sawdust reinforced high-density polyethylene laminate. It was found that the impact and flexural strengths of laminate reduced with an increment in the concentration of filler, and the flexural modulus also enhanced. Perez et al. (2012) collected that the strain at fracture, fracture toughness, and tensile strength of red pine wood fillers polypropylene laminates was reduced with an increment in the concentration of wood filer, and also the tensile modulus of the laminate was improved. The incorporation of MAPP coupling agent with the red pine wood fillers improved the tensile characteristics, but the influence on fracture toughness was insignificant. In another investigation, the tensile modulus, strength, and elongation at break of wood fillers polyvinyl chloride composites are more significant than the bamboo-based wood polymer composites because of uniform homogeneity within the matrix (Ge et al. 2004). Nikmatin et al. (2017) observed the influence of rattan nanofillers size on the mechanical, physical, and thermal characteristics of polypropylene-based laminates and compared their characteristics with the glass fiber reinforced and pure polypropylene laminates. The advanced laminate acquired an excellent surface finish and greater hardness value when compared with other laminates. The reinforcement of 40% of wood fillers in the polypropylene resin improved the tensile modulus and strength of laminate by 108% and 48%, respectively, and the elongation at break and impact strength reduced by 86% and 42%, respectively (Ichazo et al. 2001).

3.2 Thermoset Wood Polymer Laminates

Thermosetting polymers do not soften on external heating. The polyvinyl ester, epoxy, polyester, formaldehyde, phenolic, and polyurethanes are generally applied polymers to advance wood polymer composites. These resins acquire higher characteristics than the thermoplastic resins in mechanical characteristics, filler wettability, durability, thermal and electrical insulation, viscosity, molecular cross-linking, resistance from creep, and chemicals. Epoxy resin is commonly favored for laminates advancement among the abundant polymers because of many benefits such as lower viscosity, thermal stability, higher mechanical characteristics, good interfacial adhesion between the fillers and matrix, good resistance from chemicals, and after curing, lower shrinkage. Many scientists have studied the mechanical characteristics of thermosets wood laminates. Lette et al. (2018) studied the flexural and tensile characteristics of cedarwood fillers phenolic resin laminate and rice husk reinforced phenolic laminate. It was observed that 33 and 1% of higher flexural and tensile strengths of cedarwood fillers phenolic laminates were 29 and 11% higher than the flexural and tensile modulus of the rice husk phenolic laminates. Vignesh and Selvam (2015) analytically found that the tensile modulus, tensile, flexural, impact strength, and hardness of teak wood fillers polyester laminate were examined as higher when compared with the similar rubber and Sal based wood fillers polyester laminate. The flexural and tensile strength of wood shell (apple) particles epoxy laminate were improved with the incorporation of particle concentration, and the density of laminate reduced because of the lower weight of wood shell fillers (Shakuntala et al. 2014). Kumar et al. (2014) studied the flexural and tensile characteristics of wood dust (sundi) fillers epoxy laminates for seven different variations in weight% relating to three different speeds of crossheads, and the laminate with ten weight% reported higher flexural and tensile strength under crosshead speed of 1 mm/min and higher modulus under crosshead speed of 2 mm/min.

3.3 *Biopolymer Wood Polymer Laminates*

Our surroundings are getting influenced due to non-biodegradable polymers have been generally applied in different industries as packaging, furniture, automobile, and construction. The new strategies for environmental problems target to boost the application of components produced from bio-degradable materials that offer a driving load for advancing the new type of biodegradable materials, which can have a detrimental effect on the ecology environment. The new bio-based resin as green laminates is generated as a substitute material because it is recyclable, renewable, emit low greenhouse gas, and biodegradable. Biopolymers like polyester amide, polylactic acid, cellulose acetate, starch blends, polyhydroxyalkanoates, and polyvinyl alcohol and natural fillers, particles or whiskers collected from the waste of animal minerals and proteins, and plant waste are being applied as reinforcement and polymeric materials, respectively, for the advancement of green laminates (Teymoorzadeh and Rodrigue 2015). The scientists are frequently employing the sustainable improvement of bio-based wood polymer laminates. Ge et al. (2006) fabricated and mechanically characterized the bio-based bamboo wood fillers reinforced polypropylene carbonate laminate. The tensile strength of laminate was 16% greater than the pure polymer. The morphology interpreted dispersion of bamboo wood fillers in the polypropylene composite uniformly. Nagarajan et al. (2016) fabricated biochar fillers reinforced PLA-based bio-laminates and investigated the mechanical characteristics for various wood fillers' sizes. Agnantopoulou et al. (2012) measured the tensile characteristics of several types of wood fillers (general, spruce, beech, and pine) added thermoplastic starch laminates. The tensile modulus and strength of these laminates were observed from 0.85–0.95 GPa and 13.3–17.3 MPa, respectively.

3.4 *Enhanced Mechanical Characteristics of Wood Laminates by the Treatment Process*

The inappropriate interfacial adhesion between hydrophilic wood polymers incorporation and hydrophobic resins tends to the non-uniform transfer of stress by that decrement in the mechanical characteristics of wood-based polymer laminates (Hosseinaei et al. 2012). To improve the mechanical aspects of polymer-based wood polymer laminates, various kinds of surface modification techniques such as silane, maleated coupling agent, bleaching, benzoylation, acylation, etc.; physical surface modifications such as plasma, gamma, corona, etc.; compatibilizers such as metal oxides, nanoclay, etc.; and surface enhancements such as coating, roughing, and cleaning was used over the components of wood-based polymer laminates (Ichazo et al. 2001; Lee and Wang 2006). Many investigators have studied the influences of treatment techniques on the mechanical characteristics of wood polymer composites. Kord (2011) empirically evaluated the consequences of maleic anhydride polypropylene coupling agents on the impact, flexural, and tensile characteristics of sawdust reinforced polypropylene laminates. For two wt% of maleated coupling agents, the impact strength, tensile modulus, strength, flexural modulus, and laminate strength were improved by 11%, 5%, 11%, 6%, and 8%, respectively. Dairi et al. (Dairi et al. 2015) exhibited that flexural and tensile characteristics of maleated coupling agent modified pine wood fillers reinforced polyethylene-terephthalate/polypropylene laminate enhanced because of uniform interfacial bonding, which flattened the transfer of stress from fillers to the polymer resin. Kajaks et al. (2015) examined significant enhancements in impact, hardness, and flexural characteristics of pine wood fillers reinforced high-density polyethylene laminates after modified with maleated coupling agents. In another investigation, influences of chemical surface modifications such as cellulose palmitate, stearic acid, benzoylation, and maleated coupling agents on the mechanical characteristics and correlation of polypropylene depended wood laminates were performed (Danyadi et al. 2010).

The examination exhibited that incorporating nano-clay in the wood-based polymer laminate decreased the elongation at break, impact, and tensile strength but improved the modulus of laminates (Gu et al. 2010). The flexural and tensile strengths of wood fillers high-density polyethylene laminates were produced when mixed with TiO2 nanofillers, and flexural and tensile modulus were improved by 34.5% and 151%, respectively (Deka 2001). Kim et al. (2010) examined the influences of silane modifications over the Picea abies wood fillers polypropylene laminates, and Each silane modification displayed enhancements in the impact, flexural and tensile strength than the untreated wood fillers laminate. The silane modification based on VTMS exhibited higher enhancement in the impact (11%), flexural (78.4%), and tensile (85%) strength of the laminate.

3.5 *Enhanced Mechanical Characteristics of Wood Laminates from Hybridization*

The mixing of more than two wood fillers of different constituents and extents in individual polymers outcomes in a hybrid laminate counteract each kind of wood fillers' capacity. Usually, the need for hybridization reduces the demerits and boosts both kinds of wood fillers' merits. Finite published literature is found on the influences of hybridization of wood fillers on thermal, crystalline, water uptake, and wood polymer composites' mechanical properties. Saxena and Gupta (2018a, b) investigated the thermal, water uptake, and mechanical characteristics of hybrid mango and sal wood fillers reinforced epoxy laminates and examined the potential effect of hybridization mechanical attributes at the equal ration of each kind of wood fillers were combined. The hybrid shoresa and mango robust wood fillers reinforced epoxy laminate exhibited higher crystallinity than the single wood fillers reinforced epoxy laminate (Saxena and Gupta 2018a, b). Khan et al. (2019) studied the impact, flexural, microhardness, and tensile characteristics of hybrid shoresa robusta and pine fillers reinforced epoxy laminates for dissimilar weight ratios wood fillers. The hybrid laminate with an identical fraction of wood fillers exhibited higher mechanical characteristics than the similar collected for single wood filler laminates. The flexural strengths of hybrid laminate were observed 56% and 200%, tensile strengths 28% and 54%, impact strengths 14% and 18%, flexural modulus 3% and 6%, and microhardness 52% and 14% greater than the individual shorea robusta and pinewood fillers reinforced epoxy laminates, respectively.

4 Water Absorption Characteristics of Wood Laminates

The investigation of wood-based polymer laminates' water absorption characteristics is essential because the hydrophilic wood fillers are prone to water and moisture. Usually, wood polymer-based laminates were lower prone to absorption of moisture and acquired higher dimensional accuracy than the plant cellulose fiber-filled polymer laminates (Caulfield et al. 2005). The water absorption is based on various factors such as manufacturing technique, wood fillers wettability with the polymer resin, shape, size and concentration of wood fillers, wood's chemical structure, temperature, and concentration of voids (Caulfield et al. 2005). The hydrophilicity and hygroscopicity principle negatively influence the mechanical characteristics of wood polymer composite and stimulate fungal decay. The laminate initiates swelling when disclosed to moisture tend to the regional strain in polymer that outcome to micro-level cracks; subsequently, decreases the mechanical characteristics of laminates (Joseph and Thomas 1995). The wood fillers attain their original size and shape when preserved, whereas the polymer does not achieve its size and shape because of bounded strain tend to the limited debonding between the matrix and fillers. This principles a poor interfacial adhesion that damages the mechanical characteristics (Stark and Gardner 2008). Bhaskar et al. (2012) studied pine wood flour's water absorption capacity reinforced in original and recycled polypropylene laminate. The water absorption improves with an increment in the concentration of wood flour. The results found that recycled laminate exhibited 15–20% of lower water absorption capacity than the original polypropylene laminate. The maleated coupling modified recycled polypropylene laminate exhibited 15–20% lower water absorption than the original polypropylene laminate. The maleated coupling agent changed recycled polypropylene laminate exhibited decrement in water absorption up to 70% while comparing pure recycled polypropylene.

In contrast, the swelling of thickness was observed lower that further decreased to 0.43 to 0.55% with the addition of a maleated coupling agent. The water absorbed from cedar wood fillers based phenolic laminate was observed 57% higher than the rice husk phenolic laminate (Lette et al. 2018). The maleated coupling agent enhanced the resistance from water absorption of pinewood fillers based on polyethylene-terephthalate/polypropylene laminate (Dairi et al. 2015). The resistance of moisture from wood polymer composites can be enhanced through different kinds of chemical surface modifications on wood fillers as benzoylation, peroxide, esterification, acetylation, and alkaline treatment. The chemical surface modification of reinforcements changes the wood fillers' chemical constituents by eliminating the lignin, unwanted concentrations, and hydroxyl group (Kalia et al. 2011). The addition of wood fillers with polyvinyl chloride reduces the laminate's moisture-resistant characteristics (Jullianelli et al. 2010). The outcomes displayed that pine filler reinforced polypropylene laminate occurs higher resistance from moisture than the beech polypropylene laminates because of their higher lignin and hemicellulose content beech wood fillers while comparing with pinewood fillers (Ayrilmis et al. 2017). It was empirically examined that water absorbed by hybridization of charcoal and wood flour polypropylene laminate reduced with an increment in the charcoal concentration. The wood flour based polypropylene laminate soaked 38% extra moisture than the charcoal-based polypropylene laminate (Ayrilmis et al. 2015).

The mixing of TiO3 nanofillers with wood fillers reinforced high-density polyethylene laminates evaluated a 40% reduction in water absorption capacity due to the TiO2 played as a protective shield for water sorption characteristics (Deka 2001). Tazi et al. (2018) observed in their investigation that water absorption of spruce-based high-density polyethylene laminates was improved by adding the fusa bond coupling agent. The resistance to crystallinity and water absorption of beech wood fillers reinforced polypropylene laminate was potentially influenced by the organoclay insertion and maleated coupling agents (Ge et al. 2004). Danyadi et al. (2010) investigated the influences of different kinds of chemical surface modifications such as stearic acid, benzoylation, cellulose palmitate, and maleated coupling agent on the water absorption resistance of spruce wood fillers polypropylene laminate. The maleated coupling agent did not influence the resistance from water absorption of laminate, whereas the cellulose and benzoylation modifications exhibited a potential effect on laminates' water-resistance characteristics. The wood polymer-based composites have better resistance from wood.

Furthermore, the lower water absorption capacity of wood polymer composites, while comparing with the pure polymer, is the limitation of wood polymer composites that generated the decrement in mechanical characteristics and dimension instability. The hybridization of wood fillers has significantly enhanced the water-resistance features compared with the single wood filler based laminates. The investigators have applied various kinds of chemical surface modifications and compatibilizers for improving the moisture absorption resistance of laminates.

5 Thermal Characteristics of Wood Polymer Laminates

The thermal investigation contains the studies of storage and loss modulus, glass temperature transition, melting point, crystallization temperature, specific heat capacity, thermal conductivity, and damping of the laminates. The differential scanning calorimetry, dynamic mechanical analysis, thermogravimetric analysis, thermal imaging methods, and differential thermal analysis have been generally applied for the laminates' thermal properties. The dynamic automated study evaluates the laminates' viscous characteristics as the formulation of frequency and temperature and reports the loss, storage modulus, damping, and glass transition temperature. The storage modulus presents the higher quantity of energy stored in every cycle of oscillation in the laminate, whereas the loss modulus signifies the quantity of energy consumed from laminates in the conveyance as heat. The proportion of storage to loss modulus determines the damping characteristics of laminate that presents molecules' movement in the laminates. The above characteristics are necessary for the constituents applied in industrial, structural, and automotive applications, followed by dynamic loading (Sewda and Maiti 2013). Compared to the polymer and single filler incorporated laminates, a higher glass transition temperature and storage modulus were observed for hybrid shorea robusta/pine laminates. A flat cole–cole graph of hybrid laminate showed uniform and homogeneous hybrid wood fillers in the epoxy polymer (Khan et al. 2019). The dynamic mechanical analysis explained that mixing applewood fillers in the epoxy polymer appropriately increases the storage modulus, signifying the improvement in laminates' viscoelastic characteristics (Ayrilmis et al. 2015). Vimalanathan et al. (2016) examined that the damping parameter, glass transition temperature, and storage modulus of Shorea robusta/polyester laminate were enhanced significantly up to 20 wt% of the addition of fillers. The plain cole-cole curve of laminate employed uniform dispersion of wood fillers in the polyester resin. Ge et al. (2006) affirmed that the glass transition temperature and storage modulus of biodegradable bamboo wood fillers incorporated polypropylene laminate improved with bamboo reinforcement wood fillers. The thermal deterioration of wood ensued under the lower range of temperature around 200 °C (Poletto et al. 2012). Hence it is needed to forecast and examine the thermal characteristics of wood polymer composites. The thermal stability of wood polymer composites is based on the decomposition of wood fillers' structural components in terms of lignin, cellulose, hemicellulose, ashes, other extractives, and waxes (He et al. 2019). The awareness of the thermal degradation of these components assigned to their chemical structure. The hemicellulose drops their thermal stability nearly 200 to 300 °C due to their arbitrary amorphous design (John and Thomas 2008; Yang et al. 2006). The thermal decomposition of cellulosic initiates about 350 °C due to the crystalline nature (Yang et al. 2006). The lignin structure is entirely dissimilar from the cellulose and hemicellulose, which contains three kinds of propane-benzene groups that are immensely cross-linked and acquires greater molecular weight, which tends to higher stability. Hence, lignin's thermal decomposition starts near 300 to 500 °C (John and Thomas 2008; Poletto et al. 2012; Yang et al. 2006; Kim et al. 2006). Usually, spruce wood's thermal stability is greater than the pinewood due to the presence of a higher concentration of hemicellulose in the hardwood (He et al. 2019; Shen et al. 2010). Deka (2001) performed the thermal examination of TiO2 nanofillers mixed with wood fillers high-density polyethylene laminate.

The melting temperature and thermal stability of laminate were improved by 13% and 12.64%, respectively, compared with the unmixed laminate primarily because of the higher thermal diffusivity of TiO2 nanofillers. The crystallization temperature of each silane and alkali modified laminate was enhanced by approximately 5%, but the melting temperature was uninfluenced (Ichazo et al. 2001). The thermal degradation temperatures of silane (MPS, VTMS, and APTES) coupled with modified Picea abies wood fillers polypropylene laminates were enhanced for all kind of silane agent. The highest five wt% degradation temperature was examined under 308.7 °C for laminates modified with APTES agent. The similar temperature was 306.2, 306.6, and 297.5 °C for the VTMS, MTS changed and untreated laminates. Furthermore, the influence of silane agent for 50 wt% degradation temperature was observed insignificant (Kim et al. 2010).

6 Applications of Wood-Based Polymer Composites

The first profit-making application of wood polymer composite was initiated with Rolls Royce automotive (Gordon 2019). The particular utilization of wood polymer composites contains fabrication of distinct kinds of automotive components (Peaez-Samaniego et al. 2013; Ayrilmis et al. 2017) such as door panels, dashboard, rear storage, noise insulation panels, engine cover, headliner panels, bumper, wheel box, etc.; cabinets and casing of electrical and electronics equipment, interiors of buildings, packaging of other furnishings, lower price rehabilitation, and other useful household components. The wood polymer composite based components have excellent market possibility in Europe and North America with an annual expansion rate of 14% and 18%, respectively (Ashori 2008; Gupta et al. 2007). The automobile sectors explored the wood polymer laminates because of the government's higher interest towards the protection of the environment that granted the automobile sectors to use the biodegradable and recyclable constituents such as wood fillers in the advancement of automobile components. As per the European Commission regulations, around 85% of the entire automobile should be fabricated from recyclable materials (Ashori 2008).

After the successful advancements of biomass nanofillers, utilizations of their bio nanocomposites have shown interest. The bio-based nanocomposites are being applied in construction and building, electronics, packaging, cosmetics, automotive, and other film and coating industries (Sharma et al. 2019). The bio-based nanocomposites were periodically applied in biomedical utilizations like antibacterial activity, tissue engineering, drug delivery, body implants, and disease diagnostics because of their biodegradability, cytocompatibility, and hemocompatibility characteristics. A drug delivery scheme is a schedule in which an essential quantity of drugs is discharged at a particular period for the human being's specific organ. For this reason, different nanocellulose bio-based resin films were applied in the capsule form coated from different pharmaceutical industries (Jackson et al. 2011). Nontoxic behavior, porosity, lower price, network-like structure, biocompatibility, and availability are the favorable characteristics of nanocellulose, suitable for the carrier for drug delivery (Trovatti et al. 2011). The starch type of bio-based nanocomposites has been applied for producing the artificial cartilages as an implant in the osteoarthritis disease (Nasri-Nasrabadi et al. 2014).

7 Conclusion

The present chapter is stated on the studies of water absorption, thermal and mechanical characteristics, and application possibility of wood laminates. A considerable number of publications are observed on the impact of size and shape, aspect ratio, and contents of wood fillers on their polymer laminates' mechanical characteristics. It was recommended that optimum range, aspect ratio, and size attain higher mechanical characteristics. The wood-based laminates consisting of 20–30 wt% of wood fillers of smaller size exhibited better mechanical characteristics. Many methods like chemical surface modifications, incorporation of compatibilizers, and hybridization of different kinds of wood fillers have been applied to enhance wood laminates' moderate mechanical features. In the present scenario, the needs for environmental-friendly bio-based laminates are adequately improved to save our surroundings. Investigators are demanding hard to produce such components that do not affect the environment.

Furthermore, investigations on bio-based wood laminates' characterization are still restricted, which inspires the investigators for further development in this field. Bio-based nanocomposites have outstanding mechanical, barriers, and thermal characteristics, which can be examined as the future developing material. These composites can be applied in growing applications that would present the usage of waste fillers. The separation of nanofillers and their uniform dispersion in the polymer is a problem for the investigators in bio-based nanocomposites' growth. Thus, there is a better scope for producing the laminates and films of nanocellulose for new applications. The mechanical characteristics of three-dimensional printed wood laminates were observed secondary to the pure resin because of the scarcity of fabrication pressure. The lower pressure of three-dimensional printed laminates and materials restriction is still a threat, strengthening the investigators for the upcoming investigation to survive three-dimensional printing conditions. The advancement of cost-efficient manufacturing techniques for the mass fabrication of wood laminates must be performed. This technique might apply to the manufacturing of different bio-based laminate components to be used in different applications.

References

Acosta AP, Gallio E, Schulz HR, Zanatta P, Barbosa KT, Delucis RDA, Gatto DA (2020) Wood-polymer composites produced by in situ polymerization of styrene into juvenile and mature pine woods. Int Wood Prod J 11:1–7

Agnantopoulou E, Tserki V, Marras S, Philippou J, Panayiotou C (2012) Development of bodegradable composites based on wood waste flour and thermoplastic strarch. J Appl Polym Sci 126(S1):273–281

Ahmad I, Bakar DRA, Mokhilas SN, Rami A (2007) Direct usage of products of poly (ethylene terephthalate) glycolysis for manufacturing of rice husk/unsaturated polyester composite. Iran Polym J 16(4):233–239

Amar B, Salem K, Hocine D, Chadia I, Juan MJ (2011) Study and characterization of composites materials based on polypropylene loaded with olive husk flour. J Appl Polym Sci 122(2):1382–1394

Arivazhagan A, Masood S (2012) Dynamic mechanical properties of ABS material processed by fused deposition modelling. Int J Eng Res App 2:2009–2014

Arora S, Kumar M, Kumar M (2012) Flammability and thermal degradation studies of PVA/rice husk composites. J Reinf Plast Compos 31:85–93

Ashori A (2008) Wood_plastic composites as promising green-composites for automotives industries. Bioresour Technol 99:4661–4667

Ayrilmis N, Kwon JH, Han TH, Durmus A (2015) Effect of wood-derived charcoal content on properties of wood plastic composites. Mater Res 18(3):654–659

Ayrilmis N, Tasdemir M, Akbulut T (2017) Water absorption and mechanical properties of PP/HIPS hybrid composites filled with wood flour. Polym Compos 38(5):863–869

Bhaskar J, Haq S, Pandey AK, Srivastava N (2012) Evaluation of properties of polypropylene-pine wood Plastic composite. J Mater Environ Sci 3(3):605–612

Bouhamed N, Souissi S, Marechal P, Amar MB, Lenoir O, Leger R, Bergeret A (2020) Ultrasound evaluation of the mechanical properties as an investigation tool for the wood-polymer composites including olive wood flour. Mech Mater 148:103445

Castellano A, Foti P, Fraddosio A, Marzano S, Piccioni MD (2014) Mechanical characterization of CFRP composites by ultrasonic immersion tests: Experimental and numerical approaches. Compos Part B 66:299–310

Caulfield DF, Clemons C, Jacohson RE, Rowell RM (2005) Wood thermoplastic composites. In: Handbook of wood chemistry and wood composites. CRC Press, Boca Raton, pp 365–378

Chaharmahali M, Mirbagheri J, Tajvidi M (2010) Mechanical and physical properties of wood plastic composite panels. J Reinf Plast Compos 29(2):310–319

Chand N, Fahim M (2008) Wood reinforced polymer composites. In: Tribology of natural fiber polymer composites. Woodhead Publishing Limited, Cambridge, pp 180–196

Chen Z, Zheng L, Jin Q, Li X (2017) Durability study on glass fiber reinforced polymer coil nail via accelerated aging test and long-term field test. Polym Compos 38:2863–2873

Dairi B, Djidjelli H, Boukerrou A, Migneault S, Koubaa A (2015) Morphological, mechanical and physical properties of composites made with wood flour-reinforced polypropylene/recycled poly (ethylene terephthalate) blends. Polym Compos 38(8):1749–1755

Danyadi L, Moczo J, Pukanszky B (2010) Effect of various surface modification of wood flour on the properties of pp/wood composites. Compos Part a Appl Sci Manu 41:199–206

Deka BK (2001) Effect of Tio2 and nanoclay on the properties of wood polymer nanocomposite. Compos Part A Appl Sci Manu 42:2117–2125

Dong C (2018) Review of natural fibre-reinforced hybrid composites. J Reinf Plast Compos 37 (5):331–348

El-Sabbagh A, Steuernagel L, Ziegmann G (2013) Characterisation of flax polypropylene composites using ultrasonic longitudinal sound wave technique. Compos Part B 45(1):1164–1172

Espert A, Vilaplana F, Karisson S (2004) Comprison of water absorption in natural cellulosic fibres from wood and one-year crops in polypropylene composites and its influence on their mechanical properties. Compos Part A: Appl Sci Manu 35(11):1267–1276

Ge XC, Li HX, Meng ZY (2004) Tensile properties, morphology, and thermal behaviour of PVC composite containing pine flour and bamboo flour. J Appl Polym Sci 93:1804–1811

Ge XC, Zhu Q, Meng YZ (2006) Fabrication and cracterization of biodegradable poly (propylene carbonate)/wood flour composites. J Appl Polym Sci 99:782–787

Gordon JE (2019) The new science of strong materials. Princeton University Press, New Jersey

Government RM, Onukwuli OD, Azeez TO (2019) Optimization and characterization of the properties of treated avocado wood flour-linear low density polyethylene composites. Alexandria Eng J 58:891–899

Gu R, Kokta BV, Michalkova D, Dimzoski B, Fortelny I, Slouf M, Krulis Z (2010) Characteristics of wood-plastic composites reinforced with organo-nanoclays. J Reinf Plast Compos 29 (24):3566–3586

Guo J, Song K, Salmén L, Yin Y (2015) Changes of wood cell walls in response to hygro-mechanical steam treatment. Carbohydr Polym 115:207–214

Gupta BS, Reiniati I, Laborie MPG (2007) Surface properties and adhesion of wood fiber reinforced thermoplastics composites. Coll Surf A: Physicochem Eng Asp 302(1–3):388–395

Güven O, Monteiro SN, Moura EAB, Drelich JW (2016) Re-emerging field of lignocellulosic fiber – polymer composites and ionizing radiation technology in their formulation. Polym Rev 56:702–736

He Z, Wang Z, Qu L, Qian J, Yi S (2019) Gaseous decomposition products from wood degradation via thermogravimetric and fourier transform infrared analysis during thermal modification of beech and pine woods. BioResources 14(3):6883–6894

Hosseinaei O, Wang S, Enayati AA, Rials TG (2012) Effects of hemicellulose extraction on properties of wood flour and wood-pastic composites. Compos Part A: Appl Sci Manu 43:686–694

Ichazo MN, Albano C, Gonzalez J, Perera R, Candal MV (2001) Polypropylene /wood flour composites: treatments and properties. Compos Struct 54(2–3):207–214

Islam M, Hamdan D, Hassan A, Talib Z, Sobuz H (2014) The chemical modification of tropical wood polymer composites. J Compos Mater 48(7):783–789

Jackson JK, Letchford K, Wasserman BZ, Ye L, Hamad WY, Burt HM (2011) The use of nanocrystalline cellulose for the binding and controlled release of drugs. Int J Nanomed 6:321–330

John MJ, Thomas S (2008) Biofibres and biocomposites. Carbohydr Polym 71(3):343–364

Joseph K, Thomas S (1995) Effect of aging on the physical and mechanical properties of sisal-fibre-reinforced polyethylene composites. Compos Sci Technol 53(1):99–110

Kain S, Ecker JV, Musso M, Petutschnigg A (2020) Effects of the infill pattern on mechanical properties of fused layer modeling (FLM) 3D printed wood/polylactid acid (PLA) composites. Europ J Wood Wood Prod 78:65–74

Kajaks J, Zagorska A, Mezinskis A (2015) Some exploitation properties of wood plastic composites (WPC), based on high density polyethylene and timber industry waste. Mater Sci 21(3):396–399

Kalia S, Dufresne A, Cherian BM, Kaith BS, Averous L, Njugana J, Nassiopoulus E (2011) Cellulose-based bio- and nanocomposites: a review. Int J Polym Sci 2011:1–35

Kariz M, Sernek M, Obucina M, Kuzman MK (2018) Effect of wood content in FDM filament on properties of 3D printed parts. Mater Today: Comm 14:135–140

Khan MZR, Srivastava SK, Gupta MK (2019) Hybrid wood particulates composites: mechanical and thermal properties. Mater Res Exp 6:105323

Kim HS, Kim S, Kim HJ, Yang HS (2006) Thermal properties of bio-flour-filled polyolefin composites with different compatibilizing agent type and content. Thermochim Acta 451:181–188

Kim TW, Lee SY, Chun SJ, Doh GH, Paik KH (2010) Effect of silane coupling on the fundamental properties of wood flour reinforced polypropylene composites. J Compos Mater 45(5):1595–1605

Kismet Y (2017) Change of mechanical properties of powder recyclate reinforced polyolefin based on gamma radiation. Polym 9:384

Komal UK, Lila MK, Chaitanya S, Singh I (2020) Fabrication of short fiber reinforced polymer composites. In: Reinforced polymer composites. Wiley-VCH, Weinheim, pp 21–36

Kootsookos A, Mouritz AP (2004) Seawater durability of glass- and carbon-polymer composites. Compos Sci Technol 64:1503–1511

Kord B (2011) Influence of maleic anhydride on the flexural, tensile and impact characteristics of swadust flour reinforced polypropylene composites. World Appl Sci J 12(7):1014–1016

Kumar R, Kumar K, Sahoo P, Bhowmik S (2014) Study of mechanical properties of wood dust reinforced epoxy composite. Procedia Mater Sci 6:551–556

Lee SH, Wang S (2006) Biodegradable polymers/bamboo fiber biocomposite with biobased coupling agent. Compos Part A Appl Sci Manu 37(1):80–91

Lefebvre G, Wunenburger R, Valier-Brasier T (2018) Ultrasonic rheology of viscoelastic materials using shear and longitudinal waves. Appl Phys Lett 112(24):241906

Lette MJ, Ly BE, Ndiaye D, Takasaki A, Toshihiro O (2018) Evaluation of sawdust and rice husks as fillers for phenolic resin based wood-polymer composites. O J Compos Mater 8:124–137

Li X, Wu H, Southerton S (2011) Transcriptome profiling of wood maturation in Pinus radiata identifies differentially expressed genes with implications in juvenile and mature wood variation. Gene 487(1):62–71

Liang JZ (2012) Predictions of Young's modulus of short inorganic fiber reinforced polymer composites. Compos Part B 43(4):1763–1766

Lullianelli G, Tavares MB, Luetkmeyer L (2010) Water absorption behaviour and impact strength of PVC//Wood flour composites. Chem Chemi Technol 4:225–239

Martikka O, Karki T, Wu Q (2018) Mechanical properties of 3D-printed wood-plastic composites. Key Eng Mater 777:499–507

Mcdonald E, Hubert J (2002) Review of the effects of silviculture on timber quality of Sitka spruce. Forestry 75(2):107–138

Mohit H, Selvan VAM (2018) A comprehensive review on surface modification, structure interface, and bonding mechanism of plant cellulose fiber reinforced polymer based composites. Compos Interf 25(5–7):629–667

Moise V, Stanculescu I, Vasilca S, Cutrubinis M, Pincu E, Oancea P, Raducan A, Meltzer V (2019) Consolidation of very degraded cultural heritage wood artefacts using radiation curing of polyester resins. Radiat Phys Chem 156:314–319

Nagarajan V, Mohanty AK, Misra M (2016) Biocomposites with size-fractionated biocarbon: influence of the microstructure on macroscopic properties. ACS Omega 1(4):636–647

Naghmouchi I, Mutje P, Boufi S (2015) Olive stones flour as reinforcement in polypropylene composites: a step forward in the valorization of the solid waste from the olive oil industry. Ind Crops Prod 72:183–191

Nasri-Nasrabadi B, Mehrasa M, Rafienia M, Bonakdar S, Behzad T, Gavanji S (2014) Porous starch/cellulose nanofibers composite prepared by salt leaching technique for tissue engineering. Carbohydr Polym 108:232–238

Ndiaye D, Badji AM, Tidjani A (2014) Physical changes associated with gamma doses on wood/polypropylene composites. J Compos Mater 48:3063–3071

Niang B, Ndiaye D, Kholl M, Babacar L, Diallo A, Badji A (2017) Gamma radiation induced HDPE/wood composite: studies of morphological, mechanical and thermal properties of the composites. Chem Sci Int J 20:1–12

Nikmatin S, Syafiuddin A, Kuch ABH, Maddu A (2017) Physical, thermal, and mechanical properties of polypropylene composites filled with rattan nanoparticles. J Appl Res Technol 15:386–395

Nourbakhsh A, Karegarfard A, Ashori A, Nourbakhsh A (2010) Effect of particle size and coupling agent concentration on mechanical properties of particulate-filled polymer composites. J Thermoplast Compos Mater 23:169–174
Palermo P, De Figueiredo LJ, Severo E, Do Nascimento A, De Rezende M (2013) Delimitação entre os lenhos juvenil e adulto de Pinus elliottii Engelm. Revista Árvore 37(1):191–200
Peaez-Samaniego MR, Yadama V, Lowell E, Espinoza-Herrera R (2013) A review of wood thermal pretreatments to improve wood composite properties. Wood Sci Technol 47(6):1321–1322
Perez E, Fama L, Pardo SG, Abad MJ, Bernal C (2012) Tensile and fracture behaviour of PP/wood flour composites. Compos Part B Eng 43:2795–2800
Poletto M, Zattera AJ, Forte MM, Santana RM (2012) Thermal decompositon of wood: influence of wood components and cellulose crystallite size. Bioresour Technol 109:148–153
Prusty RK, Ghosh SK, Rathore DK, Ray BC (2017) Reinforcement effect of graphene oxide in glass fibre/epoxy composites at in-situ elevated temperature environments: an emphasis on graphene oxide content. Compos A 95:40–53
Quan D, Pearson RA, Ivankovic A (2018) Interaction of toughening mechanisms in ternary nanocomposites. Polym Compos 39:3482–3496
Ray BC, Rathore D (2015) Environmental damage and degradation of FRP composites: a review report. Polym Compos 36:410–423
Sawpan MA (2016) Effects of alkaline conditioning and temperature on the properties of glass fiber composite rebar. Polym Compos 37:3181–31900
Saxena M, Gupta M (2018) Mechanical, thermal, and water absorption properties of hybrid wood composites. J Mater Des Appl 233(9):1914–1922
Saxena M, Gupta MK (2018b) Hybrid mango/shorea robusta wood reinforced epoxy composites: crystalline behaviour and dynamic mechanical analysis. Mater Today: Proc 5:19359–19366
Scalerandi M, Gliozzi AS, Bruno CLE, Antonaci P (2012) Nonequilibrium phenomena in damaged media and their effects on the elastic properties. J Acoust Soc Am 131(6):4304–4315
Schneider M (2016) Polymer modified wood. Encycl Matr Sci Technol 1–4
Schwarzkopf MJ, Burnard MN (2016) Wood-plastic composites—performance and environmental impacts. In: Environmental impacts of traditional and innovative forest-based bioproducts. Springer, Singapore, pp 19–43
Sewda K, Maiti SN (2013) Dynamic mechanical properties of high density polyethylene and teak wood flour composites. Polym Bull 70(10):2657–2674
Shakuntala O, Raghavendra G, Samir KA (2014) Effect of filler loading on mechanical and tribological properties of wood apple shell reinforced epoxy composite. Adv Mater Sci Eng 2014:1–9
Sharma A, Thakur M, Bhattacharya M, Mandal T, Goswami S (2019) Commercial application of cellulose nano-composites-a review. Biotechnol Rep 21:00316
Shen DK, Gu S, Bridgwater AV (2010) The thermal perdormance of the polysaccharides extracted from hardwood: Cellulose and hemicellulose. Carbohydr Polym 82(1):39–45
Simonetti F, Cawley P, Demčenko A (2005) On the measurement of the young's modulus of small samples by acoustic interferometry. J Acoust Soc Am 118(2):832–840
Stark NM, Gardner DJ (2008) Outdoor durability of wood-polmer composites. In: Wood-polymer composites. Woodhead Publishing Limited, Cambridge, pp 142–165
Tazi M, Erchiqui F, Kaddami H (2018) Influence of softwood -fillees content on the biodegradability and morphological properties of wood -polyethylene. Polym Compos 39 (1):29–37
Teymoorzadeh H, Rodrigue D (2015) Biocomposite of wood flour and polylactic acid: processing and properties. J Biobased Mater Bioenerg 9(2):1–6
Trovatti E, Silva NHCS, Duarte IF, Rosado CF, Almeida IF, Costa P, Freire CSR, Silvestre AJD, Neto CP (2011) Biocellulose membranes as supports for dermal release of lidocaine. Biomacromol 12:4162–4168
Vignesh J, Selvam CM (2015) Expermental evaluation of wood dust particulate reinforced polymer composites. Eng Sci Technol Int J 5(4):226–229

Vimalanathan P, Venkateshwaran N, Santhanam V (2016) Mechanical, dynamic mechanical and thermal analysis of shorea robusta dispersed polyester composite. Int J Polym Anal Charact 21 (4):314–326

Wang X, Jiang M, Zhou Z, Gou J, Hui D (2017) 3D printing of polymer matrix composites: a review and propestive. Compos Part B Eng 110:442–458

Wang Z, Xian G, Zhao XL (2018) Effects of hydrothermal aging on carbon fibre/epoxy composites with different interfacial bonding strength. Constr Build Mater 161:634–648

Wolcott MP, Englund K (1999) A technology review of wood-plastic composites. In: 33rd international particleboard/composite materials symposium, Pullman, Washington

Yang H, Yan R, Chen H, Zheng C, Lee DH, Liang DT (2006) In-depth investigation of biomass pyrolysis based on three major composnents: hemicellulose, cellulose, and lignin. Energ Fuels 20:388–393

Zhang X, Fan X, Yan C, Li H, Zhu Y, Li X, Yu L (2012) Interfacial microstructure and properties of carbon fiber composites modified with graphene oxide. ACS Appl Mater Interf 4:1543–1552

Zhou Y, Fan M, Chen L, Zhuang J (2015) Lignocellulosic fibre mediated rubber composites: an overview. Compos B Eng 76:180–191

Manufacturing of Wood Polymer Composites

Neeraj Dubey, Rajesh Purohit, and R. S. Rana

Abstract Wood polymer composites (WPCs) is a category of natural fiber composites (NFCs) in which comparatively short fibers such as saw dust and wood flour are used to reinforce the plastics. Besides all the advantages of NFCs which are concern to sustainable material and environment, WPCs are easier to manufacture and economic. These qualities of WPCs have made it strong alternative for many conventional materials. This chapter presents an overview of manufacturing processes involved in the production of WPCs with their advantages and disadvantages. It also discusses over the preparation of wood fibers and suitability of manufacturing techniques for different kinds of fibers and plastics. WPC's properties such as strength, durability and external finishing highly depend upon their development process. A comparative discussion over properties of finished WPC products is included in the present chapter. It also enlighten the key factors involved in manufacturing process of WPCs such as process temperature, moisture content, heat evolved etc.

Keywords WPC (Wood Polymer Composites) · NFC (Natural Fiber Composites) · Manufacturing processes · Process parameters

1 Introduction

The concept of composite material is being utilized from the ancient civilization which is evident from the use of bricks in the ancient house building. One can achieve the combinations of the properties in the form of composites that could not be achieved with the individual materials. The concept is not a human invention, rather it also exist naturally such as, wood and bones are the examples of natural composite materials. In the domain of manmade composite materials the concept of

N. Dubey (✉)
Mechanical Engineering Department, LNCT, Bhopal, Madhya Pradesh, India

R. Purohit · R. S. Rana
Mechanical Engineering Department, MANIT, Bhopal, Madhya Pradesh, India

S. Mavinkere Rangappa et al. (eds.), *Wood Polymer Composites*,
Composites Science and Technology, https://doi.org/10.1007/978-981-16-1606-8_2

reinforcing strong fibers in the polymers results in manufacturing of many strong fibers and polymer matrix composites (PMCs) such as GFRP, CFRP composites etc. But PMCs with synthetic fibers have some disadvantages which are mainly related to environment, human health and the cost involved. Such as all synthetic fiber composites are non biodegradable, the glass fibers are dangerous to health of workers and carbon fiber and kevlar fibers are costlier. These problems related to synthetic fibers found answer in the form of natural fiber composites (NFCs) or wood polymer composites (WPCs). In the last three decades many scientists have published their work on synthesis of natural fiber composites and their characterization. The NFCs with many strong natural fibers like flax, jute, sisal, hemp coir etc. have been synthesized successfully. But to impart the value to these materials and to provide them a market it is equally necessary to mould them into the useful forms. It requires the investigation of the manufacturing processes for the NFCs. This chapter provides the study of various manufacturing processes which can be adopted for processing of NFCs.

1.1 Manufacturing Methods of WPCs and NFCs

The manufacturing process adopted for the fabrication of a NFC is a primary aspect for the final properties and performance. To manufacture a product from a traditional material, the process starts from a blank piece of material which are generally in the form of sheet, ingot and rod etc. But unlike traditional materials (such as metals, glass, wood), in case of composite materials the material and the component are being produced simultaneously. There are a broad variety of processes available for the composites manufacturer to produce cost efficient products. The choice of process depends upon the final product to be produced and the cost efficiency. This also provides the best solution to the customer. The PMCs or NFCs can be classified widely on the basis of the matrix materials i.e. thermoset composite and thermoplastic composites.

1.2 Thermoset Composites

In this type of composites the matrix material includes thermosetting resins such as polyesters, vinylesters, epoxies, bismaleimides, and polyamides. The most commonly used resins are polyester and epoxies. Although, the viscosity of these resins is low, but when they undergo chemical reaction with the hardener, the crosslinking of the polymer chains takes place. Due to this reaction the entire matrix together form a three dimensional network and this process is called Curing. The three dimensional crosslinked structures of thermosets provide high rigidity, high temperature resistance, and good resistance to solvents.

1.3 Thermoplastic Composites

This category includes the polymers as matrix material such as Polyethylene, Polystyrene, Polyamides, Nylons, Polypropylene, Polyetheretherketones (PEEK). They consists of long discrete molecules that melts to a viscous liquid at the processing temperature, typically 260 to 370°, and after forming are cooled to an amorphous semicrystalline or crystalline solid. Thermoplastics possess one or two-dimensional molecular structure and they tend to melt at an elevated temperature and show exaggerated melting point. Another benefit is that the process of softening at elevated temperatures can reversed to regain its properties during cooling. Due to which the conventional compress techniques to mould the compounds can be applied. Figure 1 illustrates the general characteristics of the thermoplastic and thermoset resins (Tenney et al. 2009).

Besides the above categories of the matrix materials another category is of Bio based polymers. This includes polymers such as poly lactic acid (PLA) (Wolf et al. 2005), Polyglycolic acid (PGA) (Maurus and Kaeding 2004), wheat gluten plastic (Wretfors et al. 2009), soya protein isolate (SPI) (Rakesh et al. 2008). This type of polymers as matrix material mainly aims to make the composites bio degradable and to develop a sustainable product. However, the compromise with strength, low resistance to solvent and high temperature is still an issue with these polymers.

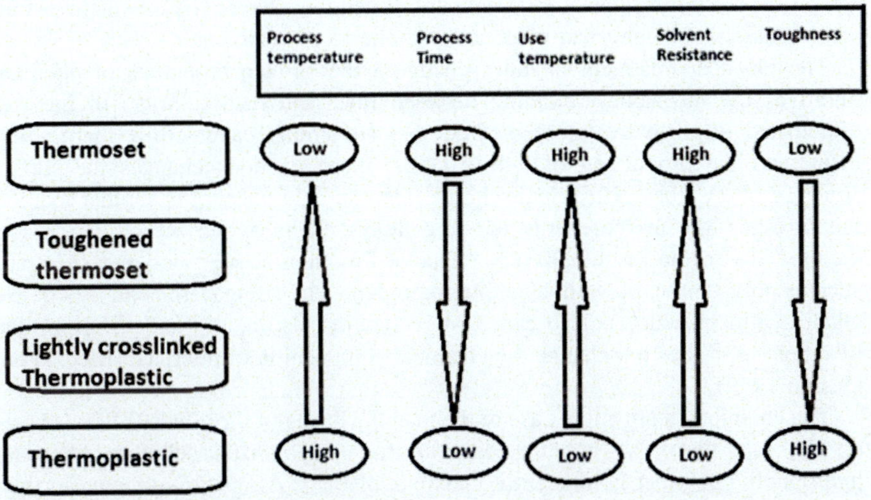

Fig. 1 General characteristics of the thermoplastic and thermoset resins

2 Preparation of Natural/Wood Fibers for Reinforcement

Natural fibers are obtained by the desiccation (drying) or Retting (controlled degradation) of the crops. These fibers are spun and processed for reinforcement into different forms. The short spun fibers are used in the manufacturing techniques such as hand layup and injection moulding. For textile processes Such as spray lamination, filament winding and pultrusion, the reinforcement fibers are processed in a continuous form. The different forms of continuous fiber are Roving, tows and yarns (Summerscales and Grove 2014).

- Roving: a bundle of continuous untwisted fibers.
- Tow: a loose strand of large number of filaments without a significant twist.
- Yarn: a product of relatively small cross-section and substantial length and with or without twist.

3 Treatments for Improving Interfacial Strength

The interfacial strength between the hydrophilic cellulosic fiber and hydrophobic matrix material (olefinic polymers) is a primary issue for overall strength of the NFCs/WPCs. To improve the interfacial bonding between the matrix and the fiber can be successfully achieved by treatment of cellulose fiber. The treatment of fiber can be classified as physical treatment and chemical treatment.

The physical treatment includes changing the surface properties of fiber and improves the mechanical bonding between fiber and matrix. Some of the well established physical treatment methods are stretching, calandering, thermotreatment, and the production of hybrid yarns, these methods changes the surface structure of the fibers. Electric discharge (corona, cold plasma) is another method; this changes the surface energy of the cellulose fibers by surface oxidation activation. This method is found very effective for "non-active" polymer substrates such as polystyrene, polyethylene, polypropylene, etc. (Belgacem et al. 1994). The alkali treatment which is also called Mercerization also comes under the physical treatment as it does not change the chemical composition of interface (Bledzki and Gassan 1999).

The chemical treatment in case of the manufacturing of the NFCs/WPCs is based on the fact that, The bonding between highly polarized cellulose fiber and hydrophobic polymer is inherently incompatible. The compatibility between them can be achieved by introducing an intermediate material between them, having intermediate properties between those of other two. The chemical treatment includes Silane treatment, Acetylation, Grafting, Oxidation, Scouring etc. (Kabir et al. 2012).

4 Manufacturing Methods for WPCs

The wood polymer composites can be developed by the manufacturing methods that are applied for other fiber reinforced polymeric composites (Salit 2014). However, on the basis of manufacturing process there are three possible ways to mix fibers into matrix to produce a composite

1. The user has to fabricate composite by processing fiber and matrix separately such as in hand lay-up process.
2. The user obtains a moulding compound by mixing fiber, matrix and additives in correct proportion. The moulding compound is then used to form a composite such as in injection moulding, extrusion, compression moulding.
3. A composite component is produced as ready to use for secondary processes such as assembly etc. for example in pultrusion and filament winding process.

Following are the manufacturing processes commonly employed for the development of the fiber reinforced plastics (FRCs) (Mayer 1993). Figure 2 shows classification of the manufacturing processes generally used for the production of polymer matrix composites (PMCs).

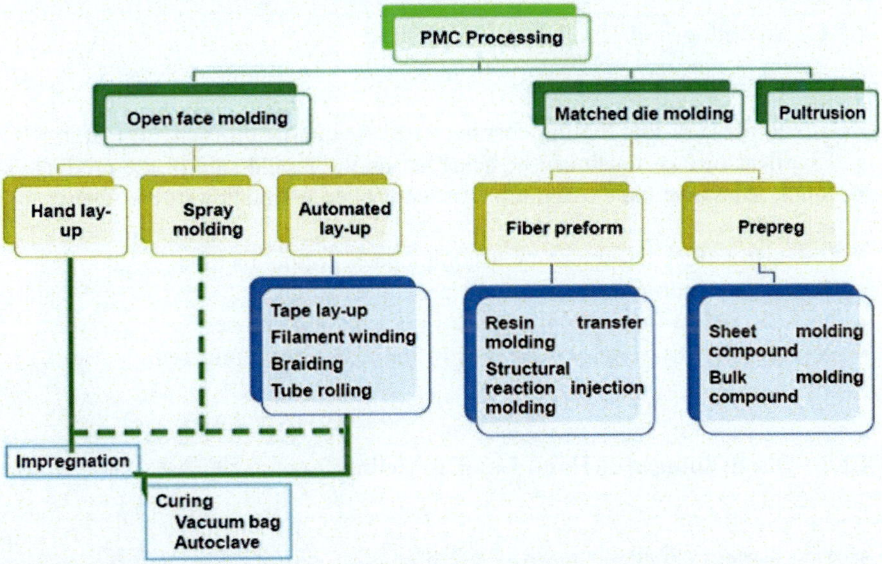

Fig. 2 Classification of manufacturing processes of PMCs

4.1 Hand Lay-Up Technique

Hand lay-up is the simplest fabrication process and is an open mould process. This process is also known as wet lay-up process. In this, the mould can be made of sheet metal, plaster, composites, wood, plastics etc. In this process the reinforcement is laid layer by layer to achieve a desired thickness. After laying a layer it is wetted with resin. After this a roller is moved to remove the air pockets and uniform distribution of resin. The curing in the hand lay-up process takes place at room temperature.

The mould is given a thin coating of release agent such as polyvinyl alcohol, silicon oil and wax. The mould is then coated with resin matrix. A pre-cut fiber cloth mat or layer of fibers is then laid over the resin layer. Another layer of resin is coated over the fiber layer. Rollers are used to press the fiber layer on the resin uniformly and also to remove the entrapped air bubbles. Alternate layer of resins and fiber layer are laid in a similar sequence until the required thickness is obtained. Now curing is done at elevated temperature. Trimming and finishing may now be taken up after removal of the composite material from mould. Sheets, auto body parts, boat- hulls, ducts building components etc. are produced by these techniques, like boat hull produced by hand lay-up method with natural fiber.

4.1.1 Advantages of Hand Lay-Up Method

- It requires very low tooling cost because no automation is involve in it.
- Excellent surface finish can be achieved on the mould side of the product.
- Since, it involve hand assembly thus, controlled orientation of the fibers can be achieved.
- Irregular surface shape can be accommodated in this method.
- There is no limit of the size of the part to be produced.
- On site production is possible because it a portable process.
- Any form (short, long, woven etc.) of the fibers can reinforced.

4.1.2 Disadvantages of Hand Lay-Up Method

- This method is labour intensive and it requires skilful operators.
- Good surface finish is achieved on only one side of the product.

4.2 Spray Lay-Up Technique

Spray lay-up process is an advanced version of hand lay-up process. This process includes a spray gun with attached chopping equipment. A continues roving of a fiber is fed through the chopper and cut to a preset length. The spray-gun sprays these chopped fibers to mix with resin stream. The mixture of fiber and resin is sprayed to deposit on the mould surface. Spray of mixture continues to achieve a desired thickness. A releasing agent (wax or gel) is coated on the surface of the mould before spraying for the easy release of part from the mould. Generally the moulds itself in the spray lay-up process are made of reinforced plastics. The inner surface of a mould takes the shape of the outer surface of the product.

Generally, the polyester resin is used in a spray gun that has a glass chopper attachment. According to the requirement of a part the layers of the composite are built up on the mould. The gun of the spray lay-up process sprays resin and catalyst in two different streams which mix with each other on the exit of the gun. However, in the spray lay-up the excess amount of resin is required in comparison to hand lay-up for the same product which is due to the over spray of the resin. However, multiple headed guns are also used. The technique is adopted for plastic articles like body parts of truck, vessels, boat hulls and lorry cabs.

4.2.1 Advantages of Spray Lay-Up Method

- In this process any materials can be used as mold.
- It is faster than hand lay-up process.
- It is continuous process.
- Error can be corrected by re-spraying

4.2.2 Disadvantages of Spray Lay-Up Method

- It is slower than other advance processes.
- It is labor intensive process thus inconsistency occurs.
- No control of fiber orientation
- Only one side (mould side) surface finishing is achieved.
- It is environmental unfriendly process due to dust of chopped fibers.

However, spray lay-up process as manufacturing process for wood polymer composites is yet not reported. This is because the spray lay-up process is mainly developed for the continuous roving reinforcement of glass fibers. In case of natural fiber the spinning short fibers into yarn and inherent low strength of yarn will increase the cost and difficulty. However, configuring spray lay-up process for wood polymer may be a new field of investigation (Summerscales and Grove 2014).

4.3 Vacuum Bagging Process

This process is basically extension of the hand lay-up technique and it is mainly used for laminated composites. In this process the composite is placed on the mould similar to the hand lay-up process, and the next layer is peel ply, this is actually a perforated release film. A breather is placed on the top of the peel ply and the layer after this is a separator. Above the breather a plastic film is sealed over the laminate and on to the tool. The sealed plastic film play the role of a vacuum bag as the air under the vacuum bagging film is extracted by a vacuum pump which allows up to one atmosphere compaction pressure to be applied to the laminate for consolidation. The compaction pressure removes the excess amount of air and resin by force. This also removes voids and increases consolidation of the laminate and results in to the production of stronger, void free and with better surface finishes composites (Biswal et al. 2020). However, this process is not suitable for polyester and vinyl ester resins because vacuum pump extract excessive styrene from resin, however it is suitable for epoxy and phenolic resins (Mohammad et al. 2018). This process is used to manufacture body parts of racing boats and racing cars. However, more advanced techniques of the vacuum bagging such as pressure molding and autoclave processes are also available (Biswal et al. 2020).

4.3.1 Advantages of Vacuum Bagging

- Production cost is low.
- Higher strength of composite as compare to hand lay-up.
- Good surface finish free from voids is obtained.
- Large variety of fiber/matrix combination can be cast.
- Reduced amount of volatile emission during the cast.

4.3.2 Disadvantages of Vacuum Bagging

- It requires frequent change of breather cloth.
- High temperature cannot be applied for fast curing.
- Pressure limited to one atmospheric only is allowed for compaction.
- It is slow process.
- There is issue of inconsistency.

4.4 Filament Winding Process

In this method a continuous length of strand, roving or woven tape of fiber is passed through a bath of resin. The excess amount of resin is squeezed out, as the strand comes out of the bath. The resin dipped filament or strand is then wound over a mandrel. Curing is then undertaken with heat. The fiber can be wound over the winding machine in a predetermined direction. The tension of the fiber and the pattern of winding are very important factors in influencing the ultimate tensile properties of the product. The winding of prepaged tapes are also possible in this method.

This technique is used for the fabrication of surfaces of revolution such as pipes, tubes, cylinders, and spheres and is frequently used for the production of large tanks and pipe work for the chemical industry. Filament winding method is based on high-speed precise lay down of continuous reinforcement in preset patterns. In this process fiber tension is the critical parameter and affects the porosity content, fraction of reinforcement of the fiber and properties of the final product. An optimum fiber tension is evaluated by considering the factors such as nature of fiber, type and geometry of the fiber as well as pattern of winding (Biswal et al. 2020).

Different commercial products by using roving of natural fiber such as kenaf fiber have been successfully manufactured by this method (Senawi 2010) The investigation of natural fiber composite with flax and hemp fibers manufactured by filament winding were also reported (Lilholt and Madsen 2012).

4.4.1 Advantages of Filament Winding Technique

- Fast lay-up speed.
- Great strength to weight ratio.
- Higher accuracy.
- Consistency.
- A huge size part can be produced.
- It is computer controlled thus less labor cost involved.

4.4.2 Disadvantages of Filament Winding

- It involve high cost for equipment such as mandrel.
- Poor surface finish is achieved.
- Limited shape of product such as solids of revolution can be produced.
- Application of heat for curing is not easy.
- Due to penetration of resin and splashing the spinning speed is limited.
- Breakage of yarn occurs.

4.5 *Pultrusion*

The word Pultrusion can be thought of as combination of pulling and extrusion. Hence, this process involves pulling of the continuous fiber and resin into a pre-defined shaped and heated die. This process is similar to extrusion process but only basic difference is that in this process the material is pulled through the dies, rather in extrusion the material is pushed through the dies.

This process is a continuous process and is used to produce thermosetting composite materials of uniform and constant cross section with rapid rate of production. In this process continuous string of reinforcing fibers are pulled. These roving of fiber pass through a liquid resin mix bath and pulled through a long machined steel die of suitable profile. The heated die initiates an exothermic reaction to polymerize the thermosetting resin matrix. The composite profile emerges from the die as a constant and hot cross-sectional material. The cured profile that cools sufficiently to be fed into a clamping and pulling mechanism which is pulled by the rollers which provides the driving force for the strands to be forced through the die. The product can then be cut to desired lengths. Examples of pultruded products include electrical insulation materials, ladders, walkway gratings, structural supports, channels and rods of different cross sections and antennas. The method is suitable for articles such as tubing or fishing rod. The natural fiber composites by using roving of fibers such as flax, hemp and kenaf are being produced by this method (Senawi 2010). Almost all types of thermosetting resins such as epoxy, polyester, phenolic and vinyl ester can be used for matrix material.

4.5.1 Advantages of Pultrusion Process

- It is a continuous reinforcement process.
- It is automated thus less labor cost involved.
- Consistency in this process is very high.
- It is capable of rapid production.
- Versatile cross section can be produced

4.5.2 Disadvantages of Pultrusion Process

- It involve expensive die.
- Only constant cross section can be produced
- It is configured only for thermosetting resins.
- A defect in die mess up whole batch of products.

4.6 Injection Molding Process

The process completes in the steps as first the polymer composite is heated to molten state, injecting the molten composite into a mould after this it is allowed to cool the composite in the mould, and lastly part is ejected. These four steps are briefly explained as follows: (Salit 2014).

1. Filling of the mould: - In this the screw rotates to move forward and force the molted composite into a closed mould. This step is mainly controlled by velocity and completed at very fast rate. When molten composite enters the mould and comes in contact with the cold walls of the mould so it freezes.
2. Packing: - In this step screw continue to push more material into the mould with pressure to ensure complete filling of the cavity of the mould. This process is to compensate shrinkage of material due to cooling. This process mainly controlled by pressure.
3. Cooling: - During this step composite is allowed to cool and hold its shape. However, screw rotates during this period but do not apply any pressure. Rather it transfers material for the next cycle.
4. Ejection: - During this step the mould is opened and the part is ejected. After this mould is again closed and prepared for the next cycle.

The injection molding machine is defined and characterized by its clamp size and its injection capacity. Clamp force ranges from a couple of tons to several thousands of tons. The most frequently used machines lie around 300 tons (Sykacek et al. 2009). Injection capacity ranges from a few grams to hundreds of kilograms. Most of the heat that is used for melting granules evolves from friction between granules, and between the barrel and the screw. The size of the screw often gradually increases. Injection molding is a very suitable procedure to process natural short fiber reinforced polymers into geometrically sophisticated structural parts. It is the primary process for producing WPCs with thermoplastic polymers such as Polyvinyl chloride (PVC) and Polypropelene (PP). However, it is also used for thermoset composites. A rough finish is indicative of a low mould temperature. Higher temperatures may be possible if the material is allowed to remain for short duration in the barrel. However, if too high a temperature is used, charring can occur. Injection molding differs from profile extrusion, in that, after the material is heated, it is pumped into a permanent mould, where it takes shape and cools. The mould is then opened and the finished part is discharged.

Different extruder systems such as single screw extruder and twin screw extruder are available for the compounding of WPCs (Schwendemann 2008). However moisture content in wood fibers is major issues in case of the pelletizing, hence wood fibers are required to dry to 12% moisture content to receive appropriate pellets (Le Baillif and Oksman 2009). It has been observed that, due to the low back pressure and poor distribution ability of the screw the agglomeration of the fibers and void inclusion in the final composite part takes place.

Usually the injection-molding screw is not designed for the distribution of particle or fiber based granulates. Rather the screw is designed to ensure the material feed-in, plasticizing of polymer, transportation of molten composite and compression of polymer composite. The pelletized material is difficult to disperse properly, because of hydrogen bonds between the fibers (Felix and Gatenholm 1991; Le Baillif and Oksman 2009), which have been formed during the process.

4.6.1 Advantages of Injection Molding Process

- Both thermoset and thermoplastic composites can be produced.
- Complex shaped product can be produced.
- It is automated process.
- Product can be produced in high volume.

4.6.2 Disadvantages of Injection Molding Process

- Instrument cost is very high.
- Only short fibers can be reinforced.
- Low strength of WPCs is obtained due to short fibers.
- Fiber agglomeration occurs.
- Voids may be present in the product.
- High Maintenance cost.

4.7 Extrusion Process

Unlike injection molding process extrusion is a continuous process and mainly suitable for thermoplastic composites. Similar to injection molding process it is also screw assisted process. The screw of the extrusion machine has three different zones: Feeding zone, compression zone and Metering zone. In Feeding zone the material is entered and heated. As the material travel along the screw, it first becomes soften and finally converted into high viscous fluid. The material are mixed and compressed before entering the metering zone, according to design of screw. After metering the material are continuously pushed toward the die.

For the production of thermoplastic WPCs by extrusion process the pellets of pure polymer are mixed or compounded with wood/natural fibers to produce composite pellets (by extrusion called compounder). These composite pellets become a raw material for the composite product. The pellets of composite can be used in both injection molding as well as in extrusion. In case of extrusion the composite pellets are feed into the extruder through a hopper. These pellets get heated to molten stage and become a viscous fluid. This viscous fluid is then compressed in compression zone and enters into melting zone. Then material is pushed through a predefined shaped die and a continuous strand of composite material is drawn out. Some cooling arrangement cools the strand. The extrusion is basically a forming process for producing product of constant cross section such as Pipes, Tubes, Sheets, Films, Strips, and Profiles.

4.7.1 Advantages of Extrusion Process

- It is a continuous process.
- High rate of production.
- It is automated thus less labor cost involved.
- Consistency in this process is very high.
- Versatile cross section can be produced.

4.7.2 Disadvantages of Pultrusion Process

- It involves expensive die and equipment.
- Due to short fiber reinforcement strength of the product is low.
- Only constant cross section can be produced.
- It is configured only for Thermoplastic polymers.
- For production of WPCs pellets of composite is required to be produced.
- Production of pellets involves high cost.
- A defect in die can messed up whole batch of products.

4.8 Resin Transfer Molding - RTM

For the production of thermoset WPCs/NFCs, RTM is the most efficient fabrication technique. In this process fibers are arranged in a pre-form according to the mould shape. To produce a pre-form the fibers are pre-pressed to match the mould shape. In case of loose fibers (wood flour) a binder material can be used to hold them together. This pre-form of fibers is then kept in the mould cavity and the second matching part of the mould is clamped over the first. After this, the mixture of resin with hardener and other additives such as colour and catalyst is injected into the cavity with pressure. Also, to support the resin to be drawn into the fibers and gaps, the vacuum can be applied to cavity. This modified process is called as Vacuum Assisted Resin Transfer Moulding (VARTM). The composite is then allowed to cure either at room temperature or elevated temperature. Resin, such as bismaleimides requires high temperature for curing. All general types of resins such as epoxy, polyester, vinylester and phenolic can be used in this process. The factors affecting RTM process are mould designing, wetting property of resin, mould temperature and mould filling. Any failure of these factors causes void inclusion, low strength and poor surface quality of the product (Chen et al. 1997). The applications include a wide range of products from general components such as door and window panels to automotive panels, and aircraft components. The products such as helmets, hockey sticks, bicycle frames have been also developed by this technique (Mazumdar 2002). In a comparative study of RTM and compression moulding process, it has been observed that the mechanical properties of Sisal/polyester composites have been found better in case of RTM. This can be attributed to the fact that, in RTM when resin flow advances through the fibers inside the mould then due to hydrostatic pressure it also pushes air out from the mould. Consequently, low voids are contained in the product and improved mechanical properties are obtained (Sreekumar et al. 2007).

4.8.1 Advantages of RTM

- It is suitable for complex shapes.
- High fiber volume (65%) laminates can be obtained with very low void contents.
- Both sides of the component have a moulded surface. Hence, the final product gets a superior surface finish.
- Possible labour reductions.
- The process is very efficient.
- Better reproducibility.
- Relatively low clamping pressure and ability to induce inserts.
- Good health, safety, and environmental control due to enclosure of resin.

4.8.2 Disadvantages of RTM

- It is only suitable for thermosets.
- It is medium volume process and large size of the product cannot be accommodated.
- It involves high tooling cost in comparison to open mould process.
- Reinforcing fiber materials are limited due to the resin saturation of fiber and flow requirement.

4.9 Compression Molding Process

It is well established process for manufacturing thermosetting and thermoplastic composites. This is the first process which was applied to replace metal parts from automotives. This technique is generally applied to manufacture larger as well as flat and moderately curved parts. This process involves two techniques which are compression and flow compression technique. Compression technique applied to manufacture thermoplastic parts whereas flow compression is used for thermosetting parts. These two techniques basically differ by the types of semi-finished product used and its cutting. In case of thermosetting material which contains a temperature initiated catalyst is compressed between heated matched die by a vertical hot or cold press. The curing temperature is kept 80–100 °C for hot pressing and 40–50 °C for cold pressing, for 1–2 h (Mohammad et al. 2018). The uncured composite materials which have high viscosity are placed into the mould area. The temperature of mould area is maintained in the range of 150–160 °C. Due to temperature the viscosity of the thermosetting composite is decreased. After this the mould is closed and a pressure of 100 psi (approximately) is applied on the mould which cause viscous composite to flow and fill cavity of the mould. The mould is opened after a period of time which depends upon the curing time of the composite. For thermoplastic material higher processing temperature and pressure is required to compress the composite pellets. Also, after pressing, the mould remains closed for the period of time which depends upon the time taken by the material to become hard.

The crucial parameters for superior properties and performance of natural fiber composites manufactured by compression molding are moisture removal and mixing (before compression) temperature, holding time and pressure. These factors have been analyzed for their optimum value and found to be interrelated to each other (Jaafar et al. 2019). In an investigation compression molding process found optimum process for manufacturing of natural fiber composites (Zampaloni et al. 2007). Moreover, in comparison to injection molding process the WPCs composites manufactured by compression molding were found to have improved mechanical properties (Liu et al. 2007; Tungjitpornkull and Sombatsompop 2009), Higher specific density and less voids (Ku et al. 2011).

4.9.1 Advantages of Compression Molding Process

- It is suitable for both thermosetting and thermoplastic materials
- Less tooling required in comparison to injection molding
- Low cost
- Complex parts can be produced
- Reproducibility and low cycle time
- Minimum waste material
- High productivity and consistency
- Flexibility

4.9.2 Disadvantages of Compression Molding Process

- Requirement of heat and pressure
- Low volume component
- Only board shape can be made

5 Key Parameters for Successful Processing of NFCs/WPCs (Summerscales and Grove 2014)

Following are the key factors which must be controlled in the manufacturing of the NFCs or WPCs. These factors affect the performance and properties of the final product of composite.

5.1 Moisture Content

Wood or natural fibers have inherent tendency to absorb moisture. This is because of the presence of –OH group and hence they absorb moisture from the atmosphere. To control moisture, it is necessary to maintain relative humidity (RH) of the surrounding of the mould shop as recommended by the manufacturer of the resins. Also, fibers must be kept at dry and clean place. It is recommended to maintain temperature 50 °C and RH < 10% for the production of NFCs or WPCs.

5.2 Thermal Transition Temperature

This factor is essential for selecting an appropriate matrix material for the reinforcement of a cellulose fiber. The important thermal transition temperatures considered for the manufacturing of WPCs or NFCs are Glass transition temperature (Tg), Peak crystalline temperature (Tc), Crystalline melting temperature (Tm) and processing temperature (range) (Tp). Tc and Tm are not applicable for amorphous materials. Generally, Tm≈Tg + 200 ± 50 °C and Tp≈Tm + 30 °C is for partially crystalline thermoplastic materials. Tg is normally similar to the post cure temperature for thermosetting resins. In case of plastics such as PVC, the decomposition temperature (Td) limits Tp. It has been observed that the dry cellulose has Tg ranges from 84 to 250 °C and Tg decreases with increase in moisture content. It has been reported that Tg of cellulose is strongly depend on the degree of crystallinity (Roig et al. 2011). Also, cellulose changes colour at 180 °C and lose approximately 70% weight between 220 to 300 °C (Szcześniak et al. 2008). The decomposition temperature (Td) of cellulose was found at ~200 °C in air. Hence, it is recommended that the Tp of cellulose should be below 200 °C.

5.3 Heat Due to Exothermic Reaction

In case of thermosetting composites heat is generated due to exothermic reaction during the curing. This heat can cause discoloring of resin and also deterioration of mechanical properties of matrix. Also, it can cause thermal damage to natural or wood fibers in case of WPCs and NFCs.

5.4 Evolution of Volatile Materials

In case of thermosetting resins volatile substances are evolved during the curing. This evolution of volatiles in the form of gases causes porosity and voids in the composite. For example styrene evolves from the polyester resin and has been studied by various researchers. Their results are shown in the Table 1.

However, It is also reported that porosity in the composite is also attributed to entry of air due to failure of sealing of the mould. Hence, it is recommended to double seal the mould (Lundstrom et al. 1992, 1993).

5.5 Rheology Behavior of Polymer

Viscosity of polymers is the main governing factor for their rheology behavior. The primary parameters which affect the viscosity of the polymers are reported as temperature and curing process. Various results for change in viscosity with respect to temperature have been published for thermosetting resins in RTM (Potter 1997; Pearce et al. 1998) and Wet lay-up/vacuum bag process (Stringer 1989) and thermoplastic material (Cogswell 1992). For thermosetting resins the curing process

Table 1 Evolution of styrene from polyester resin at different Vapor pressure and temperature

Vapour pressure (absolute vacuum)	Temperature in °C	Reference
10 mbar	40	(Lundstrom et al. 1992, 1993)
6 mbar	20	(Anon 1996)
10.3 mbar	29.9	(Dreyer et al. 1955)
18.27 mbar	39.2	
13.33 mbar	33.4	(Chaiyavech and Winkle 1959)
26.66 mbar	45.6	

includes generation of heat due to exotherm. The degree of cure and viscosity reported to increase with exotherm. The researchers developed model for curing of resins such as epoxy (White 1974), polyester and venyl ester (Kamal and Sourour 1973).

5.6 *Permeability*

In case of manufacturing of WPCs or NFCs the permeability is defined as the ease of flow of a fluid through the pore space in a reinforcement pack. Some mathematical relations have been derived to measure permeability for a reinforcement pack (Griffin et al. 1995; Kozeny 1927; Carman 1937). These relations reveal that flow rate at constant fiber volume fraction is inversely proportional to the square of the wetted perimeter or wetted area. In an investigation on jute fibers it has been found that the permeability increases with increase in porosity this also confirmed an inverse relationship with fiber volume fraction (Francucci et al. 2010). However, permeability measures differ for unsaturated flow and saturated flow for a reinforcement pack.

5.7 *Shrinkage*

Like other materials, polymers also shrink on cooling according to their thermal coefficient of expansion. This parameter is important because many manufacturing techniques take place at elevated temperature. Change in volume due to shrinkage takes place in a thermoplastic system due to the chemical reactions whereas in case of thermoset it is due to the cross linking during curing. The defects such as distortion, matrix cracking and surface defects can occur in case of thermosetting system because of thermal residual stresses arise due to the contraction.

6 Conclusions and Future Perspective

This chapter presents a review on different manufacturing processes of conventional composite materials. These processes have successfully adopted for production of polymer composites with reinforcement of natural or wood fibers. All processes of manufacturing of WPCs have their merits and limitations. The choice of appropriate process for a WPC product mainly depends upon the material type, size and volume of product and also the production rate. Incompatibility of natural fibers with polymers is the primary issue which affects the performance of WPC product. However, there are many physical and chemical treatments of natural fiber have been investigated to improve the performance of WPC product. These treatments and preparation of natural fibers for reinforcement involve a cost of production but it can be compromised with the advantage of sustainability. The present chapter also summarizes the parameters such as moisture content, transition temperature, permeability which must be considered for the manufacturing of WPCs. For the environment concern it is essential to replace the parts made up of conventional material with WPCs in many relevant applications. This requires the modification of the conventional manufacturing processes of composite materials to adopt the production of WPCs.

Manufacturing industries of WPCs aim to produce quality products with high rate and at low cost. Three basic steps as mixing of materials, bonding of materials and heating are involved in the WPC's production. In view of this, extrusion and IM are the most appropriate manufacturing methods for the WPCs. Among these, extrusion is a continuous process and limited to produce linear profile. In this process the manufacturing line, before the extruder need to be refined to feed a constant input of material. Whereas, IM is not a continuous process, due which production rate is comparatively slow. The hybridization of manufacturing process has been suggested to improve the production rate. In case of thermoplastic based WPC, to achieve a uniform distribution of fibers in compounding process is still a critical issue. Adoption of technology of additive manufacturing for the production of WPCs has extended the application areas and it has also provided the solution to limitation of size and complexity of product.

References

Anon (1996) European Organisation of Reinforced Plastics Composite Materials, Brussels, UP-resin handling guide. DuPont Magazine circa 1994

Belgacem MN, Bataille P, Sapieha S (1994) Effect of corona modification on the mechanical properties of polypropylene/cellulose composites. J Appl Polym Sci 53:379–385

Biswal T, Badjena SK, Pradhan D (2020) Synthesis of polymer composite materials and their biomedical applications. Mater Today: Proc. https://doi.org/10.1016/j.matpr.2020.01.567

Bledzki AK, Gassan J (1999) Composites reinforced with cellulose based fibers. Prog Polym Sci 24:221–274

Carman PC (1937) Fluid flow through a granular bed. Trans Inst Chem Eng (London) 15:150–166

Chaiyavech P, van Winkle M (1959) Styrene-ethylbenzene vapor-liquid equilibria at reduced pressures. J Chem Eng Data 4(1):53–56

Chen YF, Stelson KA, Voller VR (1997) Prediction of filling time and vent locations for resin transfer mould. J Compos Mater 37(11):1141–1161

Cogswell FN (1992) Thermoplastic Aromatic Polymer Composites. Butterworth-Heinemann, Oxford. ISBN 0-7506-1986-7. Now Woodhead Publishing, Cambridge. ISBN 978-1-85573-086-1

Dreyer R, Martin W, Weber U von (1955) Die s ä ttigungsdampfdrucke von benzol, toluol, ä thylbenzol, styrol, cumol und brombenzol zwischen 10 und 760 torr.Journal für Praktische Chemie. 1(5–6):324–328

Felix JM, Gatenholm P (1991) The nature of adhesion in composites of modified cellulose fibers and polypropylene. J Appl Polym Sci 42(3):609–620

Francucci G, Rodríguez ES, Vázquez A (2010) Study of saturated and unsaturated permeability in natural fiber fabrics. Compos a Appl Sci Manuf 41(1):16–21

Griffin PR, Grove SM, Russell P, Short D, Summerscales J, Guild FJ, Taylor E (1995) The effect of reinforcement architecture on the long-range flow in fibrous reinforcements. Compos Manuf 6(3–4):221–235

Jaafar J, Siregar JP, Tezara C (2019) A review of important considerations in the compression molding process of short natural fiber composites. Int J Adv Manuf Technol 105:3437–3450

Kabir MM, Wang H, Lau KT, Cardona F (2012) Chemical treatments on plant based natural fiber reinforced polymer composites: an overview. Compos B Eng 43(7):2883–2892

Kamal MR, Sourour S (1973) Kinetics and thermal characterization of thermoset cure. Polym Eng Sci 13(1):59–64

Kozeny J (1927) Uber die kapillare Leitung des Wassers in Boden, Sitzungsberichte Akademie der Wissenschaft Wien Math-naturw. 136(kl.Abt.IIa):271–306

Ku H, Wang H, Pattarachaiyakoop N, Trada M (2011) A review on the tensile properties of natural fiber reinforced polymer composites. Compos B Eng 42(4):856–873

Le Baillif M, Oksman K (2009) The effect of processing on fiber dispersion, fiber length, and thermal degradation of bleached sulfite cellulose fiber polypropylene composites. J Thermoplast Compos Mater 22(2):115–133

Lilholt H, Madsen B (2012) Properties of flax and hemp composites. In: Reux F, Verpoest I (eds) Flax and Hemp Fibers: A Natural Solution to the Composite Industry. JEC Composites, Paris, pp 119–140

Liu W, Drzal LT, Mohanty AK, Misra M (2007) Influence of processing methods and fiber length on physical properties of kenaf fiber reinforced soy based biocomposites. Compos B Eng 38 (3):352–359

Lundstrom TS, Gebart BR, Lundemo CY (1992) Void formation in RTM. In: Proceedings of the 47th annual conference, composites institute. The Society of the Plastics Industry, pp 1–5

Lundstrom TS, Gebart BR, Lundemo CY (1993) Void formation in RTM. J Reinf Plast Compos 12:1339–1349

Maurus PB, Kaeding CC (2004) Bioabsorbable implant material review. Operative Tech Sports Med 12(3):158–160

Mayer RM (1993) Design with reinforced plastics: a guide for engineers and designers. Design Council, London. Springer. ISBN 978-94-011-2210-8

Mazumdar SK (2002) Composites Manufacturing: Materials, Product, and Process Engineering. CRC Press, Boca Raton

Mohammad RMJ, Mohammad SAM, Khasri A (2018) Natural lightweight hybrid composites for aircraft structural applications. In: Sustainable composites for aerospace applications. Series in composites science and engineering. Woodhead Publishing, pp 155–170

Pearce NRL, Guild FJ, Summerscales J (1998) An investigation into the effects of fabric architecture on the processing and properties of fiber reinforced composites produced by resin transfer moulding. Compos A Appl Sci Manuf A29(1):19–27

Potter K (1997) Resin Transfer Moulding. Chapman & Hall, London. ISBN 0-412-72570-3

Rakesh K, Veena C, Saroj M, Varma IK (2008) Banana fiber-reinforced biodegradable soy protein composites. Front Chem China 3(3):243–250

Roig F, Dantras E, Dandurand J, Lacabanne C (2011) Influence of hydrogen bonds on glass transition and dielectric relaxations of cellulose. J Phys D: Appl Phys 44(4):045403

Salit MS (2014) Manufacturing techniques of tropical natural fiber composites. In: Tropical natural fiber composites. Engineering materials. Springer, Singapore

Schwendemann D (2008) Manufacturing technologies for wood-polymer composites. In: Oksman K, Sain M (eds) Wood-polymer-composites. Woodhead Publishing and Maney Publishing, Boca Raton, pp 72–100

Senawi R (2010) Innovative Pultrusion Sdn Bhd. No. 43, jalan sj 8/1, taman seremban jaya, 70450, negeri sembilan, malaysia, 70450, Malaysia

Sreekumar PA, Kuruvilla J, Unnikrishnan G, Sabu T (2007) A comparative study on mechanical properties of sisal-leaf fiber-reinforced polyester composites prepared by resin transfer and compression moulding techniques. Compos Sci Technol 67:453–461

Stringer LG (1989) Optimization of the wet lay-up/vacuum bag process for the fabrication of carbon fiber epoxy composites with high fiber fraction and low void content. Composites 20 (5):441–452

Summerscales J, Grove S (2014) Manufacturing methods for natural fiber composites. In: Natural fiber composites: materials, processes and applications, pp 176–215. https://doi.org/10.1533/9780857099228.2.176

Sykacek E, Hrabalova M, Frech H, Mundigler N (2009) Extrusion of five biopolymers reinforced with increasing wood flour concentration on a production machine, injection moulding and mechanical performance. Compos A Appl Sci Manuf 40(8):1272–1282

Szcześniak L, Rachocki A, Tritt-Goc J (2008) Glass transition temperature and thermal decomposition of cellulose powder. Cellulose 15(3):445–451

Tenney DR, Davis JG, Pipes BR, Johnston N (2009) NASA composite materials development: lessons learned and future challenges. In: NATO RTO AVT-164 Workshop on Support of Composite Systems, Bonn, Germany, 19 October 2009–22 October 2009

Tungjitpornkull S, Sombatsompop N (2009) Processing technique and fiber orientation angle affecting the mechanical properties of E-glass fiber reinforced wood/PVC composites. J Mater Proc Technol 209(6):3079–3088

White RP (1974) Time-temperature superpositioning of viscosity-time profiles of three high temperature epoxy resins. Polym Eng Sci 14(1):50–57

Wolf O, Crank M, Patel M, Marscheider-Weidemann F, Schleich J, AngereBH (2005) Techno-economic feasibility of large-scale production of bio-based polymers in Europe. European Commission Director-General Joint Research Centre Technical report EUR 22103 EN. ISBN 92-79-01230-4

Wretfors C, Hedenqvist MS, Marttila S, Nimmermark S, Johansson E (2009) Use of industrial hemp fibers to reinforce wheat gluten plastics. J Polym Environ 17:259–266

Zampaloni M, Pourboghrat F, Yankovich S, Rodgers B, Moore J, Drzal L, Mohanty A, Misra M (2007) Kenaf natural fiber reinforced polypropylene composites: a discussion on manufacturing problems and solutions. Compos A Appl Sci Manuf 38(6):1569–1580

Wood Treatments and Interfacial Bonding in Wood-Plastic Composites

Matheus de Prá Andrade and Matheus Poletto

Abstract This chapter presents an overview of different physical, chemical and biological methods used to improve the mechanical and thermal properties of wood and wood plastic composites. The physical methods addressed contemplate plasma, corona, ultraviolet (UV) radiation, ultrasound, heat treatments, fibre beating and electron radiation. The chemical methods discussed in this work were alkali, benzyl, acetyl, acryl, silane, permanganate, among others. Biological treatments, such as enzyme and fungal treatment were also reviewed. Methods used to impregnation and chemical modification of wood using sustainable reagents were also presented. The furfurylation of wood, the impregnation and polymerization of lactic acid directly in wood structure and the usage of citric acid are also discussed as natural compounds that can be used to promote wood chemical modification. The usage of coupling agents or compatibilizers from renewable and non-renewable sources and their impact on the thermo-mechanical properties of the wood-plastic composites will be discussed. Moreover, the chemistry and the mechanism of compatibility between the wood and polymer matrix will be properly evaluated.

Keywords Composites · Polymer · Wood · Interface · Mechanical properties

1 Introduction

The growing interest in wood-plastic composites for several industrial applications may be explained due to the potential of these composites to replace synthetic fibers reinforced plastics. Wood fibers can be obtained with low cost, when compared to glass fibers, and also comes from a renewable source which contributes to improve

M. de Prá Andrade · M. Poletto (✉)
Postgraduate Program in Engineering of Processes and Technologies (PGEPROTEC), University of Caxias do Sul (UCS), Francisco Getúlio Vargas, 1130, Caxias do Sul 95070-560, Brazil
e-mail: mpolett1@ucs.br

S. Mavinkere Rangappa et al. (eds.), *Wood Polymer Composites*,
Composites Science and Technology, https://doi.org/10.1007/978-981-16-1606-8_3

composite sustainability. The main advantages and disadvantages of wood-plastic composites are summarised in Table 1.

Table 1 Main advantages and disadvantages of wood-plastic composites

Advantages	Disadvantages
Wood comes from renewable source	Lower durability than inorganic fiber composites
During growth wood remove CO_2 from the environment and return O_2	Greater variability of properties due to wood heterogeneity
Low density, high specific strength and stiffness	High moisture absorption
Wood fibers have lower cost than inorganic fibers	Lower impact strength when compared to synthetic fibers
Low hazard manufacturing process	Lower processing temperatures
Less abrasive damage during processing when compared to inorganic fibers	Lower interfacial adhesion with several polymers
Wood is biodegradable	Wood need to be previous treated to avoid fungi attack and UV degradation
Low emission of toxic fumes during incineration at end of life	

The influence of chemical and physical treatments on mechanical and thermal properties of wood and its composites with thermoplastic polymers will be the bases for the further discussions done in this work.

2 Interfacial Adhesion in Wood-Plastic Composites

The performance of a wood-plastic composite is intrinsically associated with the interfacial adhesion between polymer and fiber. Higher interfacial adhesion results in composites with higher mechanical properties since the strength can be transferred from the matrix to the reinforcement fibers. On the other hand, composites with higher thermal properties can be also obtained when the interfacial bounding is achieved, because the fibers are covered by the matrix and probably will initiate a thermal degradation in a higher temperature range, when compared with composites without interfacial adhesion. Thus, interfacial bonding between polymer matrix and wood fiber plays a vital role, determining the mechanical and thermal properties of wood-plastic composites (Pickering et al. 2016; Elamin et al. 2020). However, wood-plastic composites normally present weak interfacial bonding due to the limited interactions that occurs between hydrophilic wood fibers and hydrophobic polymer matrix. As a result, the mechanical and thermal performance of the composite material is reduced. When fiber and matrix are put into intimate contact,

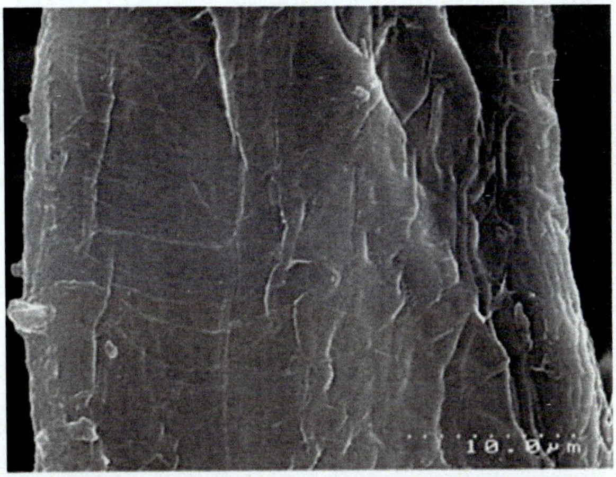

Fig. 1 Surface morphology of a rough wood fiber (Reproduced with permission (Pickering et al. 2016))

the wettability can be improved, and bonding can be obtained. However, insufficient fiber wetting can generate interfacial defects, such as voids and gaps between filler and matrix, which can act as stress concentrators. To avoid poor wettability chemical and physical treatments are used. These treatments are responsible to improve the compatibility between fiber and matrix and also cause an improvement in tensile and flexural composite properties (Mohit and Arul Mozhi Selvan 2018; Rezaee Niaraki and Krause 2020).

According to Matthews and Rawlings (1999), interfacial bonding can be promoted by means of inter-diffusion bonding, chemical bonding, electrostatic bonding and mechanical interlocking. When the wood fiber surface is rough the mechanical interlocking may be favoured. The fiber roughness contributes to generate sites to polymer deposition, as can be seen in Fig. 1. As a consequence of mechanical interlocking, the interfacial shear strength increases.

The attraction between opposite charges is responsible to create electrostatic bonding. This type of bonding is not so common in wood-plastic composites. However, when metallic interfaces are involved the electrostatic bonding has a significant influence. The mechanism that involves chemical bonding is associated with chemical groups presented in the matrix and in the fiber surface that may react to form chemical bonds, in this case the interfacial strength is deeply associated with the type and density of the bonds formed (Pickering et al. 2016). Generally, to promote chemical bonding in wood-plastic composites coupling agents are used. This type of chemical agent is normally composed by a carbonic chain that may interact with the polymer matrix and a polar group that may form chemical bonds with the hydroxyl groups presented in the wood surface. As previous reported (Poletto et al. 2011, 2014; Liu et al. 2016; Quiles-Carrillo et al. 2018), coupling agents acts as bridges between filler and matrix.

When the atoms and molecules presented on matrix and fiber interact at the interface the inter-diffusion bonding occurs. The entanglements involved the polymer chains at the interface is the main mechanism observed for polymer composites. The number of chains per unit area, the degree of entanglement and length of chains are the main factors that influence the inter-diffusion bonding in polymeric interfaces (Pickering et al. 2016). The type of bonding formed is mainly associated with the materials used in composite development. In some occasions, several types of bonding can occurs at the interface at the same time (Pickering et al. 2016).

2.1 *Wood Physical Treatments to Improve Interfacial Bonding in Composites*

The physical methods used to treat wood and other cellulosic and lignocellulosic fibers may include plasma, corona, ultraviolet (UV) radiation, ultrasound, heat treatments, fibre beating and electron radiation. Plasma and corona treatments are similar in some aspects. During plasma treatment gas is continuously supplied to a vacuum chamber to maintain the pressure and gas composition adopted for fiber modification (Pickering et al. 2016; Mantia and Morreale 2011; Koohestani et al. 2019). On the other hand, corona treatment also uses plasma, but generated by the application of a high voltage to sharp electrode tips separated by quartz at low temperature and atmospheric pressure (Pickering et al. 2016; Koohestani et al. 2019). The corona treatment is normally adopted to species that contain oxygen in its structure, such as wood and others natural fibers. Both treatments increase the fiber surface roughness which contribute to mechanical interlocking between fiber and matrix and improves the interfacial adhesion. However, plasma treatment commonly generates hydrophobicity at fiber surface, while corona treatment increases surface polarity due the formation of carboxyl and hydroxyl groups at fiber surface. As previously discussed, both methods presented some similarities, but one point should be highlighted. While plasma treatment increases the fiber hydrophobic character the corona treatment acts in the opposite way and reduces the material hydrophobicity.

Several studies used corona, plasma and both treatments to change the properties of natural fibers (Liu et al. 2010; Acda et al. 2012; Podgorski et al. 2000; Fazeli et al. 2019; de Almeida Mesquita et al. 2017; Islam et al. 2017; Ragoubi et al. 2010). Dong et al. (1993), used corona treatment to modify a hardwood fiber and after prepared a composite with linear low-density polyethylene (LLDPE). The authors observed that the treatment affect the fiber polarity but reduces the composite strength probably due to the poor adhesion at polymer/fiber interface. When the LLDPE was treated by corona discharge, the authors observed that the compatibility between filler and matrix improved and the mechanical properties increased. Islam et al. (2017) combined plasma and corona treatments by means of

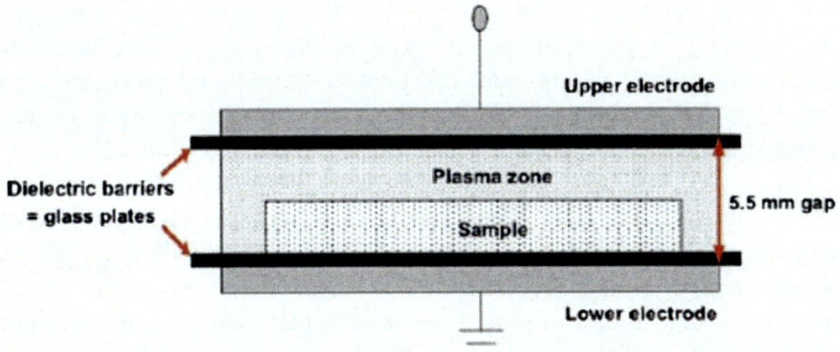

Fig. 2 Corona discharge system (Reproduced with permission (Ragoubi et al. 2010))

atmospheric pressure corona-based weakly ionized plasma to modify the surface of pine wood. Figure 2 shows the principle of corona discharge system.

The authors observed that treated wood contain on their surface nano and micro-nodules of plasma-polymerized acetylene which can interact with polymer matrix and improve the fiber matrix adhesion in composite materials. According to the authors corona-based plasma method has potential to become cost-effective when compared with others commercially plasma systems. Cademartori et al. (2017) also used atmospheric pressure cold plasma to deposit a fluorocarbon film on the surface of white spruce and Brazilian cedar wood. The authors observed that fluorocarbon deposition increases the wood roughness and also creates a hydrophobic surface for both wood species evaluated. Liu et al. (2010) used air plasma to treated wood/polyethylene composites. The results showed that the polar groups, such as hydroxyl, carbonyl and carboxyl, formed at the composites surface improved the composite adhesion. The shear bonding strength test showed that the adhesion properties of the composites studied improved effectively after plasma treatment.

The UV radiation generally promotes the depolymerization of some wood components, such as lignin and in minor extension the extractives may also be affected by UV light. Todaro et al. (2015) observed that UV treatment causes irreversible changes in chemical composition and morphological aspects of Turkey oak wood after 4000 min exposed to UV-C irradiation. However, in short time of exposition to UV treatment wood presented the formation of carbonyl groups at the surface, probably associated with lignin depolymerization. Poletto (2017) also observed that UV-B irradiation causes severe lignin degradation after exposure *Eucalyptus grandis* and *Dipteryx odorata* wood species to 6000 min of UV treatment.

When wood is exposed to heat treatment, the fibers are heated to temperatures close to promote fiber degradation (Pickering et al. 2016; Tomak et al. 2018). During treatment chemical, physical and mechanical wood properties can be affected. As a result, fiber morphology and strength, cellulose crystallinity and depolymerisation of wood components can occurs. The effect of heat treatment, also occurs for corona, plasma and UV treatments, is based on temperature, time and type of the gases used during the treatment (Koohestani et al. 2019; Venkatachalam et al. 2016). Yin et al. (2011) submitted spruce wood to steam treatment at 140, 160 and 180 °C during a total time of 120 min. The authors observed a decrease of indentation modulus and hardness of wood with the temperature increases which is associated with the hemicelluloses degradation. Hosseinihashemi et al. (2016) used saturated steam to treat beech wood at 120, 150 and 180 °C during 30 or 120 min. After the thermal treatment, wood was used in the development of polypropylene composites. The composites produced with treated wood at temperature of 180 °C for 120 min presented the lowest water absorption. Aydemir et al. (2015) observed that exposing wood to 212 °C for 8 h, was capable to improve the thermal stability and crystallinity of pine and maple wood species. The authors also observed that flexural modulus of elasticity increased by 101 and 82% with the addition of 30 wt% of heat-treated pine and 20 wt% of heat treat maple in wood-nylon 6 composites.

Electron radiation basically consists in treating the sample with a radiation beam forming radicals in the sample. Based on the irradiation dose, two major reaction are possible: degradation and crosslinking (Huber et al. 2010). Which reaction is dominant depends on several factors: type of material (type of polymer and/or type of fiber), additives, temperature, pressure, dose, dose rate, morphology, crystallinity and the surface/volume ratio (Huber et al. 2010). Huber et al. (2010) observed that electron radiation can improve interfacial bonding due to the formation of crosslinking between polypropylene and the radicals formed at the ramie fiber surface. Takinami et al. (2010) also verified a similar behavior when gelatin/piassava composites were treated by electron beam radiation.

Fiber beating is a mechanical treatment used in defibrillation and consequently increases the fiber surface area. The treatment causes generally three main effects on fibers (Ahmad et al. 2019; Chen et al. 2012). The first is internal fibrillation that increases the flexibility of fibers by the breakdown of fiber walls into separate lamellae (Ahmad et al. 2019). The second is external fibrillation described as the creation and/or exposure of fibrils on the surface of the fibers (Ahmad et al. 2019). The last one is the generation of fines from fibers when they are no longer (Ahmad et al. 2019). Beg and Pickering (Beg and Pickering 2008) verified that fiber beating causes a 10% increase in strength of Kraft fiber reinforced polypropylene composites due the fiber defibrillation and increase of surface area, which can result an improve of mechanical interlocking.

2.2 *Wood Chemical Treatments to Improve Interfacial Bonding in Composites*

The chemical methods adopted to modify the properties of wood and others natural fiber includes alkali, benzyl, acetyl, acryl, silane, permanganate, isocyanate, peroxide, zirconate, titanat, acrylonitrile, etherification and epoxi treatments. The usage of maleated anhydride grafted coupling agents is also very common.

The chemical compounds used to provide the chemical fiber modification normally acts improving the surface and mechanical characteristics of natural fibers. Some of them can significantly improve the mechanical properties of lignocellulosic fibers just with the removal of impurities and changing the fiber crystalline structure. The most common treatments are alkali (mercerization), silane, acetyl and graft polymerization. These most popular chemical treatments will be briefly commented in this section.

The alkali treatment consists in the removal of oils, wax, pectin and other extractives from the fiber surface together with certain amounts of hemicelluloses and lignin using sodium hydroxide. The alkaline treatment increases surface fiber roughness and area, which contribute to improve the interfacial adhesion, but exposes the cellulose fibers and can also affect the cellulose crystallinity. Several studies related improvements in mechanical properties of composites when alkali treated fibers were used (Saiful Islam et al. 2012; Amroune et al. 2015; Essabir et al. 2015; Gwon et al. 2010; Jiang et al. 2018). The increase in composite thermal stability was also reported in the literature (Rosa et al. 2009; Saini et al. 2009).

Silane treatment has been extensively used to promote interfacial bonding in natural fiber composite material (Xie et al. 2010a; Goriparthi et al. 2012; Liu et al. 2019). The silane molecule has bifunctional groups in its structure which generates the possibility of react with the hydroxyl groups presented in the fiber surface forming hydrogen bonds and also interact with the polymer chains of the matrix phase. The reaction with wood fibers involves hydrolysis of alkoxy groups on silane with water to form silanol groups that can after react with the hydroxyl groups on the wood fiber surface, as can be seen in Fig. 3.

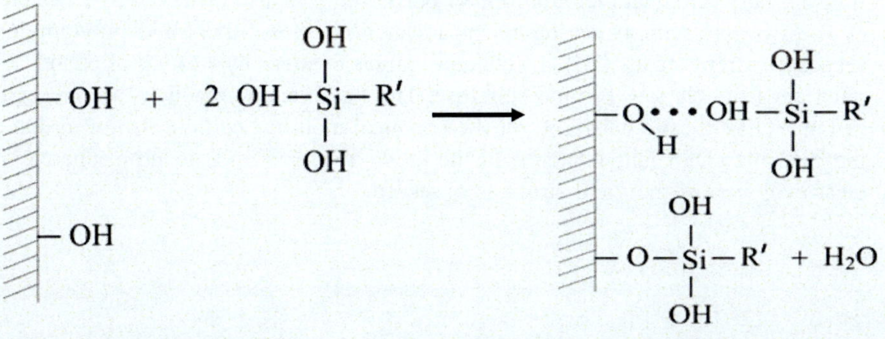

Fig. 3 Schematic representation of reaction of silane and wood fiber (Reproduced with permission (Pickering et al. 2016))

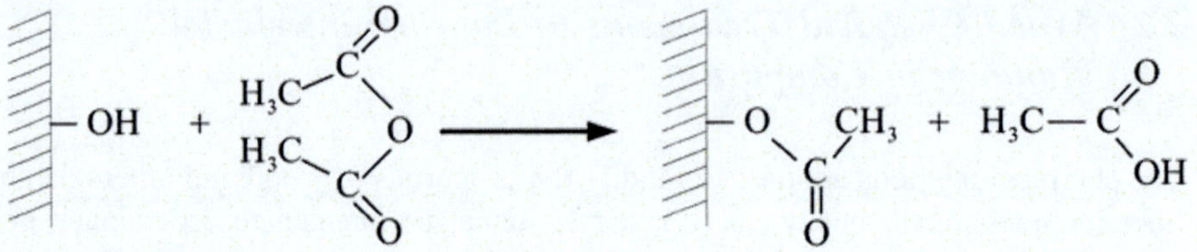

Fig. 4 Reaction of acetic anhydride with hydroxyl groups of wood fiber (Reproduced with permission (Pickering et al. 2016))

Reddy and Simon (2010) performed a surface wood functionalization using bi-functional organo-silane 7-octenyldimethylchlorosilane and mono-functional organo-silane n-octyldimethylchlorosilane. The modified wood fiber with organo-silane were used as reinforcement in polypropylene composites. The authors observed that vinyl functionality in wood promotes good fiber coverage with polypropylene after in situ polymerization.

The acetylation treatment is used to increase the hydrophobicity at the wood fiber surface (Tserki et al. 2005; Hung et al. 2016). The esterification reaction occurs by the reaction of acetyl groups with hydroxyl groups on the wood fiber surface, as shown in Fig. 4.

According to the literature (Tserki et al. 2005; Bledzki et al. 2008; Ali et al. 2016) the composites developed with treated fibers presented an improvement in interfacial bonding which results in composites with better mechanical properties when compared with composites prepared with non-treated fibers. In addition, composite thermal stability and resistance to fungal attack also increases (Tserki et al. 2005; Bledzki et al. 2008; Ali et al. 2016), probably due to the better fiber coating by the polymer matrix.

Freire et al. (2008) filled low-density polyethylene with unmodified and fatty acid (hexonic, dodecanoic, octadecanoic and docosanoic acids) esterified cellulose. The authors observed that the chemical modification proposed to cellulose fibers improved the interfacial bonding with polymer matrix. As result, the composite mechanical properties increased and water absorption decreased. The authors concluded that the composite performance was affected by the degree of substitution and fatty acid chain length. Similar behaviour was also observed by Fernandes et al. (2011) in composites based on acylated cellulose fibers and poly(ethylene sebacate). Crépy et al. (2011) obtained cellulose esters by linking aliphatic acid chlorides from C8 to C18 onto cellulose. The esterification reactions are presented in Fig. 5. The tensile behaviour of the chemical modified cellulose showed that as the carbonic chain length increased the strain at break tends to drop whereas no clear evolution of the yield stress as observed.

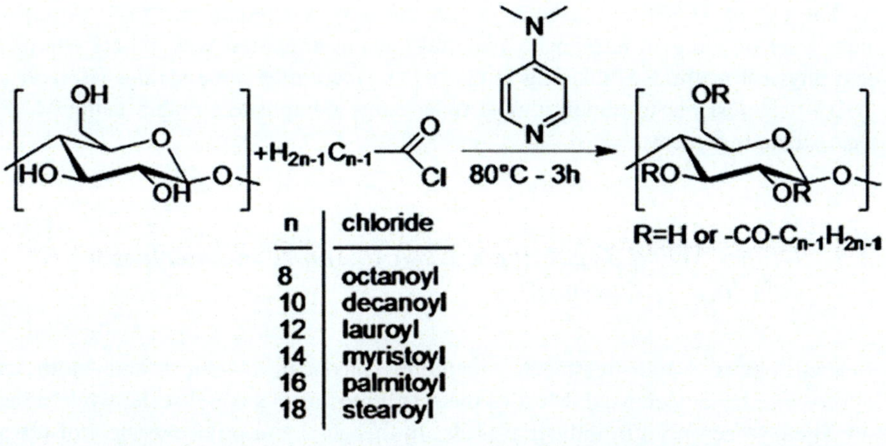

Fig. 5 Esterification of cellulose using aliphatic acid chlorides (Reproduced with permission (Crépy et al. 2011))

Maleated coupling agents are widely used to improve composite properties. Maleic anhydride is commonly grafted to the polymer, who is used as composite matrix, ensuring the compatibility between the polymer matrix and coupling agent. The most commonly maleated coupling agent is polypropylene grafted with maleic anhydride (MAPP). This coupling agent acts as a bridge between filler and matrix. MAPP can react with the hydroxyl groups at the wood fiber surface forming hydrogen or covalent bond. On the other hand, the PP chains from the matrix may form entanglements with the polymer chains from the MAPP which result in higher interfacial bonding. A representation of the interactions between MAPP and wood fibers in a polymer composite is illustrated in Fig. 6.

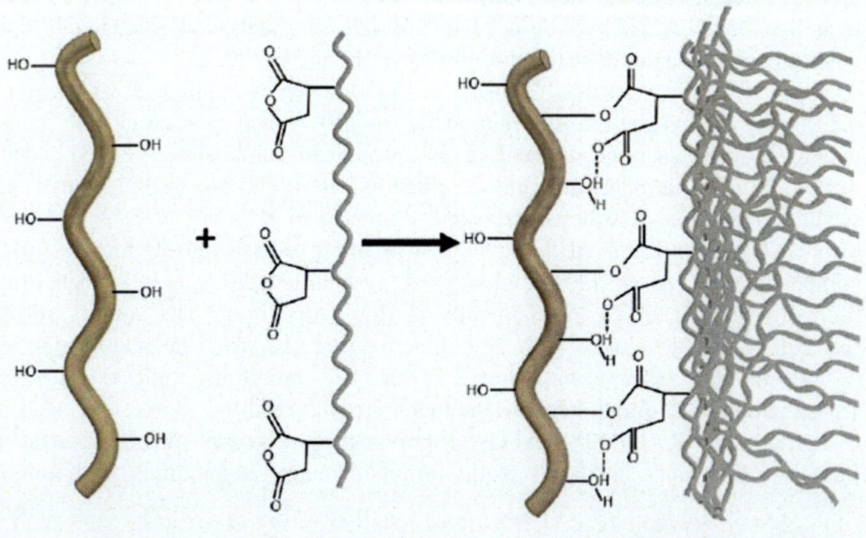

Fig. 6 Interfacial bonding in wood polymer composite using MAPP as coupling agent. (Reproduced with permission (Belgacem et al. 2020))

The use of MAPP could be regarded the most successful method to improve interfacial bonding in composite materials when compared with others chemical and physical methods (Pickering et al. 2016). According some studies (Bera et al. 2010) it is possible to obtain almost twice more composite strength using MAPP than silane treatment.

2.3 Others Wood Treatments Used to Improve Interfacial Bonding in Composites

Several others wood treatments, including physical, chemical and biological approaches have been used. Most of them still remain in a small scale, while others are already been used in industrial scale. In this section a brief overview of others wood treatments will be explored.

Biological treatments, such as enzyme treatment, are increasing due to their environmental friendliness. The treatment consists in using enzymes for selectively removing hydrophilic components to reduce the hydrophilicity of cellulose fibers. The microorganisms and bacteria are used to degrade the main linkages between lignin and cellulose or lignin and hemicelluloses (Zadeh et al. 2020). However, some drawbacks of using biological treatments, such as incubation time and overall efficiency require improvement to make this process an alternative to others chemical and physic methods (Vasco-Correa et al. 2016).

Fungal treatment is used to remove noncellulosic components, such as lignin, wax or pectin from the lignocellulosic fiber surface (Ferreira et al. 2019). Gulati and Sain (2006) observed that after fungi (*Ophiostoma ulmi*) treatment the unsaturated polyester resin composites reinforced with treated hemp fibers presented an increase of 22% in flexural strength when compared to the composites obtained with untreated fibers. The addition of bacterial cellulose onto lignocellulosic fiber surface is another method that can be used to improve the interfacial adhesion in composites based on natural fibers and polymers (Ferreira et al. 2019).

Xie et al. (2010b) used glutaraldehyde and 1,3-dimethylol-4,5-dihydroxyethyleneurea to chemically modify wood particles. The treated wood was used as reinforcement in polypropylene composites. The composites treated with glutaraldehyde and 1,3-dimethylol-4,5-dihydroxyethyleneurea presented a reduction in water uptake and dimensional swelling of 39% and 46%, respectively. Treatments of wood flour also improved the tensile strength of the composites. Ou et al. (2014) also used glutaraldehyde and 1,3-dimethylol-4, 5-dihydroxyethyleneurea to chemically modify wood flour. The authors verified that composites developed with glutaraldehyde treated wood decreased melt viscosity, moduli, and shear stress, while no modifying effects of using 1,3-dimethylol-4,5-dihydroxyethyleneurea were observed.

Spiridon et al. (2012) developed polypropylene composites reinforced by toluene diisocyanate modified wood. Spruce wood was modified with toluene

2,4-diisocyanate (TDI). The addition of 15 wt% of treated wood causes an increase of 19% in composite tensile strength. The wood chemical modification was observed by the formation of a urethane bond between wood hydroxyl groups and phenyl isocyanate, evidenced by FTIR analysis, according to the reaction:

$$\mathrm{Wood} - \mathrm{OH} + \mathrm{R} - \mathrm{N} = \mathrm{C} = \mathrm{O} \rightarrow \mathrm{Wood} - \mathrm{O} - \mathrm{C}(=\mathrm{O}) - \mathrm{NH} - \mathrm{R}$$

Islam et al. (2014) impregnated wood with acrylonitrile monomer solution thought vacuum-pressure method. The wood polymer composites developed with modified wood showed an increase of 23% in compressive modulus compared to the composites reinforced by raw wood, followed by a reduction in water absorption due the better interfacial bonding.

Xu et al. (2014) prepared wood flour/chitosan/poly(vinyl chloride) composites using a three step modification process. Firstly, a vacuum-pressure treatment of sodium montmorillonite was done, followed by inner intercalation replacement of organically modified montmorillonite, and finally surface grafting of glycidyl methacrylate. The results reveal that flexural and tensile strengths were increased by 11 and 12%, respectively, while water absorption rate was decreased by approximately 62%, due the better in interfacial compatibility between polymer and matrix.

Kord et al. (2020) used ionic liquids to improve interfacial adhesion in wood/high density polyethylene composites. Wood flour was pretreated with two ionic liquids, named 1-(3-trimethoxysilylpropyl)-3-methylimidazolium chloride and 1-(3-trimethoxysilylpropyl)-3-methylimidazolium thiocyanate. The results indicate strong interfacial adhesion between filler and matrix after wood pretreatment. A remarkable increase in composite thermal stability was observed when the ionic liquids were used. The temperature associated with 10% of weight loss has an increase at almost 30 °C, when compared to the composite without the addition of ionic liquids.

3 Wood Impregnation and Chemical Modification Using Sustainable Reagents

The environmental impact associated with the wood modification is the driving force to the development of green modified wood (Dong et al. 2020). The usage of renewable agents to promote wood impregnation and chemical modification has been explored in research and industrial fields. Natural products, biological-based chemical materials and biomass-derived compounds are generally used as renewable agents (Dong et al. 2020; Colmenares et al. 2016). To develop a sustainable wood modification process three main aspects should be observed: use of renewable modifying agents, environmentally friendly processes and nontoxic and biodegradable resultant products (Dong et al. 2020). In this context, this section contemplates some friendly process used to wood modification.

The furfurylation of wood is a process that uses furfuryl alcohol (FA) solutions with the addition of a catalyst at elevated temperature to promote the polymerization of FA (Dong et al. 2020; Mantanis 2017). The hydrolysis of pentosan-rich biomass, such as agricultural residues and wood results in the formation of furfural and after chemical conversion process FA is obtained (Gong et al. 2017). Some studies reveal that lignin present in wood could form covalent bonds with poly (furfuryl alcohol) (PFA) (Gong et al. 2017; Nordstierna et al. 2008). Barsberg and Thygesen (2017) observed that condensation of FA with lignin formed a lignin-PFA copolymer localized especially in the lignin rich areas of the wood cell wall. However, furfurylated wood could be degraded by oxidative and enzymatic degradation (Gong et al. 2017). In addition, the formation of a stiff cross-linked PFA in wood structure promotes the embrittlement of wood, resulting in a decrease in its impact strength and modulus of rupture (Gong et al. 2017; Xie et al. 2013). Dong et al. (2015) used nano-SiO_2 mixed with FA water solution to impregnate poplar wood. The results indicated that the nanoparticles can improve the modulus of rupture of furfurylated wood.

The furfurylation is an attractive method to chemically modify wood trough a sustainable process. The drawback is the embrittlement of wood caused by furfurylation. However, controlling the polymerization of FA in the wood structure combined with the use of reinforcements, such as nanoparticles could be an effective and environmental friendly method to chemical modified wood (Gong et al. 2017).

In general, the impregnation of wood with natural compounds extracted from lignocellulosic materials has become a more attractive method to wood modification (Gong et al. 2017; Broda 2020). Compounds present in wood, such as, oils, extractives, resins and waxes can act as wood modifiers. The water repellence of wood and also better interactions with hydrophobic polymers may be obtained using beeswax, paraffin, montan and carnauba waxes (Gong et al. 2017).

The wood impregnation with biodegradable polymers is another method to develop sustainable wood polymer composites. The impregnation and polymerization of lactic acid directly in wood structure is an example to produce wood composite based on sustainable reagents and also in sustainable processes. Noël et al. (Noël et al. 2009) impregnated wood with lactic acid and observed that the hardness of wood was improved without strongly affect the composite mechanical properties. Other types of polyesters were also used for wood impregnation such as polybutylene adipate (PBA), polyglycolic acid (PGA) and polybutylene succinate (PBS) (Gong et al. 2017).

Citric acid is another natural compound that can be used to promote wood chemical modification. This organic acid is crosslinking with wood by means of two esterification steps in the presence of a catalyst (Gong et al. 2017). As previously reported, the modification of wood with citric acid increase wood dimensional stability (Gong et al. 2017), resistance against fungi (Despot et al. 2008) and compression strength (Šefc et al. 2012).

The usage of natural compounds to promote wood chemical modification exhibits great potential to replace the non-renewable compounds. However, more studies must be done to better understand the chemical reactions and the mechanism that can generates a greatly improve in wood properties.

4 Wood Composites Developed with Coupling Agents and Compatibilizers from Renewable Sources

The reduction of fossil fuels associated with environmental concerns is increasingly driving the development of materials from renewable sources. The usage of polymers obtained from biomass has become a reality in composite industry. However, the search for additives from renewable sources that can replace the compounds derived from petroleum remains a challenge. This section will present some of these studies that use natural compounds as coupling agents or compatibilizers in natural fiber composites.

The usage of coupling agents or compatibilizers from natural sources has increased in the last decade. Several natural oils and compounds chemically modified or not have been used to promote interfacial bonding in composites reinforced by natural fibers (Liminana et al. 2018; Poletto 2018; Castro et al. 2017; Husseinsyah et al. 2016). Some studies also related that these compounds may act as flow agents, due to this low molar mass, helping during the composite processing by extrusion and injection moulding (Castro et al. 2017; Poletto 2019). The main components of vegetable oils are triglycerides. These compounds are renewable, abundant when vastly investigated alternative feedstock for polymeric materials (Zhang et al. 2017; Sawpan 2018). These natural oils are used to develop polymer and monomers to be used as an alternative to the petrochemical resources.

Tran et al. (2006) developed biocomposites using epoxidized and maleated soybean oil as polymer matrixes in composites reinforced by natural fibers. The authors observed that composites tensile strength greatly enhanced when soy-based resins were used. Dai et al. (2014) used epoxidized soybean oil to improve the toughness of microcrystalline cellulose reinforced polylactide acid composites. When 10 wt% of epoxidized soybean oil were added to the composites an increase of 38.5% in composite elongation was observed. The oil chemical structure promoted a plasticizing effect which also resulted in a slightly decrease in the composite Tg. Meng et al. (2018) also used epoxidized soybean oil to improve toughness of PLA/nanocellulose composites. The authors observed that the addition of epoxidized soybean oil can improve the ductility of the nanocomposites until tenfold with a slight decreased in mechanical properties.

Castro et al. (2017) reinforced high density biopolyethylene with curauá fibers. The authors tested castor and canola oils as potential compatibilizers in the composite. The results indicated that the addition of oils improved the mechanical properties of the composites and also lead an improvement in composite thermal stability. Castor oils presented the best results when compared with canola oils. The authors also concluded that the mechanical properties of the composites developed with natural oils can be modified by the viscosity of the oils chosen, as well as from the level of incorporation of the oil in composite to balance the action of the oils as plasticizer and compatibilizer. Poletto (2019) reacted soybean oil with maleic anhydride to obtain maleated soybean oil and compared the performance of soybean oil and maleated soybean as potential coupling agents in polypropylene wood flour composites. As previous observed by Castro et al. (2017), both oils act as plasticizers and reduce the composite viscosity, which have a positive effect during the processing steps. The addition of soybean oil reduced the flexural strength of the composite, but causes an increase of 10% in impact strength, when compared with the composite without coupling agent. However, maleated soybean oil causes an increase of 10% in composite flexural strength, also when compared with the composite without coupling agent. The usage of maleated soybean oil also improves the composite thermal stability.

Poletto et al. (2014) used natural oils as coupling agents to improve the interfacial adhesion in recycled polypropylene wood flour composites. Hexanoic, octanoic, decanoic and dodecanoic acid were manually mixture with wood flour, without further modification. Composites treated with octanoic acid showed higher thermal stability and interfacial adhesion, than the composites treated with the others natural oils tested. Figure 7 shows the micrographs from the fracture surfaces of non-treated (a) and composites treated (b) with octanoic acid. According to the authors, the improvement caused by octanoic acid addition is similar to that obtained by the addition of MAPP. The authors verified that the boiling temperature and chain length of the acids tested directly affect the interfacial adhesion of the composites.

Martins and Santana (2016) tested carboxylic acids as compatibilizers on polypropylene/thermoplastic starch. The myristic (C14), palmitic (C16) and stearic (C18) acids were used as natural compatibilizers in the blend studied. When compared with the blend without compatibilizers, the addition of C18, C16 and C14

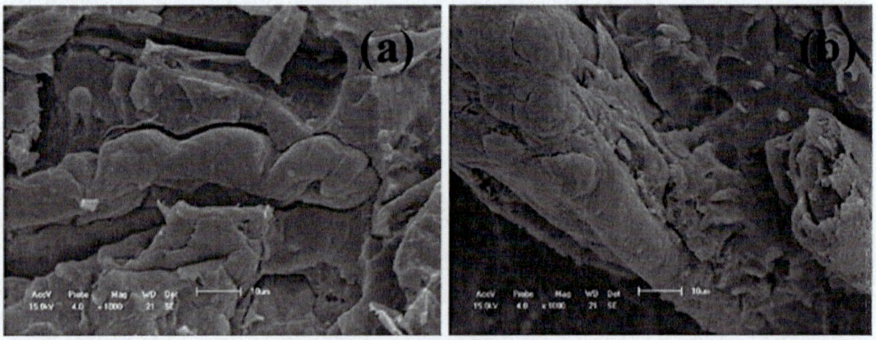

Fig. 7 Morphological aspect of the composites without (a) and with addition of octanoic acid (b) (Reproduced with permission (Poletto et al. 2014))

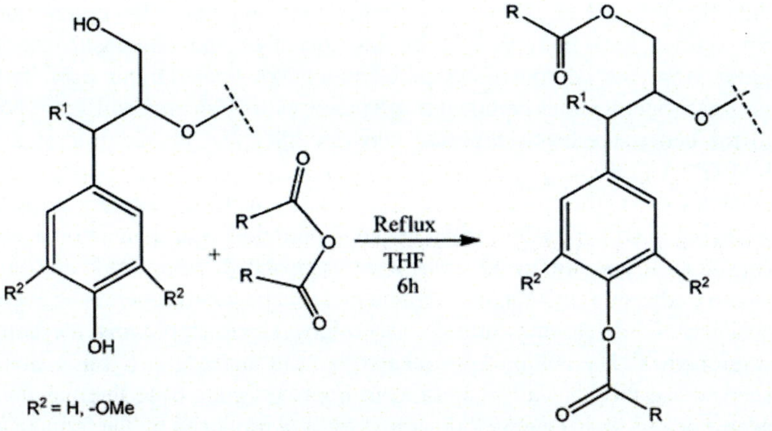

Fig. 8 Lignin esterification reaction (Reproduced with permission (Chatterjee et al. 2014))

presented an improvement of 25, 10 and 17% in the tensile strength and the addition of C14 also promote an improvement of 54% in the impact strength.

Lignin is the second most abundant biopolymer in nature after cellulose (Chatterjee et al. 2014). This biopolymer is generally used as a reinforcement agent in composite formulations due to its three dimensional structure consisting of randomly cross-linked propane building blocks formatted by p-hydroxyphenyl, guaiacly and syringyl units (Chatterjee et al. 2014; Naseem et al. 2016). However, is difficult blend lignin with others polymers and biopolymers due to the chemical incompatibility of lignin with these materials (Thakur et al. 2014; Kun and Pukánszky 2017). So, lignin is chemically modified by esterification reactions using normally dicarboxylic acid anhydrides that can react with hydroxyl groups presented in lignin surface, as illustrated in Fig. 8. Acetic, phthalic, succinic and maleic anhydride are commonly used in esterification reactions. Lignin chemical modification also includes graft copolymerization, functionalization, among others (Naseem et al. 2016; Gordobil et al. 2015; Wei et al. 2006).

Lignin and chemically modified lignin have also been used as compatibilizers in polymer composites (Thakur et al. 2014; Fernandes et al. 2014; Wood et al. 2011). Wood et al. (Wood et al. 2011) used lignin as a compatibilizer in epoxy composites reinforced with hemp mats. The authors observed an improvement in mechanical properties of the composites developed with lignin. However, at higher loadings (10 wt%) the lignin particles can increase the resin viscosity and prevent the complete resin infusion across the hemp mat which results in composites with lower mechanical properties. Tay et al. (2011a) used chemically modified Alcell lignin as a compatibilizer in polypropylene reinforced oil palm empty fruit bunch composites. The lignin was chemically modified by glycidyl methacrylate (GMA) and allyl glycidyl ether (AGE) (Tay et al. 2011a, 2011b). The authors used 1, 3 and 5% of modified lignin based on the weight of filler used to prepare the composites. The authors verified that both treated lignins increased the composite mechanical properties. However, composites prepared with GMA-treated lignin exhibited better mechanical properties that composites prepared with AGE-treated lignin, probably due to the better interfacial adhesion between filler and matrix promoted by the usage of GMA-treated lignin.

Fernandes et al. (Fernandes et al. 2014) developed high density polyethylene reinforced with cork powder using benzoyl peroxide was a initiator agent, and suberin (2 wt%) or lignin (2 and 4 wt%) isolated from cork to enhanced filler-matrix adhesion and improve composite mechanical properties. Suberin acts as plasticizer, while lignin works as a coupling agent, improving the composite tensile properties. The authors also related that the increase in lignin content does not improve the mechanical properties but improves composite thermal stability.

Itaconic acid is an unsaturated dicarboxylic acid produced by the fermentation of glucose (Rose and Palkovits 2011; Klement and Büchs 2013). The acid contains two carboxy functionalities and is commonly used to production of polyesters, wood modification and graft polymerization (Sabaa and Mokhtar 2002; Moncada et al. 2006; Sailaja and Seetharamu 2008). Poletto (Poletto 2017) developed composites of recycled polypropylene and wood flour using 2 wt% of polypropylene grafted itaconic acid as coupling agent. The flexural strength improved 29% with addition of polypropylene grafted with itaconic acid when compared with the wood plastic composites without coupling agent. The incorporation of polypropylene grafted with itaconic acid does not change the impact strength values significantly. The author also observed an increased in composite thermal stability with the addition of polypropylene grafted with itaconic acid. Han (Han et al. 2020) prepared wood composites by the impregnation of itaconic acid solution in wood followed by in situ polymerization of styrene, as can be seen in Fig. 9. The authors observed a reduction in the water uptake, an improvement in dimensional stability and also an enhancement in mechanical and thermal properties of the treated composites.

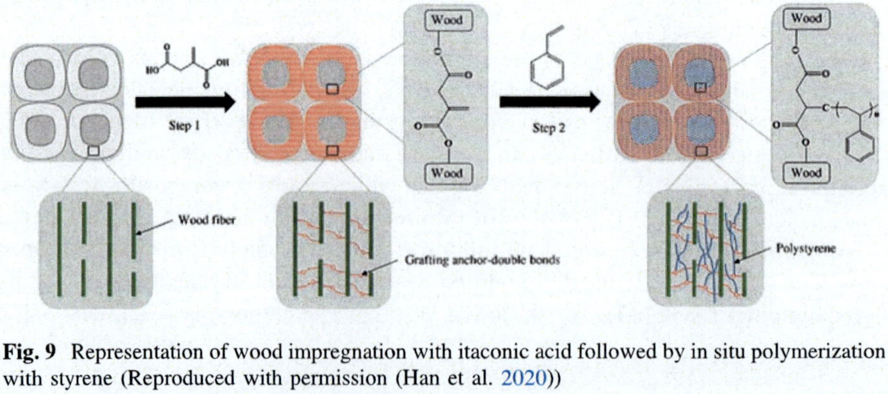

Fig. 9 Representation of wood impregnation with itaconic acid followed by in situ polymerization with styrene (Reproduced with permission (Han et al. 2020))

5 Conclusions and Future Perspectives

The application of wood and others types of natural fibers in several industrial fields continues to growth due to its sustainable origin associated with environmental and economic benefits when compared with non-renewable materials. These lignocellulosic fibers presented high specific properties, lower density and high performance when used as reinforcements in composite materials. However, one the main drawbacks of use natural fibers in polymer composites is the incompatibility between the hydrophilic fibers and the hydrophobic character of the polymer matrix. To solve this problem the fibers were exposed to chemical, physical or biological treatments, as previous described in this chapter. In addition, compatibilizers and/or coupling agents were also used. These treatments were essential to promote fiber compatibility with polymeric materials. The challenge still remains in found effectively treatments that can improve the lignocellulosic fiber properties and also can improve the composite performance for many industrial applications. The addition of natural compounds to improve the fiber or composite properties is also growing based on the environmental concerns. The usage of bio-renewable feedstock is a potential driving force to developed novel lignocellulosic fiber composites. The driving force for future developments based on wood modification, coupling agents and compatibilizers for wood composites must be in agreement with sustainable products and processes. The challenge for use friendly compounds and processes to obtain better interfacial adhesion in wood polymer composites still remains.

Acknowledgements The author Matheus Poletto thanks the Conselho Nacional de Desenvolvimento Científico e Tecnológico (CNPq/Brazil), grant number 435110/2018-7, for financial support.

References

Acda MN, Devera EE, Cabangon RJ, Ramos HJ (2012) Effects of plasma modification on adhesion properties of wood. Int J Adhes Adhes 32:70–75. https://doi.org/10.1016/j.ijadhadh.2011.10.003

Ahmad R, Hamid R, Osman SA (2019) Physical and chemical modifications of plant fibres for reinforcement in cementitious composites. Adv Civ Eng 2019:5185806. https://doi.org/10.1155/2019/5185806

Ali A, Shaker K, Nawab Y et al (2016) Hydrophobic treatment of natural fibers and their composites—a review. J Ind Text 47:2153–2183. https://doi.org/10.1177/1528083716654468

Amroune S, Bezazi A, Belaadi A et al (2015) Tensile mechanical properties and surface chemical sensitivity of technical fibres from date palm fruit branches (Phoenix dactylifera L.). Compos Part A Appl Sci Manuf 71:95–106. https://doi.org/10.1016/j.compositesa.2014.12.011

Aydemir D, Kiziltas A, Erbas Kiziltas E et al (2015) Heat treated wood–nylon 6 composites. Compos Part B Eng 68:414–423. https://doi.org/10.1016/j.compositesb.2014.08.040

Barsberg ST, Thygesen LG (2017) A combined theoretical and FT-IR spectroscopy study of a hybrid poly(furfuryl alcohol) – lignin material: basic chemistry of a sustainable wood protection method. ChemistrySelect 2:10818–10827. https://doi.org/10.1002/slct.201702104

Beg MDH, Pickering KL (2008) Mechanical performance of Kraft fibre reinforced polypropylene composites: Influence of fibre length, fibre beating and hygrothermal ageing. Compos Part A Appl Sci Manuf 39:1748–1755. https://doi.org/10.1016/j.compositesa.2008.08.003

Belgacem C, Tarrés Q, Espinach F et al (2020) High-yield lignocellulosic fibers from date palm biomass as reinforcement in polypropylene composites: effect of fiber treatment on composite properties. Polymers 12:1423. https://doi.org/10.3390/polym12061423

Bera M, Alagirusamy R, Das A (2010) A study on interfacial properties of jute-PP composites. J Reinf Plast Compos 29:3155–3161. https://doi.org/10.1177/0731684410369723

Bledzki AK, Mamun AA, Lucka-Gabor M, Gutowski VS (2008) The effects of acetylation on properties of flax fibre and its polypropylene composites. Express Polym Lett. https://doi.org/10.3144/expresspolymlett.2008.50

Broda M (2020) Natural compounds for wood protection against fungi—a review. Molecules 25:3538

Castro DO, Passador F, Ruvolo-Filho A, Frollini E (2017) Use of castor and canola oils in "biopolyethylene" curauá fiber composites. Compos Part A Appl Sci Manuf 95:22–30. https://doi.org/10.1016/j.compositesa.2016.12.024

Chatterjee S, Clingenpeel A, McKenna A et al (2014) Synthesis and characterization of lignin-based carbon materials with tunable microstructure. RSC Adv 4:4743–4753. https://doi.org/10.1039/C3RA46928J

Chen Y, Wan J, Zhang X et al (2012) Effect of beating on recycled properties of unbleached eucalyptus cellulose fiber. Carbohydr Polym 87:730–736. https://doi.org/10.1016/j.carbpol.2011.08.051

Colmenares JC, Varma RS, Lisowski P (2016) Sustainable hybrid photocatalysts: titania immobilized on carbon materials derived from renewable and biodegradable resources. Green Chem 18:5736–5750. https://doi.org/10.1039/C6GC02477G

Crépy L, Miri V, Joly N et al (2011) Effect of side chain length on structure and thermomechanical properties of fully substituted cellulose fatty esters. Carbohydr Polym 83:1812–1820. https://doi.org/10.1016/j.carbpol.2010.10.045

Dai X, Xiong Z, Na H, Zhu J (2014) How does epoxidized soybean oil improve the toughness of microcrystalline cellulose filled polylactide acid composites? Compos Sci Technol 90:9–15. https://doi.org/10.1016/j.compscitech.2013.10.009

de Almeida Mesquita RG, da Silva César AA, Mendes RF et al (2017) Polyester composites reinforced with corona-treated fibers from pine, eucalyptus and sugarcane bagasse. J Polym Environ 25:800–811. https://doi.org/10.1007/s10924-016-0864-6

de Cademartori PHG, Stafford L, Blanchet P et al (2017) Enhancing the water repellency of wood surfaces by atmospheric pressure cold plasma deposition of fluorocarbon film. RSC Adv. 7 (46):29159–29169. https://doi.org/10.1039/c7ra03334f

Despot R, Hasan M, Jug M, Šefc B (2008) Biological durability of wood modified by citric acid. Drv Ind

Dong S, Sapieha S, Schreiber HP (1993) Mechanical properties of corona-modified cellulose/polyethylene composites. Polym Eng Sci 33:343–346. https://doi.org/10.1002/pen.760330606

Dong Y, Yan Y, Zhang S et al (2015) Flammability and physical–mechanical properties assessment of wood treated with furfuryl alcohol and nano-SiO2. Eur J Wood Wood Prod. https://doi.org/10.1007/s00107-015-0896-y

Dong Y, Wang K, Li J et al (2020) Environmentally benign wood modifications: a review. ACS Sustain Chem Eng 8:3532–3540. https://doi.org/10.1021/acssuschemeng.0c00342

Elamin MAM, Li SX, Osman ZA, Otitoju TA (2020) Preparation and characterization of wood-plastic composite by utilizing a hybrid compatibilizer system. Ind Crops Prod 154:112659. https://doi.org/10.1016/j.indcrop.2020.112659

Essabir H, Achaby MEI, Hilali EM et al (2015) Morphological, structural, thermal and tensile properties of high density polyethylene composites reinforced with treated argan nut shell particles. J Bionic Eng. 12(1):129–141. https://doi.org/10.1016/S1672-6529(14)60107-4

Fazeli M, Florez JP, Simão RA (2019) Improvement in adhesion of cellulose fibers to the thermoplastic starch matrix by plasma treatment modification. Compos Part B Eng 163:207–216. https://doi.org/10.1016/j.compositesb.2018.11.048

Fernandes TF, Trovatti E, Freire CSR et al (2011) Preparation and characterization of novel biodegradable composites based on acylated cellulose fibers and poly(ethylene sebacate). Compos Sci Technol 71:1908–1913. https://doi.org/10.1016/j.compscitech.2011.09.005

Fernandes EM, Aroso IM, Mano JF et al (2014) Functionalized cork-polymer composites (CPC) by reactive extrusion using suberin and lignin from cork as coupling agents. Compos Part B Eng 67:371–380. https://doi.org/10.1016/j.compositesb.2014.07.028

Ferreira DP, Cruz J, Fangueiro R (2019) Surface modification of natural fibers in polymer composites (chapter 1). In: Koronis G, Silva ABT-GC for AA (eds) Woodhead publishing series in composites science and engineering. Woodhead Publishing, pp 3–41

Freire CSR, Silvestre AJD, Neto CP et al (2008) Composites based on acylated cellulose fibers and low-density polyethylene: Effect of the fiber content, degree of substitution and fatty acid chain length on final properties. Compos Sci Technol 68:3358–3364. https://doi.org/10.1016/j.compscitech.2008.09.008

Gong W, Chen C, Zhang Y et al (2017) Efficient synthesis of furfuryl alcohol from H2-hydrogenation/transfer hydrogenation of furfural using sulfonate group modified Cu catalyst. ACS Sustain Chem Eng 5:2172–2180. https://doi.org/10.1021/acssuschemeng.6b02343

Gordobil O, Delucis R, Egüés I, Labidi J (2015) Kraft lignin as filler in PLA to improve ductility and thermal properties. Ind Crops Prod 72:46–53. https://doi.org/10.1016/j.indcrop.2015.01.055

Goriparthi BK, Suman KNS, Mohan Rao N (2012) Effect of fiber surface treatments on mechanical and abrasive wear performance of polylactide/jute composites. Compos Part A Appl Sci Manuf 43:1800–1808. https://doi.org/10.1016/j.compositesa.2012.05.007

Gulati D, Sain M (2006) Fungal-modification of natural fibers: a novel method of treating natural fibers for composite reinforcement. J Polym Environ 14:347–352. https://doi.org/10.1007/s10924-006-0030-7

Gwon JG, Lee SY, Chun SJ et al (2010) Effects of chemical treatments of hybrid fillers on the physical and thermal properties of wood plastic composites. Compos Part A Appl Sci Manuf 41:1491–1497. https://doi.org/10.1016/j.compositesa.2010.06.011

Han X, Wang Z, Zhang Q, Pu J (2020) An effective technique for constructing wood composite with superior dimensional stability. Holzforschung 74:435–443. https://doi.org/10.1515/hf-2019-0176

Hosseinihashemi SK, Arwinfar F, Najafi A et al (2016) Long-term water absorption behavior of thermoplastic composites produced with thermally treated wood. Measurement 86:202–208. https://doi.org/10.1016/j.measurement.2016.02.058

Huber T, Biedermann U, Müssig J (2010) Enhancing the fibre matrix adhesion of natural fibre reinforced polypropylene by electron radiation analyzed with the single fibre fragmentation test. Compos Interfaces 17:371–381. https://doi.org/10.1163/092764410X495270

Hung K-C, Wu T-L, Chen Y-L, Wu J-H (2016) Assessing the effect of wood acetylation on mechanical properties and extended creep behavior of wood/recycled-polypropylene composites. Constr Build Mater 108:139–145. https://doi.org/10.1016/j.conbuildmat.2016.01.039

Husseinsyah S, Seong Chun K, Hadi A, Ahmad R (2016) Effect of filler loading and coconut oil coupling agent on properties of low-density polyethylene and palm kernel shell eco-composites. J Vinyl Addit Technol 22:200–205. https://doi.org/10.1002/vnl.21423

Islam M, Hamdan S, Hassan A et al (2014) The chemical modification of tropical wood polymer composites. J Compos Mater 48. https://doi.org/10.1177/0021998313477894

Islam R, Xie S, Lekobou W et al (2017) Surface modifications of wood materials using atmospheric pressure corona-based weakly ionized plasma. J Thermoplast Compos Mater 31:089270571772901. https://doi.org/10.1177/0892705717729019

Jiang X, Wang J, Wu G et al (2018) Significant reinforcement of polypropylene/wood flour composites by high extent of interfacial interaction. J Thermoplast Compos Mater 32:577–592. https://doi.org/10.1177/0892705718761476

Klement T, Büchs J (2013) Itaconic acid–a biotechnological process in change. Bioresour Technol 135:422–431. https://doi.org/10.1016/j.biortech.2012.11.141

Koohestani B, Darban AK, Mokhtari P et al (2019) Comparison of different natural fiber treatments: a literature review. Int J Environ Sci Technol 16:629–642. https://doi.org/10.1007/s13762-018-1890-9

Kord B, Movahedi F, Adlnasab L, Masrouri H (2020) Influence of eco-friendly pretreatment of lignocellulosic biomass using ionic liquids on the interface adhesion and characteristics of polymer composite boards. J Compos Mater 0021998320918345. https://doi.org/10.1177/0021998320918345

Kun D, Pukánszky B (2017) Polymer/lignin blends: Interactions, properties, applications. Eur Polym J 93:618–641. https://doi.org/10.1016/j.eurpolymj.2017.04.035

La Mantia FP, Morreale M (2011) Green composites: a brief review. Compos Part A Appl Sci Manuf 42:579–588. https://doi.org/10.1016/j.compositesa.2011.01.017

Liminana P, Garcia-Sanoguera D, Quiles-Carrillo L et al (2018) Development and characterization of environmentally friendly composites from poly(butylene succinate) (PBS) and almond shell flour with different compatibilizers. Compos Part B Eng. https://doi.org/10.1016/j.compositesb.2018.02.031

Liu Y, Tao Y, Lv X et al (2010) Study on the surface properties of wood/polyethylene composites treated under plasma. Appl Surf Sci 257:1112–1118. https://doi.org/10.1016/j.apsusc.2010.08.032

Liu W, Xie T, Qiu R (2016) Improvement of properties for biobased composites from modified soybean oil and hemp fibers: Dual role of diisocyanate. Compos Part A Appl Sci Manuf 90:278–285. https://doi.org/10.1016/j.compositesa.2016.07.018

Liu Y, Xie J, Wu N et al (2019) Influence of silane treatment on the mechanical, tribological and morphological properties of corn stalk fiber reinforced polymer composites. Tribol Int 131:398–405. https://doi.org/10.1016/j.triboint.2018.11.004

Mantanis GI (2017) Chemical modification of wood by acetylation or furfurylation: a review of the present scaled-up technologies. Bioresour 12(2):4478–4489

Martins AB, Santana RMC (2016) Effect of carboxylic acids as compatibilizer agent on mechanical properties of thermoplastic starch and polypropylene blends. Carbohydr Polym. https://doi.org/10.1016/j.carbpol.2015.08.074

Matthews FL, Rawlings RDB (1999) Composite Materials: Engineering and Science. Woodhead Publishing, Sawston

Meng X, Bocharova V, Tekinalp H et al (2018) Toughening of nanocelluose/PLA composites via bio-epoxy interaction: mechanistic study. Mater Des 139:188–197. https://doi.org/10.1016/j.matdes.2017.11.012

Mohit H, Arul Mozhi Selvan V (2018) A comprehensive review on surface modification, structure interface and bonding mechanism of plant cellulose fiber reinforced polymer based composites. Compos Interfaces 25:629–667. https://doi.org/10.1080/09276440.2018.1444832

Moncada E, Quijada R, Lieberwirth I, Yazdani-Pedram M (2006) Use of PP grafted with itaconic acid as a new compatibilizer for PP/Clay nanocomposites. Macromol Chem Phys 207:1376–1386. https://doi.org/10.1002/macp.200600150

Naseem A, Tabasum S, Zia KM et al (2016) Lignin-derivatives based polymers, blends and composites: a review. Int J Biol Macromol 93:296–313. https://doi.org/10.1016/j.ijbiomac.2016.08.030

Noël M, Mougel E, Fredon E et al (2009) Lactic acid/wood-based composite material. Part 2: physical and mechanical performance. Bioresour Technol 100:4717–4722. https://doi.org/10.1016/j.biortech.2009.04.042

Nordstierna L, Lande S, Westin M et al (2008) Towards novel wood-based materials: chemical bonds between lignin-like model molecules and poly(furfuryl alcohol) studied by NMR. Holzforschung 62:709–713. https://doi.org/10.1515/HF.2008.110

Ou R, Wang Q, Wolcott M et al (2014) Effects of chemical modification of wood flour on the rheological properties of high-density polyethylene blends. J Appl Polym Sci 131. https://doi.org/10.1002/app.41200

Pickering KL, Efendy MGA, Le TM (2016) A review of recent developments in natural fibre composites and their mechanical performance. Compos Part A Appl Sci Manuf 83:98–112. https://doi.org/10.1016/j.compositesa.2015.08.038

Podgorski L, Chevet B, Onic L, Merlin A (2000) Modification of wood wettability by plasma and corona treatments. Int J Adhes Adhes 20:103–111. https://doi.org/10.1016/S0143-7496(99)00043-3

Poletto M (2017) Polypropylene-based wood-plastic composites: effect of using a coupling agent derived from a renewable resource. Maderas Cienc Tecnol 19:265–272

Poletto M (2018) Maleated soybean oil as coupling agent in recycled polypropylene/wood flour composites. J Thermoplast Compos Mater. https://doi.org/10.1177/0892705718785707

Poletto M (2019) Natural oils as coupling agents in recycled polypropylene wood flour composites: mechanical, thermal and morphological properties. Polym Polym Compos. https://doi.org/10.1177/0967391119886941

Poletto M, Zeni M, Zattera AJ (2011) Effects of wood flour addition and coupling agent content on mechanical properties of recycled polystyrene/wood flour composites. J Thermoplast Compos Mater 25:821–833. https://doi.org/10.1177/0892705711413627

Poletto M, Zattera AJ, Santana RMC (2014) Effect of natural oils on the thermal stability and degradation kinetics of recycled polypropylene wood flour composites. Polym Compos. https://doi.org/10.1002/pc.22852

Quiles-Carrillo L, Montanes N, Sammon C et al (2018) Compatibilization of highly sustainable polylactide/almond shell flour composites by reactive extrusion with maleinized linseed oil. Ind Crops Prod. https://doi.org/10.1016/j.indcrop.2017.10.062

Ragoubi M, Bienaimé D, Molina S et al (2010) Impact of corona treated hemp fibres onto mechanical properties of polypropylene composites made thereof. Ind Crops Prod 31:344–349. https://doi.org/10.1016/j.indcrop.2009.12.004

Reddy CR, Simon LC (2010) Surface modification of wood fiber and preparation of a wood fiber-polypropylene hybrid by in situ polymerization. Macromol Mater Eng 295:906–914. https://doi.org/10.1002/mame.201000210

Rezaee Niaraki P, Krause A (2020) Correlation between physical bonding and mechanical properties of wood plastic composites: part 1: interaction of chemical and mechanical treatments on physical properties. J Adhes Sci Technol 34:744–755. https://doi.org/10.1080/01694243.2019.1683325

Rosa MF, Chiou B, Medeiros ES et al (2009) Effect of fiber treatments on tensile and thermal properties of starch/ethylene vinyl alcohol copolymers/coir biocomposites. Bioresour Technol 100:5196–5202. https://doi.org/10.1016/j.biortech.2009.03.085

Rose M, Palkovits R (2011) Cellulose-based sustainable polymers: state of the art and future trends. Macromol Rapid Commun 32:1299–1311. https://doi.org/10.1002/marc.201100230

Sabaa MW, Mokhtar SM (2002) Chemically induced graft copolymerization of itaconic acid onto cellulose fibers. Polym Test 21:337–343. https://doi.org/10.1016/S0142-9418(01)00094-0

Saiful Islam M, Hamdan S, Jusoh I et al (2012) The effect of alkali pretreatment on mechanical and morphological properties of tropical wood polymer composites. Mater Des 33:419–424. https://doi.org/10.1016/j.matdes.2011.04.044

Sailaja RRN, Seetharamu S (2008) Itaconic acid – grafted – LDPE as compatibilizer for LDPE – plasticized Tapioca starch blends. React Funct Polym 68:831–841. https://doi.org/10.1016/j.reactfunctpolym.2007.12.003

Saini G, Narula AK, Choudhary V, Bhardwaj R (2009) Effect of particle size and alkali treatment of sugarcane bagasse on thermal, mechanical, and morphological properties of PVC-bagasse composites. J Reinf Plast Compos 29:731–740. https://doi.org/10.1177/0731684408100693

Sawpan MA (2018) Polyurethanes from vegetable oils and applications: a review. J Polym Res 25:184. https://doi.org/10.1007/s10965-018-1578-3

Šefc B, Trajković J, Sinković T et al (2012) Compression Strength of Fir and Beech Wood Modified by Citric Acid. Drv Ind. https://doi.org/10.5552/drind.2012.1123

Spiridon I, Darie RN, Bodîrlău R et al (2012) Polypropylene-based composites reinforced by toluene diisocyanate modified wood. J Compos Mater 47:3451–3464. https://doi.org/10.1177/0021998312466906

Takinami PYI, Shimazaki K, Colombo MA et al (2010) Gelatin/piassava composites treated by electron beam radiation. Matéria (Rio Janeiro) 15:303–308

Tay GS, Shannon-Ong SH, Goh SW, Rozman HD (2011a) Thermoplastic–lignocellulose composites enhanced by chemically treated Alcell lignin as compatibilizer. J Thermoplast Compos Mater 26:733–746. https://doi.org/10.1177/0892705711428660

Tay GS, Shannon-Ong SH, Goh SW, Rozman HD (2011b) Enhancement of tensile and impact properties of thermoplastic lignocellulose composites by incorporation of chemically treated alcell lignin as compatibilizer. Polym Plast Technol Eng 50:160–167. https://doi.org/10.1080/03602559.2010.531423

Thakur VK, Thakur MK, Raghavan P, Kessler MR (2014) Progress in green polymer composites from lignin for multifunctional applications: a review. ACS Sustain Chem Eng 2:1072–1092. https://doi.org/10.1021/sc500087z

Todaro L, D'Auria M, Langerame F et al (2015) Surface characterization of untreated and hydro-thermally pre-treated Turkey oak woods after UV-C irradiation. Surf Interface Anal 47:206–215

Tomak ED, Ustaomer D, Ermeydan MA, Yildiz S (2018) An investigation of surface properties of thermally modified wood during natural weathering for 48 months. Measurement 127:187–197. https://doi.org/10.1016/j.measurement.2018.05.102

Tran P, Graiver D, Narayan R (2006) Biocomposites synthesized from chemically modified soy oil and biofibers. J Appl Polym Sci 102:69–75. https://doi.org/10.1002/app.22265

Tserki V, Zafeiropoulos NE, Simon F, Panayiotou C (2005) A study of the effect of acetylation and propionylation surface treatments on natural fibres. Compos Part A Appl Sci Manuf 36:1110–1118. https://doi.org/10.1016/j.compositesa.2005.01.004

Vasco-Correa J, Ge X, Li Y (2016) Biological pretreatment of lignocellulosic biomass (chapter 24). In: Mussatto SIBT-BFT for a LFBB (ed). Elsevier, Amsterdam, pp 561–585

Venkatachalam N, Navaneethakrishnan P, Rajsekar R, Shankar S (2016) Effect of pretreatment methods on properties of natural fiber composites: a review. Polym Polym Compos 24:555–566. https://doi.org/10.1177/096739111602400715

Wei M, Fan L, Huang J, Chen Y (2006) Role of star-like hydroxylpropyl lignin in soy-protein plastics. Macromol Mater Eng 291:524–530. https://doi.org/10.1002/mame.200500372

Wood BM, Coles SR, Maggs S et al (2011) Use of lignin as a compatibiliser in hemp/epoxy composites. Compos Sci Technol 71:1804–1810. https://doi.org/10.1016/j.compscitech.2011.06.005

Xie Y, Hill CAS, Xiao Z et al (2010a) Silane coupling agents used for natural fiber/polymer composites: a review. Compos Part A Appl Sci Manuf 41:806–819. https://doi.org/10.1016/j.compositesa.2010.03.005

Xie Y, Xiao Z, Grüneberg T et al (2010b) Effects of chemical modification of wood particles with glutaraldehyde and 1,3-dimethylol-4,5-dihydroxyethyleneurea on properties of the resulting polypropylene composites. Compos Sci Technol 70:2003–2011. https://doi.org/10.1016/j.compscitech.2010.07.024

Xie Y, Fu Q, Wang Q et al (2013) Effects of chemical modification on the mechanical properties of wood. Eur J Wood Wood Prod 71:401–416. https://doi.org/10.1007/s00107-013-0693-4

Xu K, Li K, Tu D et al (2014) Reinforcement on the mechanical-, thermal-, and water-resistance properties of the wood flour/chitosan/poly(vinyl chloride) composites by physical and chemical modification. J Appl Polym Sci 131. https://doi.org/10.1002/app.40757

Yin Y, Berglund L, Salmén L (2011) effect of steam treatment on the properties of wood cell walls. Biomacromol 12:194–202. https://doi.org/10.1021/bm101144m

Zadeh ZE, Abdulkhani A, Aboelazayem O, Saha B (2020) Recent insights into lignocellulosic biomass pyrolysis: a critical review on pretreatment, characterization, and products upgrading. Processes

Zhang C, Garrison TF, Madbouly SA, Kessler MR (2017) Recent advances in vegetable oil-based polymers and their composites. Prog Polym Sci 71:91–143. https://doi.org/10.1016/j.progpolymsci.2016.12.009

Property Improvements of Wood and Wood-Plastic Composites

Md. Minhaz-Ul Haque

Abstract Several factors such as type of wood species e.g. soft wood, hard wood, wood flour (WF) particle size, particle shape, particle aspect ratio, type of polymer matrices, dispersion of wood flour in polymer matrix as well as interfacial interaction i.e. interfacial adhesion between wood flour particle and polymer matrix affect the properties of Wood plastic-composite (WPC). Among the above factors, the most significant and important factor is the interfacial adhesion between WF and polymer matrix that affects the properties of WPC materials largely. WPC material obtained by normal blending of wood flour and polymer matrix, due to their incompatibility, does not exhibit improved properties. The poor adhesion or incompatibilization at the interface between WF and polymer matrix decreases the mechanical properties of WPC materials that limits the use of WF as filler in polymer matrices. Hence, compatibilization between wood flour and polymer matrices is the main focus of this chapter. By reading this chapter reader will learn the chemistry and mechanism of compatibility of different compatibilizing systems (such as coupling agents as well as silane treatments, acetylation and benzylation of wood flour, etc.) and their effect on the properties of WPC materials. After reading this chapter reader will also know the effect of thermal treatment of wood flour, effect of nanofillers addition as well as wet pulverization of WF on the properties of WPC materials.

Keywords Wood-plastic composites · Interface modification · Chemical treatments · Wood flour · Wet pulverization

Abbreviations

ABS	Acrylonitrile butadiene styrene
APM2ES	γ-Aminopropylmethyldiethoxysilane
APTES	γ -Aminopropyltriethoxysilane
E-GMA	Ethylene-glycidyl methacrylate copolymer
EPDM-MA	Maleic anhydride graft ethylene propylene diene monomer

Md. Minhaz-Ul Haque (✉)
Applied Chemistry and Chemical Engineering, Isamic University, Kushtia, Bangladesh
e-mail: minhaz@acce.iu.ac.bd

S. Mavinkere Rangappa et al. (eds.), *Wood Polymer Composites*,
Composites Science and Technology, https://doi.org/10.1007/978-981-16-1606-8_4

EVA	Ethylene vinyl acetate
HDPE	High density polyethylene
LLDPE	Linear low density polyethylene
MAPE	Maleic anhydride graft polyethylene
MAPP	Maleic anhydride graft polypropylene
MPTES	γ-Methacryloxypropyltrimethoxysilane
NAPTES	N-2(aminoethyl)-3-aminopropyltrimethoxysilane
PE	Polyethylene
PMPPIC	Polymethylene polyphenylisocyanate
PP	Polypropylene
PS	Polystyrene
PVC	Polyvinyl chloride
SANMA	Maleic anhydride graft styrene acrylonitrile
SEBS-MA	Maleic anhydride graft Styrene Ethylene Butylene Styrene
SMA	Styrene maleic anhydride
VTES	Vinyltriethoxysilane
VTMS	Vinyltrimethoxysilane
V2MES	Vinyltri(2-methoxyethoxy)silane
WPC	Wood plastic composite
WF	Wood flour

1 Introduction

Wood plastic composite (WPC), a partially green and renewable composite, is made of wood-based fillers e.g. wood flour (WF) or wood fibres which are green materials, and petroleum based thermosets such as epoxies, phenolics etc. resins or thermoplastic polymer matrices such as polypropylene (PP), polyethylene (PE), polyvinyl chloride (PVC), polystyrene (PS) etc. Of course, WPC can be a fully green and renewable composite material if the used polymer matrix is biodegradable and it is obtained from green resources (Chan et al. 2018). However, applications of renewable and biodegradable polymer matrices in WPC is still insufficient because of higher cost, poor moisture resistance and processing limitation of the biodegradable polymer matrices (Reddy et al. 2013). In WPC industry, wood flours are commonly used because of their easier processing compared with wood fibres. Wood flour micro-particles are usually obtained by milling hard wood or soft wood directly. WF can also be obtained by milling sawdust of sawmills. Wood flour obtained after milling has a very broad particles size distribution as shown in Fig. 1. Thus, wood flour obtained are comparatively cheap, therefore, from economic point of view, incorporation of wood fillers into polymer matrices not only contribute to make it green but also reduce the cost a great extent of the wood plastic composites. Simultaneously, wood flours can impart their physical and mechanical properties such as high stiffness, high tensile strength etc. to WPC materials. However, WPC materials obtained by normal blending of WF and polymer matrix, due to their incompatibility, does not display improvement in mechanical properties as much as expect from it. The properties of WPC materials depend on several factors such as the type of wood species e.g. soft wood, hard wood (Berger and Stark 1997), wood flour particle size, particle shape, particle aspect ratio, (Delviawan et al. 2019; Khonsari et al. 2015; Stark and Berger 1997; Seo and Kim 2014), type of polymer matrices, dispersion of wood flour in polymer matrix as well as interfacial interaction i.e. interfacial adhesion between wood flour particle and polymer matrix (Dányádi et al. 2010; Zhu et al. 2014). Among the above factors, the most significant factor is the interfacial adhesion between WF and matrix, which has received the attention very much by academic researchers and the WPC industries. The poor dispersion of WF in polymer matrix and the insufficient adhesion at the interface between WF and polymer matrix limit the use of WF as filler in polymer matrices.

Wood flour is a hydrophilic in character whereas polymer matrix is a hydrophobic material. As a result, WF is not compatible with hydrophobic polymer matrix and their dispersion in hydrophobic polymer become inhomogeneous. Therefore, the surface modification of wood flour particles i.e. the necessity of development of compatibilization of hydrophilic WF and hydrophobic polymer matrix become a great and challenging issue in WPC for obtaining improved mechanical properties of WPC.

The major constituents of wood flour are cellulose, hemicellulose and lignin. Wood flour is often called as a lignocellulosic material. The chemical structure of cellulose is shown in Fig. 2. D-glucose units are connected through $\beta(1 \rightarrow 4)$ linked to form the linear chain of cellulose. Cellulose is a semicrystalline material. Wood flour strength and stiffness are mainly due to the hydrogen bonds and crystalline cellulose parts.

Fig. 1 A scanning electron microscopic image of pine wood flour (Haque et al. 2019a)

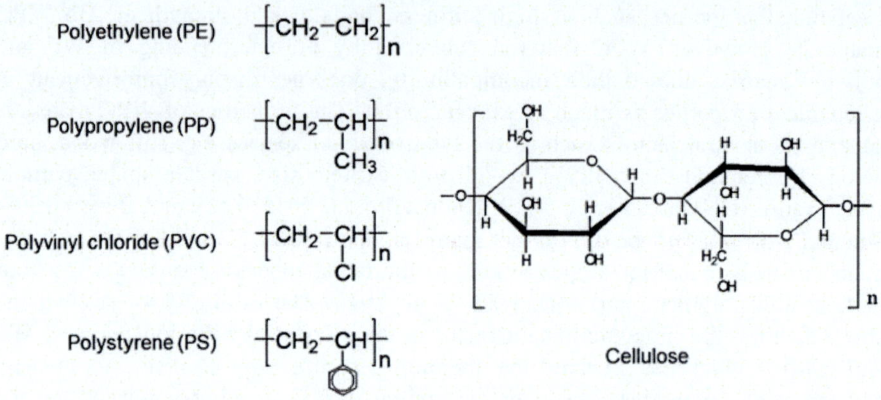

Fig. 2 Chemical structure of PE, PP, PVC, PS polymers and cellulose

In the structure of cellulose it can be noticed that each repeating glucose unit has a primary and two secondary hydroxyl groups. These hydroxyl groups are chemically reactive. Therefore, most of the developed compatibilizing processes are based on the available hydroxyl groups present on cellulose chain of wood flour. To improve interfacial adhesion between WF and polymer matrices various chemical and physical interface modification processes are used such as use of maleic anhydride grafted polypropylene (MAPP) or maleic anhydride grafted polyethylene (MAPE) (Delviawan et al. 2019; Zhu et al. 2014; Kim et al. 2007), surface modification of wood flour with silanes (Kim et al. 2011; Geng and Laborie 2010; Fang et al. 2014; Koohestani et al. 2017), benzylation of WF (Hon and Chao 1993; Dominkovics et al. 2007), acetylation of WF (Mbarek et al. 2013; Özmen et al. 2013; Kurimoto and Sasaki 2013), etc. as well as thermal treatment (Luo et al. 2014; Askanian et al. 2015; Källbom et al. 2020) etc. A synergic effect on the properties of WPC can also be achieved by using modified wood flour and coupling agent together (Farsi 2010).

The mechanical properties of wood–plastic composite can also be one step enhanced by wet pulverization of wood flours. Pulverization of wood flour can bring changes on WF particles surfaces and WF particles aspect ratio (Niu et al. 2016), the ratio of the length to the width of particles, which has a large effect on the mechanical properties of WPC materials. Since the particle aspect ratio is changed after pulverization and the properties of WPC largely depends on WF particle aspect ratio (Delviawan et al. 2019; Khonsari et al. 2015; Stark and Berger 1997; Seo and Kim 2014; Peng et al. 2015; Soccalingame et al. 2015), the physical and mechanical properties of WPC materials are significantly affected by the wet pulverized WF (Niu et al. 2016; Haque et al. 2019; Haque et al. 2019a; Haque et al. 2019b).

The mechanical properties, thermal stability, moisture barrier properties etc. of WPC can also be improved significantly by adding a very small portion (1–3 wt%) of nanofiller such as nanoclay (Fayzullin et al. 2016; Hafshejani et al. 2019; Yadav and Yusoh 2019), carbon nanotubes (Kaymakci et al. 2019; Peng et al. 2020), nanoparticles of SiO_2 (Tabar et al. 2015; Ma et al. 2020) and TiO_2 (Kaymakci 2019). Nanofillers, at least one dimension 1–100 nm, filled polymer composites are particular type of composite because of higher interfacial interaction between polymer and nanoparticles compared with traditional microparticles. Different compatibilizing systems with their effect on the properties of different WPC materials such as wood flour reinforced polypropylene (PP), polyethylene (PE), polyvinyl chloride (PVC), polystyrene (PS) etc. matrices are reported in Table 1 as well as the chemical structure of the above polymers is also shown in Fig. 2.

In this chapter, the chemistry and mechanism of compatibility of different compatibilizing systems are discussed in detail. In addition, recently developed processes i.e. the effect of nanofillers as well as wet pulverization of WF in WPC are also discussed in detail.

Table 1 Property improvements of WPC materials by different treatments or modifications

Type of treatments or modifications	Matrices	Type of wood flours	Type of property improvements	References
MAPP	PP	Particles size, 20 and 40 mesh	Tensile and flexural strength, unnotched impact strength	(Myers et al. 1991)
MAPP, EPDM-MA, SEBS-MA	PP	Pine, particles size, 420 μm	Tensile strength, impact strength, interfacial adhesion	(Oksman and Clemons 1998)
MAPP	PP	Particles size, 80–100 mesh	Tensile strength, creep resistance	(Lee et al. 2004)
MAPP	PP	Particles size, 110 μm	Tensile, flexural and impact strengths, thermal stability, interfacial adhesion	(Kim et al. 2007)
MAPP	HDPE	Particles size, 75, 225, and 900 μm	Mechanical properties	(Cui et al. 2008)
MAPP	PP	Particles size, 20 and 40 mesh	Tensile strength and modulus, water resistance	(Ichazo et al. 2001)
SMA, SANMA	ABS	Particles size, 177 and 210 μm	Tensile strength	(Dányádi et al. 2010; Yeh et al. 2009)
E-GMA	HDPE/PET	Pine, particles size, 40 mesh	Tensile and flexural strengths, interfacial adhesion	(Lei and Wu 2010)
MAPP	PP	Poplar, particles size, 100 mesh	Interfacial compatibility	(Zhu et al. 2014)
MAPP	PP	Particles size, 40–60 mesh	Bonding strength, water reistance, swelling resistance	(Malakani et al. 2015)

(continued)

Table 1 (continued)

Type of treatments or modifications	Matrices	Type of wood flours	Type of property improvements	References
MAPP	HDPE	Pine, particles size, 500 μm	Flexural properties, water resistance	(Kajaks et al. 2015)
MAPP	PP	Beech wood flour	Tensile and flexural strength	(Farsi 2010)
MAPE	HDPE	Particles size, 354 μm (45 mesh)	Tensile and flexural properties	(Mengeloğlu and Karakuş 2008)
MAPE	HDPE	Paulownia, particles size, 75 μm	Tensile and flexural properties, dimensional stability	(Tisserat et al. 2013)
MAPE		Teak, particles size, 180 μm	Dynamic storage modulus	(Sewda and Maiti 2013)
VTES	PP	Beech wood flour	Tensile and flexural strength	(Farsi 2010)
V2MES, MPTES, APTES	PP, PS	Particles size, 500–900 μm, 20–60 mesh	Tensile strength	(Kokta et al. 1989; Maldas et al. 1988)
V2MES	PP	Particles size, 20 and 40 mesh	Interacial adhesion and dispersion	(Ichazo et al. 2001)
V2MES	HDPE	Particles size, 75, 225, 450 and 900 μm	Mechanical properties	(Cui et al. 2008)
APTES	PVC	Particles size, 50–150 μm	Tensile properties, starin%, impact strength, water resistance	(Müller et al. 2012)
APTES	PP	Birch, particles size, 630 μm	Flexural properties, inter facial adhesion	(Khonsari et al. 2015)
APTES, MPTES, VTMS	PP	Particles size, 70–150 μm	Tensile, flexural, and impact strengths, water resistance, interfacial adhesion	(Kim et al. 2011)
VTMS	HDPE	Pine, particles size, 40 mesh	Flexural strength and elongation at break	(Bengtsson et al. 2006)
VTMS	HDPE	Spruce, pine, particles size, 200–400 μm	Toughness, impact strength, creep resistance	(Bengtsson et al. 2006)
APM2ES	PVC	Pine, particles size, 100 mesh	Tensile and impact strengths	(Zhao et al. 2006)
VTMS	LLDPE	Pine, particles size, 40 mesh	Dynamic modulus	(Geng and Laborie 2010)
VTMS	HDPE	Particles size, 300 μm	Shear strength, storage modulus, interfacial adhesion, water resistance, dimensional stability	(Fang et al. 2014)

(continued)

Table 1 (continued)

Type of treatments or modifications	Matrices	Type of wood flours	Type of property improvements	References
NAPTES	PVC	Particles size, 70–150 μm	Tensile strength, water resistance	(Yim and Kim 2012)
Acetylation	PP	Particles size, 40–60 mesh	water reistance, swelling resistance	(Malakani et al. 2015)
Acetylation	PVC	Particles size, 50–150 μm	Tensile properties, strain%, impact strength, water resistance	(Müller et al. 2012)
Acetylation	HDPE	Particles size, 220 μm	Interfacial adhesion	(Mbarek et al. 2013)
Acetylation	HDPE	Pine, particles size, 250–425 μm	Tensile and flexural strengths, thermal stability	(Özmen et al. 2013)
Acetylation	PP	Particles size, 1.0–2.0 mm mesh sieves	Tensile and bending strength, dimesional stability	(Kurimoto and Sasaki 2013; Kurimoto and Sasaki 2013a)
Acetylation	PP	Particles size, 40 mesh (420 μm),	Long term durability	(Ibach and Clemons 2017)
Acetylation	HDPE	Particles size, 16 and 24 mesh	Dimensional stability, tensile strength, weathering properties	(Chao et al. 2020)
Acetylation	PP	Particles size, 30–60 mesh (250–595 μm)	Creep resistance	(Hung et al. 2016)
Acetylation	HDPE	Particles size, 16 and 24 mesh	Interfacial adhesion	(Hung and Wu 2010)
Acetylation	PS	Particles size, 100 mesh	Thermal stability	(Lisperguer et al. 2007)
Benzyalation	PP	Particles size, 210 μm	Water resistance, processability	(Dányádi et al. 2010)
Benzyalation	PS	Wood powder	Plasticizing effect and compatibility	(Hon and Chao 1993)
Benzyalation	PVC	Particles size, 100 μm	Electric insulation	(Djidjelli et al. 2007)
Benzyalation	PP	Particles size, 210 μm	Processability and surface appearance	(Dominkovics et al. 2007)
Thermal	PP	Poplar, Particle size, 10–60 mesh	Interfacial bonding	(Luo et al. 2014)
Thermal	PP	Particle size, 400 μm and 40 mesh	Thermal stability	(Askanian et al. 2015; Kaboorani 2009)
Thermal	HDPE	–	Creep behaviour	(Alrubaie et al. 2020)

(continued)

Table 1 (continued)

Type of treatments or modifications	Matrices	Type of wood flours	Type of property improvements	References
Thermal	PP	Particles size, 30–60 mesh (200–630 μm)	Water resistance, dimensional stability	(Källbom, et al. 2020)
Thermal	LDPE	Pparticles size, 40–100 mesh (400–150 μm)	Thermal stability, tensile modulus	(Lafia-Araga et al. 2012)
Thermal	HDPE	Particles size, 40 mesh	Tensile and flexural modulus	(Kaboorani et al. 2008)
Ballmiling	PP	Pine, cedar, particles size, 2 mm, 170 μm	Tensile strength	(Delviawan et al. 2019; Isa et al. 2014)
Ballmiling	PP	Pine, particles size, 2 mm	Tensile strength, impact strength	(Murayama et al. 2019)
Disc milling	PP	Cypress, pine, particles size, 165–379 μm	Tensile strength, impact strength, Fatigue behaviour,	(Haque et al. 2019; Haque et al. 2019a; Haque et al. 2019b)
Pan miling	PP	Particles size, 250 μm	interfacial bonding, storage modulus	(Niu et al. 2016)
Montmorillonite	PVC	Pine, particles size, 100 mesh	Fire flame retardancy and smoke suppression	(Zhao et al. 2006)
Cloisite 30B nanoclay	PVC	Particles size, 70–150 μm	Tensile strength, water resistance	(Yim and Kim 2012)
Cloisite 15A nanoclay, carbon nanotubes	PP	Particles size, 60, 170, 410 and 500 μm	Elastic modulus	(Fayzullin et al. 2016)
Cloisite 30B nanoclay	HDPE	Spruce, particles size, 40–50 mesh	Toughness, tensile strength and modulus	(Hafshejani et al. 2019)
Cloisite 20A nanoclay	PP	–	Hardness, elastic modulus, creep resistance	(Yadav and Yusoh 2019)
Carbon nanotube	PP	Pine, particles size, 60 mesh (250 μm)	Scratch hardness	(Kaymakci et al. 2019)
Carbon nanotube	PP	Poplar, particles size, 500 μm	Flexural and impact properties, color stability	(Peng et al. 2020)
SiO_2 nano powder	HDPE	Poplar, particles size, 40–60 mesh	Tensile and flexural properties	(Tabar et al. 2015)
SiO_2 nano particles	PP	Particles size, 80 mesh	Tensile and flexural strengths, thermal stability	(Ma et al. 2020)

(continued)

Table 1 (continued)

Type of treatments or modifications	Matrices	Type of wood flours	Type of property improvements	References
TiO_2 nano particles	PP	Particles size, 40–80 mesh	Flexural and tensile properties	(Kaymakci 2019)
PMPPIC	PP, PS	Particles size, 500–900 μm, 20–60 mesh	Tensile strength, tensile modulus	(Kokta et al. 1989; Maldas et al. 1988)
Microencapsulation	PP	Particles size, 30–60 μm	Storage modulus	(Ye et al. 2018)
EVA	HDPE	Poplar, particles size, 100 mesh	Interfacial adhesion, thermal stability	(Hu et al. 2020)
Phosphate-based fire retardant	PP	Spruce, aspect ratio 5	Fire resistance	(Hämäläinen and Kärki 2014)
Silane treated mineral	HDPE	Maple, particles size, $\geq$ 100 μm	Tensile and flexural strengths	(Koohestani et al. 2017)
Chitin and chitosan	PVC	Maple, particles size, 40 mesh	Flexural strength and modulus	(Shah et al. 2005)
Cumyl peroxide	LDPE	Particles size, 87 μm	Strain-hardening behavior	(Harnnarongchai et al. 2012)

2 Property Improvements by Wood Flour/Polymer Interface Modifications

Interface modifications of wood flour/polymer can be carried out and broadly categorised as chemical treatments and physical treatments.

2.1 Chemical Treatments

We know that wood flour, due to hydrophilic character, is incompatible with hydrophobic polymer matrices. This incompatibility causes insufficient interfacial adhesion between WF and polymer matrix as well as poor dispersion of WF into polymer matrix. Consequently, WPC materials without compatibilization show poor mechanical properties. Therefore, the main objective of the chemical modifications or treatments is to convert the surfaces of hydrophilic WF to hydrophobic. Besides, addition of a third compound, coupling agent, which is able to react with the hydroxyl groups of cellulose of WF and to be compatible with the polymer matrix, can also reduce this incompatibility greatly. Different chemical treatments used in WPC are discussed in the following sections.

2.1.1 Addition of Coupling Agents

Coupling agents are very much similar in their action with emulsifying agents. Emulsifying agents, used to make emulsion (a stable mixture of two immiscible liquids), have two parts a hydrophobic part and a hydrophilic part. Similarly, coupling agents, monomeric or polymeric, have two parts. A polar part which is able to react with the hydoxyl groups of cellulose of WF and a nonpolar hydrophobic part, which is miscible or compatible with the hydrophobic polyer matrix because of similar chemical nature. In this way, copuling agents make a bridge between wood flour particle and polymer matrix. The most common used coupling agent in WPC is maleic anhydride graft polypropylene (MAPP), which is obtained by free radical grafting reaction of maleic anhydride onto PP chain. Maleic anhydride grafted polyethylene (MAPE), maleic anhydride grafted styrene-ethylene-butylene-styrene (SEBS-MA) etc. are also used as coupling agents in WPC. The nonpolar hydrophobic polymer part of a maleic anhydride grafted polymer is miscible or compatible with the hydrophobic polymer matrix due to their similar chemical structure, and the polar part i.e. the anhydride group of the maleated polyolefin polymer makes ester bonds with the hydroxyl groups on the wood flour surfaces. Thus, maleated polyolefin polymer acts as an emulsifying agent and create a strong interphase (WF-coupling agent-polymer matrix). A reaction scheme of a maleic anhydride graft polyolefin polymer is shown in Fig. 3. Coupling agent reduces the surface tension of WF particle surfaces by converting the hydrophilic surface of WF to hydrophobic that improve wetting and adhesion.

Figure 4 shows an effect of MAPP coupling agent on the tensile properties of 40 wt% wood flour reinforced PP composite. Small amount of coupling agent e.g. about 2 wt% MAPP is sufficient to improve the interfacial bonding between wood flour and polypropylene matrix. WPC materials developers should be cautious that higher weight percentage of MAPP is not beneficial to improve internal compatibility of WF/PP composite. Even in stead of improvement higher weight percentage

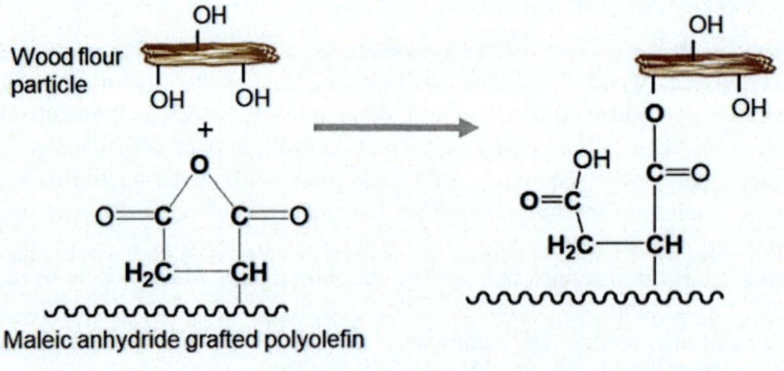

Fig. 3 A reaction scheme of maleic anhydride graft polyolefin polymer with hydroxyl group of WF

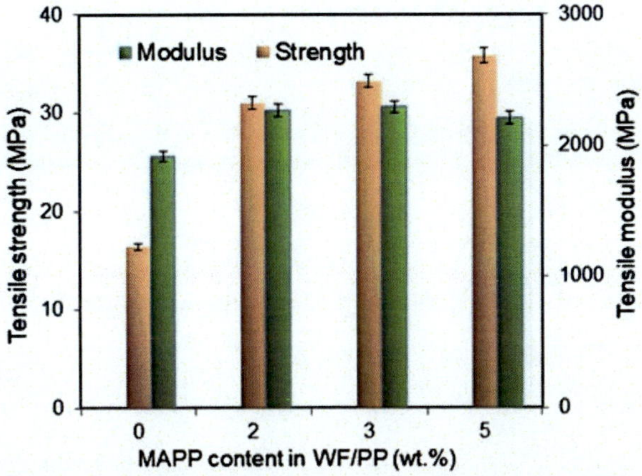

Fig. 4 Effect of MAPP on the tensile properties of 40 wt % wood flour reinforced PP composite (Lee and Kim 2009)

of coupling agent can deteriorate the mechanical properties of wood-plastic composites (Zhu et al. 2014). The mechanical properties such as tensile, impact and flexural strengths of WPC are increased by MAPP compared with untreated WPCs in one hand, but the extent of improvements in properties of WPCs depend on the molecular weight of MAPP as well as maleic anhydride graft percentage of MAPP (Kim et al. 2007).

2.1.2 Silane Treatments of Wood Flour

Silane treatments are widely investigated and found to be very effective to improve interfacial bonding of wood flour and polymer matrices. Silanes are low molecular weight organosilicon compounds which are highly hydrophobic in nature. Each molecule of silane has a polar part and a nonpolar hydrophobic part like maleic anhydride graft polyolefin coupling agents. Silane coupling agents have a generic chemical structure $R_{(4-n)}$-Si-(R°X)$_n$ (n = 1, 2) where R° is alkoxy, X represents an organofunctionality, and R° is an alkyl bridge (or alkyl spacer) connecting the silicon atom and the organofunctionality (Xie et al. 2010). Different organofunctionality such as amino, vinyl, methacryl etc. of trialkoxysilanes are mostly used silanes for wood-plastic composites. The simplified generic chemical structure for trialkoxysilanes is R_3-Si-(R°X). Mainly, a substituted silyl group (R_3Si-) is introduced onto wood flour particle surfaces by silane treatment that increase the hydrophobicity of wood flour. Since a chemical or physical compatibility between polymer matrix and organofunctionality of a silane is expected after silane treatment of WF, a silane is selected for wood flour treatment considering the interaction mode between the organofunctionality (such as amino, vinyl, methacryl etc.) of the silane and targeted polymer matrix. Before silane treatments wood flour is usually treated with NaOH, since silane treatment is more effective to NaOH treated WF. Usually, silane, 1–3 wt% of WF, is mixed with ethanol–water mixture. Firstly, the alkoxy groups of silane compound are hydrolyzed by water to form silanol (Si–OH) groups. Then in the second step, the silanol groups are reacted with the hydroxyl

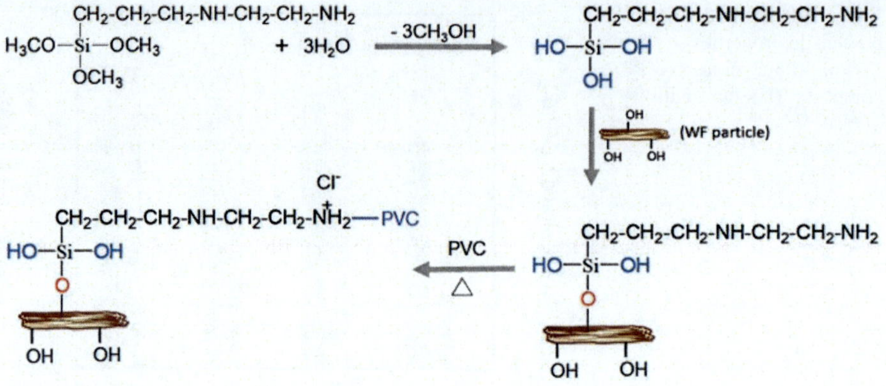

Fig. 5 A reaction scheme for silylation of WF with aminosilane and possible interaction between PVC and silane treated WF (Yim and Kim 2012)

groups of cellulose of wood flour to form silyl ether groups, which are covalently bonded to the wood flour particle surfaces (Xie et al. 2010).

A reaction scheme for silylation of WF with N-2(aminoethyl)-3-aminopropyltrimethoxysilane and a possible interaction between PVC and silane treated WF is shown in Fig. 5. The interaction between amino-silane treated WF and polyvinyl chloride mainly follows Lewis acid–base theory. Ionic bonds are formed between the negatively charged chlorine atoms of the PVC and the positively charged amino groups of treated wood flour and the PVC is bonded with amino group covalently (Matuana et al. 1998). Figure 6 shows the effect of aminosilane treatment of WF on the tensile and flexural strength of 30% WF filled PP and 10% WF filled PVC composites. The higher tensile and flexural strength of WPC composites with the silane-treated wood flour is obtained due to the improved interfacial adhesion between silane treated wood flour and polymer matrices (Kim et al. 2011; Yim and Kim 2012).

2.1.3 Acetylation of Wood Flour

Acetylation of wood flour means introduction of acetyl groups onto wood flour particle surfaces by a chemical reaction. The purpose of acetylation of WF is to convert the hydrophilic wood particle surfaces to hydrophobic surfaces. The acetylated WF or hydrophobic WF improve the interfacial adhesion with hydrophobic polymer matrices. The acetyl groups introduced onto WF surfaces can also cause plasticization of WF. The acetylated WF increasing the interfacial adhesion or compatibility with hydrophobic polymer matrices can improve the mechanical properties of WPC materials. The acetylation of WF can be carried out by acetic anhydride or vinyl acetate (Özmen et al. 2013). Before acetylation, wood flours are also treated with dilute NaOH solution to activate the OH groups of cellulose of wood flours. In the case of acetylation of wood flour by acetic anhydride, the wood flour is impregnated with acetic anhydride and then reacted at an elevated temperature. H2SO_4 is also used as catalyst. Acetyl group of acetic

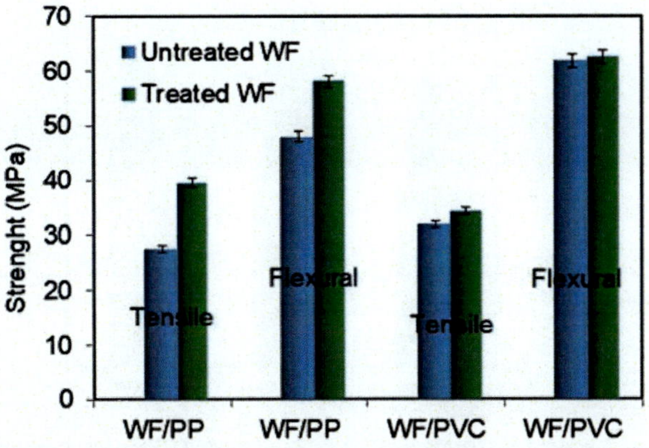

Fig. 6 Effect of silane treatment of WF on the tensile and flexural strength of 30% WF filled PP and 10% WF filled PVC composites (Kim et al. 2011; Yim and Kim 2012)

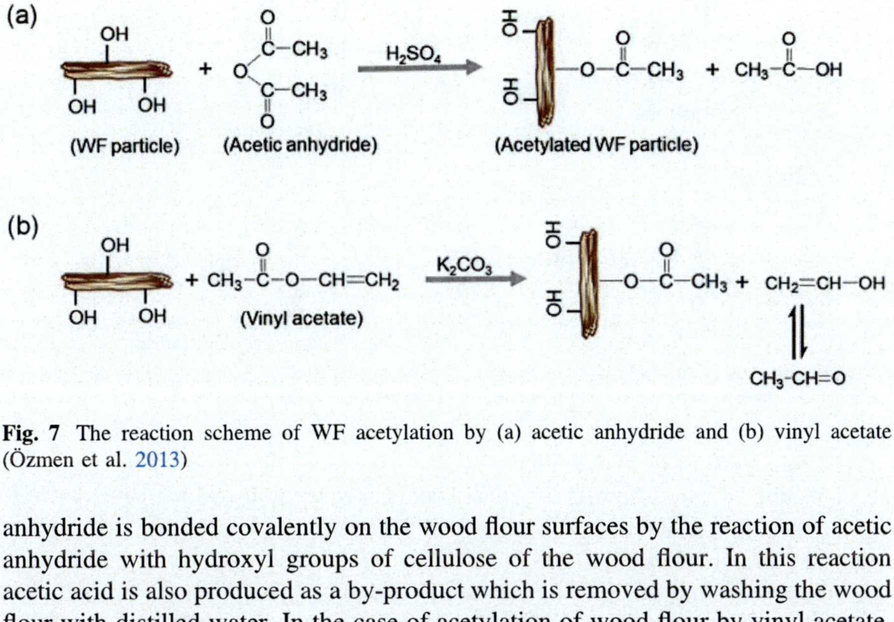

Fig. 7 The reaction scheme of WF acetylation by (a) acetic anhydride and (b) vinyl acetate (Özmen et al. 2013)

anhydride is bonded covalently on the wood flour surfaces by the reaction of acetic anhydride with hydroxyl groups of cellulose of the wood flour. In this reaction acetic acid is also produced as a by-product which is removed by washing the wood flour with distilled water. In the case of acetylation of wood flour by vinyl acetate, K_2CO_3 is used as catalyst and a transesterification reaction mechanism is followed (Jebrane and Sebe 2007). Figure 7 shows reactions of acetylation of wood flour by acetic anhydride and vinyl acetate. Tensile and bending strengths of 20 wt% untreated WF, acetic anhydride treated WF (AAWF) and vinyl acetate treated WF (VAWF) filled HDPE composites are shown in Fig. 8 (Özmen et al. 2013). The effect of acetylated wood flour on the properties of WPC composites is obvious in this figure. Comparing the improvements in mechanical properties of WPC the acetylated WF by vinyl acetate is better than that of acetic anhydride.

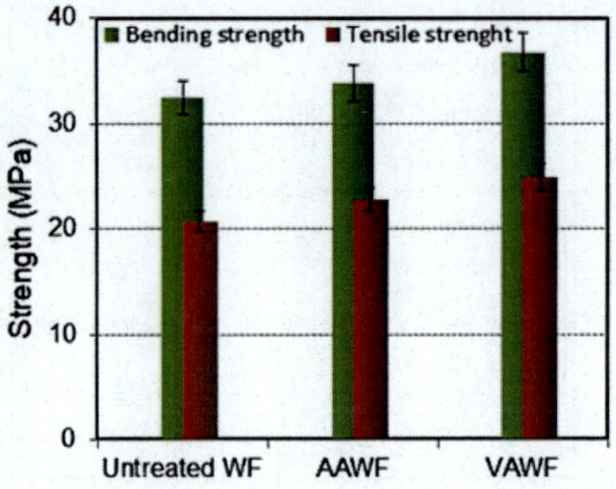

Fig. 8 Tensile and bending strengths of 20 wt% untreated WF, acetic anhydride treated WF (AAWF) and vinyl acetate treated WF (VAWF) filled HDPE composites (Özmen et al. 2013)

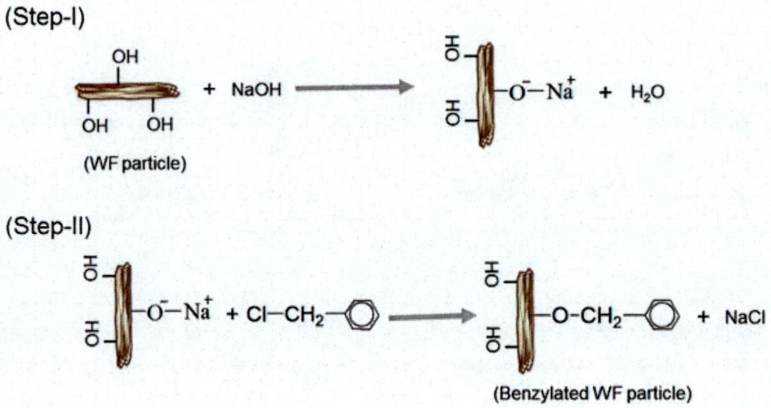

Fig. 9 A reaction scheme of WF benzylation (Hon and Ou 1989)

2.1.4. Benzylation of Wood Flour

Benzylation of wood flour is a typical chemical reaction by which benzyl (C_6H_5-CH_2-) groups are introduced onto wood flour surfaces by treating with benzyl chloride. In benzylation process NaOH is necessary and extent of benzylation of wood flour depends on NaOH concentration (Hon and Ou 1989). Therefore, wood flour must be treated with NaOH aqueous solution before benzylation of WF. Reaction mechanism of WF benzylation is shown in Fig. 9. Benzylation of WF can change its suprastructure significantly and wood flour can be converted into thermoplastic by benzylation. Since benzyl functional groups decrease the hydrophilicity of the wood flour surfaces, therefore, after benzylation the hydrophilicity of wood flour particle surfaces is changed into hydrophobic surfaces which enhance the compatibility of WF with hydrophobic polymer matrices. Use of benzylated wood flour can also enhance processability of polymer matrices as thermoplasticity of wood flour is obtained by benzylation of wood flour. Since benzylated wood flour particle surfaces contain phenyl structure similar to that of polystyrene, benzylated WF achieve a very good dispersion in polystyrene matrix. Benzylated wood also act as good plasticizer for polystyrene that improve dynamic processability of polystyrene for extrusion and compression molding (Hon et al. 1993). Bezylation of WF can also reduce the surface tension of WF particles that decrease the water absorption of WPC materials (Dominkovics et al. 2007).

2.1.5 Miscellaneous Chemical Treatments

The properties of wood-plastic composite may also be improved by using some other chemical treatments such as peroxide treatment (Harnnarongchai et al. 2012; Borah and Kim 2016), stearic acid (Dányádi et al. 2010; Gregorova et al. 2009), isocyanate treatment (Kokta et al. 1989), NaOH, cyanoethylation of WF by acrylonitrile (Kallakas et al. 2015) etc. In the case of peroxide treatment, wood flours are usually treated with benzoyl peroxide or dicumyl peroxide. A solution of peroxide is prepared in an organic solvent and then WF are immersed in it for few hours. Finally, the treated WF are washed and dried in an oven under vacuum. Organic peroxide such as RO-OR is dissociated and free radicals are generated as 2RO• in the processing of WPC. The generated free radicals can generate new free radicals on wood flour surfaces and the polymer chain abstracting hydrogen atom from wood flour and polymer chain. As a result, chemical bond is formed between wood flour and polymer matrix. Hence, compatibilization of wood flour and polymer matrix is achieved by grafting polymer chain on wood flour surfaces.

2.2 *Physical Treatments*

Physical treatments of wood flour include mainly thermal treatments and solid-state shear pulverization or milling. The thermal treatment of wood is usually done at temperature of 150–240 °C, whcich higher than those of a convensional drying temperature. However, above 200 °C wood is thermally unstable. Pulverization of wood flour is carried out at a very high speed of rotor of pulverizer such as ball mill, pan mill, disc mill etc.

2.2.1. Thermal Treatment

Thermal treatment or heat treatment of wood is an alternative to chemical treatments. Thermal treatment is carried out to reduce hygroscopicity of solid wood. It also improves dimensional stability, durability and biological resistance of wood (Tuong and Li 2010; Kamdem et al. 2002; Kocaefe et al. 2008). Heat treatment of wood not only removes water, but also causes significant changes in the wood chemical compostions such as hemicelluloses of wood are degraded and the number of free hydroxyl groups of wood are reduced by the thermal treatment. Consequently, the thermally treated wood shows lower mositure uptake, since moisture properties of wood strongly depends on the chemical constituents of wood (Niemz et al. 2010). The changes of chemical compositions of wood by heat treatment also change the polarity of wood. Thus, treated wood can be a more compatible material with non-polar hydrophobic matrix. This character of treated wood hints higher performance of WPC compared with untreated wood. Thermally treated wood flour is found to be compatible with hydrophobic polymer matrix because of lower hygroscopicity of treated wood compared with untreated ones that results high quality and thermally stable wood plastic composites (Kaboorani 2009; Kaboorani et al. 2008). The thermal stability, dimensional stability, flexural modulus and strength, moisture resistance of WPC materials are also improved by using heat treated wood (Lafia-Araga et al. 2012; Ayrilmis et al. 2011). Although, the properties of WPC are improved by heat treatment of wood, but too high temperature and long time treatment of wood may result in mechanical loss of wood and more brittle wood, which largely affect the properties of WPC. Since the improvement in properties of WPC is significantly affected by the heat treatment temperature, WPC developers should determine the extent of degradation during treatment and composites fabrication by thermal analysis. Figure 10 shows the effect of heat treatment temperature on the tensile properties of wood flour filled LDPE composites.

2.2.2 Solid-State Shear Pulverization

Generally, commercial wood flours which are used as filler in WPC, are mixture of different particle sizes and have low aspect ratio (Peng et al. 2015). Several study on WPC particularly on wood flour reinforced polypropylene composites reported that solid-state shear milling or pulverization of wood flour with and without water significantly affected the physical and mechanical properties of WF/PP composites. It is also reported that wet pulverized wood flour (i.e. pulverization of wood flour with water) improved the performance of WPC materials (Delviawan et al. 2019; Niu et al. 2016; Haue et al. 2019; Haue et al. 2019a; Haue et al. 2019b; Isa et al. 2014; Murayama et al. 2019). Wet pulverization of WF is a nonhazardous and simple mechanical treatment. It is a very suitable approach compared with any other

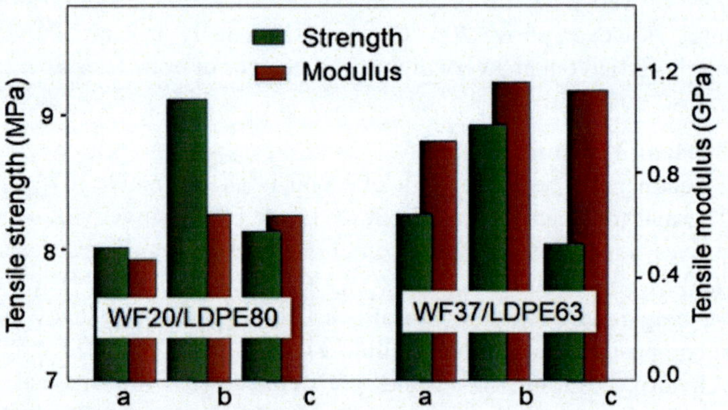

Fig. 10 Tensile strength and modulus of 20 wt% and 37 wt% wood flour filled LDPE composites: (a) untreated WF, (b) heat treated WF at 180 °C, (c) heat treated WF at 200 °C (Lafia-Araga et al. 2012;).

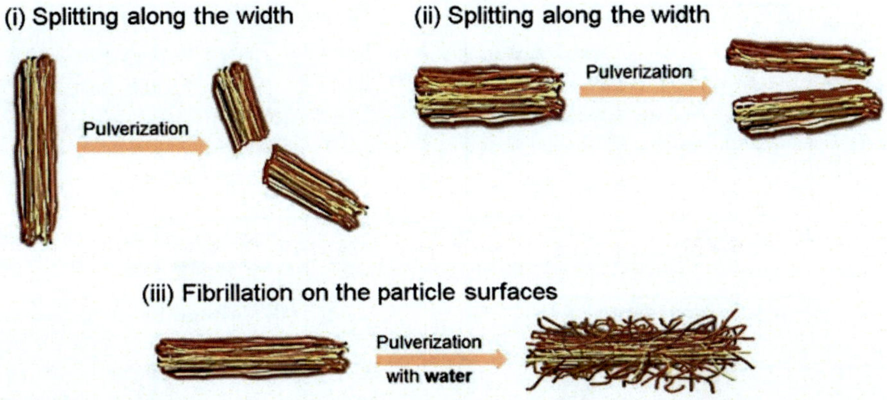

Fig. 11 Scheme represents three mechanisms of wood flour particle splitting by pulverization (Haque et al. 2019)

existing treatment of wood flour. Now aquestion is raised, what happen if wood flours are wet pulverized? If wood flours are pulverized, wood flour particles may be splitted by following the three mechanisms as shown in Fig. 11, (i) splitting of longer particles along the width, (ii) splitting of short particles along the length and, (iii) making fibrillation on the particle surfaces.

It can be assumed that wood flour particles are splitted by following the first and second mechanisms mostly when wood flours are dry pulverized. On the other hand, wood flour particles are splitted by following the third mechanism too when wood flours are wet pulverization. Splitting of wood flour particles by the first mechanism can increase the surface area of WF only, but it decrease aspect ratio of WF. On the other hand, splitting of wood particles by the second mechanism can increase both the surface area and aspect ratio of WF. However, splitting of wood particles by the third mechanism can create fibrillation on the wood flour particle surfaces. As a result surface roughness of WF is increased. The fibrils on WF particle surfaces can behave like as anchor in polymer matrices that results more stress transfer from polymer matrix to WF. In wet pulverization of wood flour, water acts as lubricant as well as water makes weak the inter fibrillar bonds through the interaction with them. If diameter of the fibrils are in nanometer scale then the fibrillated wood particles may exhibit significant effect on the properties of WPC like as nanofillers as discussed in the next section. Although, the three mechanisms of wood flour splitting have different effects on the characteristics of wood flour, but in real, pulverization of wood flour does not split particle by a single mechanism as mentioned in Fig. 11 rather particle splitting is accomplished by the combination of the three mechanisms. Therefore, the resultant effect of the three mechanisms followed in pulverization dominates the physical and mechanical properties of WPC materials (Haque et al. 2019). Melt-viscosity, tensile strength, impact strength, fatigue life etc. of WPCs can be improved by using wet pulverized WF as shown in Fig. 12 (Delviawan et al. 2019; Haque et al. 2019; Haque et al. 2019a).

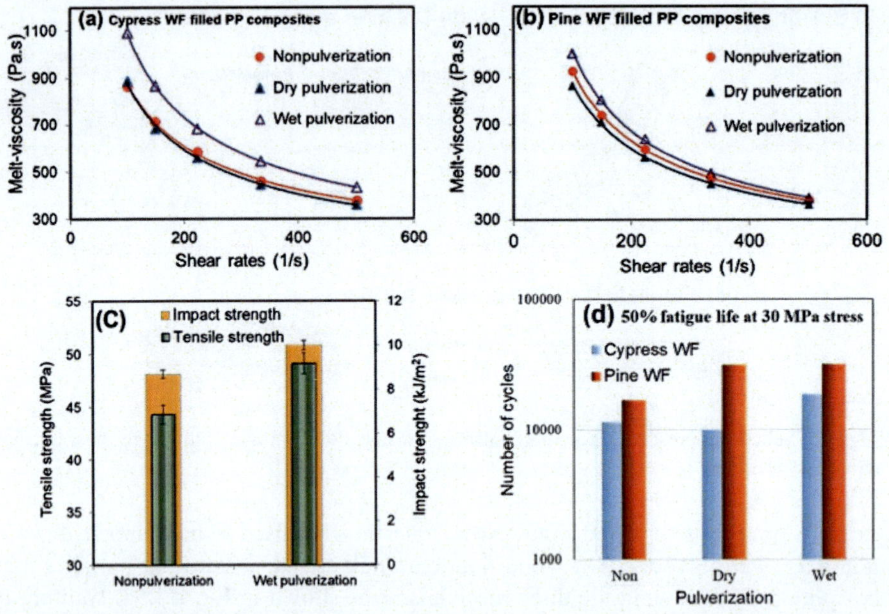

Fig. 12 Pulverization effect on the (a, b) melt-viscosity, (c) tensile strength and impact strength, (d) fatigue life of 25 wt% wood flour filled PP composites (Delviawan et al. 2019; Haque et al. 2019; Haque et al. 2019a)

3 Property Improvements by Incorporating Nanofillers

Nanofiller or nanomaterial is such a material whose one dimension (at least) such as length, width, diameter, thickness etc. is in the range of 1–100 nm. Nanomaterials have a high surface area. This particular characteristic of nanofillers can increase interfacial interaction with polymer matrices manifold compared with microfillers. Of course the higher interaction depend on the homogeneous dispersion of nanoparticles within polymer matrices. Incorporation of little amount of nanomaterials or nanofillers such as nanoclay, nanosilica, carbon nanotube, TiO_2 nanoparticles etc., in WPC can improve the thermal stability, mechanical properties, and moisture barrier properties of WPC materials. However, the performance of nanofller filled WPC materials is largely affected by the processing methods of nanofiller filled WPC materials. Single step or direct blending of polymer matrix, coupling agent, wood flour and nanofillers is not effective to disperse nanofillers homogeneously. Therefore, nanofillers incorporated wood plastic composites are prepared in two steps. Firstly nanofillers are blended with coupling agent and polymer matrix as a masterbatch. Then in the second step the masterbatch is blended with wood flour. Thus the dispersion of nanofillers in WPC materials is facilitated (Faruk and Matuana 2008; Turku and Kärki 2014). The performance of nanofllers filled WPC materials is also affected by the chemical compatibility i.e.

interaction of nanofillers with wood flour and polymer matrices. Among the various nanofillers such as nanoclay, nanosilica, carbon nanotube, TiO_2 nanoparticles etc. the most commonly used nanofillers are nanoclay as for example montmorillonite (MMT) nanoclay. The chemical composition of MMT nanoclay is $M_x(Al_{4-x}Mg_x)Si_8O_{20}(OH)_4$ (Turku and Kärki 2014; Alexandre and Dubois 2000). This pristine MMT nanoclay is a layered structure crystalline material in which each silicate layer attracts other layers by van der Waals forces.

The pristine MMT, due to hydroxyl groups on silicate layers, has hydrophilic surface which is incompatible with hydrophobic polymer matrix. Therefore, coupling agents such MAPP is used as compatibilizer whose polar part i.e. anhydride groups is reacted with the hydroxyl groups of silicate layers of nanocaly. Alternatively, the surfaces of silicate layers are modified by surfactants to convert the hydrophilic surfaces of nanoclay to hydrophobic. Thus, enhancing interaction of nanoclay with hydrophobic polymer matrices the dispersion of nanoclay is improved. Based on the degree of dispersion of nanoclay layers, WPC materials are classified as intercalated or exfoliated as shown in Fig. 13 (Alexandre and Dubois 2000). Exfoliated systems shows better performance compared with intercalated nanofillers filled WPC materials. As described above, pristine nanoclay is chemically modified to enhance the compatibility with the hydrophobic polymer matrices. Similarly, the surfaces of carbon nanotube, silica nanoparticles, TiO_2 nanoparticles etc. are also modified by various chemicals to improve the interfacial interaction of these nanofillers with WPC. CNTs are hydroxylated as the OH groups of CNT-OH

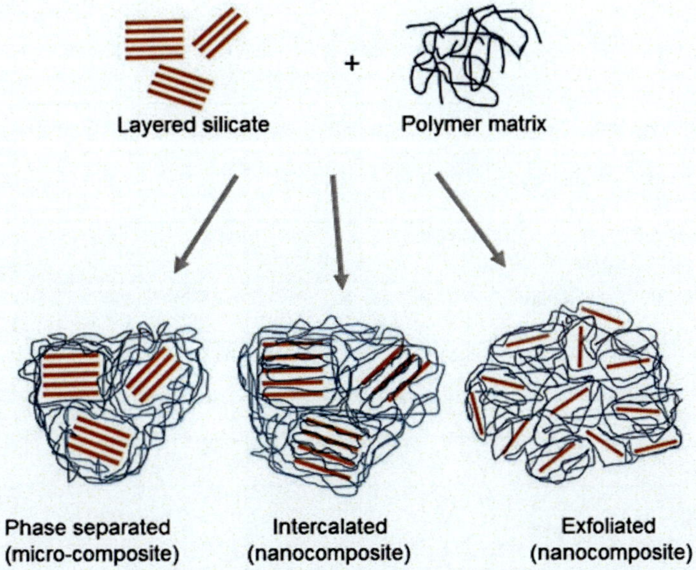

Fig. 13 A schematic representation of the intercalated and exfoliated nanocomposites (Alexandre and Dubois 2000)

have good compatibility with strong polar wood flour surfaces. In addition, hydroxyl groups of CNT-OH may also interact with maleated polypropylene (Fu et al. 2010). Silica nanoparticles and TiO_2 nanoparticles are usually modified by surfactants to enhance hydrophobicity. The -OH groups on these metal oxides surfaces interact with the functional groups of WF and maleated polypropylene, leading to strong interfacial interaction of these nanofillers with wood flour and polymer matrices in WPC materials (Deka and Maji 2011, 2012). The effect of different nanofillers on the tensile and flexural properties of different WPC materials is shown in Table 2.

Table 2 Tensile and flexural properties of WPC materials (with and without nanofillers)

Wood plastic composites	Type of nanofillers	Tensile		Flexural		References
		Strength (MPa)	Modulus (GPa)	Strength (MPa)	Modulus (GPa)	
WF40/MAPE5/PB55	–	20.3	0.27	19.3	3.9	(Deka and Maji 2012)
WF40/MAPE5/PB52	Nanosilica (3%)	33.8	0.60	26.5	4.8	(Deka and Maji 2012)
WF40/MAPE5/PB55	–	17.1	0.26	16.7	3.8	(Deka and Maji 2011)
WF40/MAPE5/PB51	Nanoclay (3%), TiO_2 (1%)	33.5	0.60	28.6	4.8	(Deka and Maji 2011)
WF40/MAPE5/PB49	Nanoclay (3%), TiO_2 (6%)	36.0	0.65	33.9	5.1	(Deka and Maji 2011)
WF40/PP60	–	16.4	1.9	–	–	(Lee and Kim 2009)
WF40/MAPP3/PP57	–	33.2	2.2	–	–	(Lee and Kim 2009)
WF40/MAPP3/PP57	Nanoclay (1 phr)	33.7	2.4	–	–	(Lee and Kim 2009)
WF10/PVC90	–	31.9	–	61.8	2.54	(Yim and Kim 2012)
WF10/PVC90	Nanoclay (3 phr)	37.0	–	63.5	2.57	(Yim and Kim 2012)
SWF10/PVC90	Nanoclay (3 phr)	41.6	–	70.1	–	(Yim and Kim 2012)
WF50/MAPE4/HDPE46	–	12.6	1.1	–	–	(Hafshejani et al. 2019)
WF50/MAPE4/HDPE46	Nanoclay (2.5 phr)	15.7	1.18	–	–	(Hafshejani et al. 2019)

(continued)

Table 2 (continued)

Wood plastic composites	Type of nanofillers	Tensile		Flexural		References
		Strength (MPa)	Modulus (GPa)	Strength (MPa)	Modulus (GPa)	
WF40/ MAPE4/ HDPE56	–	17.3	2.15	–	–	(Tabar et al. 2015)
WF40/ MAPE4/ HDPE54	Nanosilica (2%)	18.3	2.24	–	–	(Tabar et al. 2015)
WF50/ LDPE50/ MAPE 3phr	–	–	–	63.0	1.4	(Kordkheili et al. 2013)
WF50/ LDPE50/ MAPE 3phr	Carbon nanotube (3 phr)	–	–	156.0	4.5	(Kordkheili et al. 2013)

Note: *PB (polymer blend)- blend of LDPE, HDPE, PP and PVC as 1:1:1:0.5 ratio; SWF-silane treated wood flour*

4 Conclusion and Future Directions

The properties of wood-plastic composites depend on several factors such as type of wood species, size of wood flour particles, type of polymer matrices, dispersion of wood flour in polymer matrix, interfacial interaction of wood particle and polymer matrix etc. However, interfacial interaction or adhesion of wood particle and polymer matrix play important role on the improvement in properties of WPC. Incorporation of neat wood flour in polymer matrices do not improve most of the properties of WPC because of the incompatibility of wood flour and polymer matrices. Therefore, a compatibilization technique is necessary to improve the properties of WPC. The incombatibility between wood flour and polymer matrices can be removed by either modification of wood flour surfaces or by adding a third component or coupling agent which is compatible with both wood flour and polymer matrix. Addition of coupling agent can be the most effective technique. It can create a strong interphase (WF-coupling agent-polymer matrix) by making a strong bond between wood flour particle and polymer matrix. There are number of wood flour modification processes such as silane treatment, acetylation, benzylation, peroxide treatment, thermal treatment etc. and coupling agents e.g. maleic anhydride graft polyolefins are available to overcome the incompatibility between wood flour and polymer matrix. Maleated polyolefins can create a strong interphase by making covalent bonds, through the reaction of polar anhydride functional groups with the hydroxyl groups of wood flour, on wood flour surfaces, and making chain entanglements with polymer matrix. Synergistic effect on the properties of WPC can also be achieved by combining two or more modification processes. However, more research need to be carried out on it. Overall WPC fabrication cost should also be considered when combining various modification processes for achieving synergistic effect. Recently, developed processes such as wet pulverization of wood flour and nanofillers addition also have significant effect on the improvement in properties of wood plastic composites. Wet pulverization of wood flour can create fibrillation on the WF particle surfaces and mechanical properties such as tensile strength, impact strength and fatigue life of WPC are improved by the wet pulverized WF. More details study on the fatigue life of WPC is necessary, if WPCs are particularly made for structural application. Small amount of nanofillers such as nanoclay, silica nanoparticles, TiO_2 nanoparticles, carbon nanotubes etc. dispersion in WPC improved the properties of wood-plastic composites significantly. However, homogeneous dispersion of nanofillers into WPC is a challenging task that need to be developed more.

References

Alexandre M, Dubois P (2000) Polymer-layered silicate nanocomposites: preparation, properties and uses of a new class of materials. Mater Sci Eng R Rep 28(1–2):1–63

Alrubaie MAA, Lopez-Anido RA, Gardner DJ (2020) Flexural creep behavior of high-density polyethylene lumber and wood plastic composite lumber made from thermally modified wood. Polymers 12(2):262

Askanian H, Verney V, Commereuc S, Guyonnet R, Massardier V (2015) Wood polypropylene composites prepared by thermally modified fibers at two extrusion speeds: mechanical and viscoelastic properties. Holzforschung 69(3):313–319

Ayrilmis N, Jarusombuti S, Fueangvivat V, Bauchongkol P (2011) Effect of thermal-treatment of wood fibres on properties of flat-pressed wood plastic composites. Polym Degrad Stab 96 (5):818–822

Borah JS, Kim DS (2016) Recent development in thermoplastic/wood composites and nanocomposites: a review. Korean J Chem Eng 33(11):3035–3049

Bengtsson M, Oksman K, Stark NM (2006) Profile extrusion and mechanical properties of crosslinked wood–thermoplastic composites. Polym Compos 27(2):184–194

Bengtsson M, Oksman K (2006) Silane crosslinked wood plastic composites: processing and properties. Compos Sci Technol 66(13):2177–2186

Berger MJ, Stark NM (1997) Investigations of species effects in an injection-molding-grade, wood-filled polypropylene. In The fourth international conference on woodfiber-plastic composites, pp. 19–25, May 1997

Chan CM, Vandi LJ, Pratt S, Halley P, Richardson D, Werker A, Laycock B (2018) Composites of wood and biodegradable thermoplastics: a review. Polym Rev 58(3):444–494

Chao YY, Hung KC, Xu JW, Wu TL, Wu JH (2020) Effects of acetylated veneer on the natural weathering properties of adhesive-free veneer overlaid wood-plastic composites. Polymers 12 (3):513

Cui Y, Lee S, Noruziaan B, Cheung M, Tao J (2008) Fabrication and interfacial modification of wood/recycled plastic composite materials. Compos A Appl Sci Manuf 39(4):655–661

Dányádi L, Móczó J, Pukánszky B (2010) Effect of various surface modifications of wood flour on the properties of PP/wood composites. Compos A Appl Sci Manuf 41(2):199–206

Deka BK, Maji TK (2011) Effect of TiO_2 and nanoclay on the properties of wood polymer nanocomposite. Compos A Appl Sci Manuf 42(12):2117–2125

Deka BK, Maji TK (2012) Effect of silica nanopowder on the properties of wood flour/polymer composite. Polym Eng Sci 52(7):1516–1523

Delviawan A, Kojima Y, Kobori H, Suzuki S, Aoki K, Ogoe S (2019) The effect of wood particle size distribution on the mechanical properties of wood–plastic composite. J Wood Sci 65(1):1–11
Djidjelli H, Benachour D, Boukerrou A, Zefouni O, Martinez-Véga J, Farenc J, Kaci M (2007) Thermal, dielectric and mechanical study of poly (vinyl chloride)/olive pomace composites. Express Polym Lett 1(12):846–852
Dominkovics Z, Dányádi L, Pukanszky B (2007) Surface modification of wood flour and its effect on the properties of PP/wood composites. Compos A Appl Sci Manuf 38(8):1893–1901
Fang L, Chang L, Guo WJ, Chen Y, Wang Z (2014) Influence of silane surface modification of veneer on interfacial adhesion of wood–plastic plywood. Appl Surf Sci 288:682–689
Faruk O, Matuana LM (2008) Nanoclay reinforced HDPE as a matrix for wood-plastic composites. Compos Sci Technol 68(9):2073–2077
Farsi M (2010) Wood-plastic composites: influence of wood flour chemical modification on the mechanical performance. J Reinf Plast Compos 29(24):3587–3592
Fayzullin IZ, Volfson SI, Musin IN, Fayzullin AZ, Nikiforov AA. (2016) Influence of the type of wood flour and nanoadditives on the structure and mechanical properties of polypropylene-based wood-polymer composites. In: AIP conference proceedings, vol 1785, no 1, p 040098. AIP Publishing LLC, November 2016
Fu S, Song P, Yang H, Jin Y, Lu F, Ye J, Wu Q (2010) Effects of carbon nanotubes and its functionalization on the thermal and flammability properties of polypropylene/wood flour composites. J Mater Sci 45(13):3520–3528
Geng Y, Laborie MPG (2010) The impact of silane chemistry conditions on the properties of wood plastic composites with low density polyethylene and high wood content. Polym Compos 31 (5):897–905
Gregorova A, Wimmer R, Hrabalova M, Koller M, Ters T, Mundigler N (2009) Effect of surface modification of beech wood flour on mechanical and thermal properties of poly (3-hydroxybutyrate)/wood flour composites. Holzforschung 63(5):565–570
Hafshejani KT, Khorasani SN, Jahadi M, Hafshejani MS, Neisiany RE (2019) Improving mechanical and thermal properties of high-density polyethylene/wood flour nanocomposites. J Therm Anal Calorim 137(1):175–183
Hämäläinen K, Kärki T (2014) Effects of wood flour modification on the fire retardancy of wood–plastic composites. Eur J Wood Wood Prod 72(6):703–711
Harnnarongchai W, Kaschta J, Schubert DW, Sombatsompop N (2012) Shear and elongational flow properties of peroxide-modified wood/low-density polyethylene composite melts. Polym Compos 33(11):2084–2094
Haque MMU, Goda K, Ito H, Ogoe S, Okamot M, Ema T, Nogami H (2019) Melt-viscosity and mechanical behaviour of polypropylene (PP)/wood flour composites: effect of pulverization of wood flour with and without water. Adv Ind Eng Polymer Res 2(1):42–50
Haque MMU, Goda K, Ito H, Ogoe S, Okamoto M, Ema T, Nogami H (2019a) Fatigue performance of wet and dry pulverized wood flour reinforced PP composites. J Comp Sci 3 (1):20
Haque MMU, Goda K, Ogoe S, Sunaga Y (2019b) Fatigue analysis and fatigue reliability of polypropylene/wood flour composites. Adv Ind Eng Polymer Res 2(3):136–142
Hon DNS, Ou NH (1989) Thermoplasticization of wood. I. Benzylation of wood. J Polym Sci A Polym Chem 27(7):2457–2482
Hon DNS, Chao WY (1993) Composites from benzylated wood and polystyrenes: their processability and viscoelastic properties. J Appl Polym Sci 50(1):7–11
Hung KC, Wu TL, Chen YL, Wu JH (2016) Assessing the effect of wood acetylation on mechanical properties and extended creep behavior of wood/recycled-polypropylene composites. Constr Build Mater 108:139–145
Hung KC, Wu JH (2010) Mechanical and interfacial properties of plastic composite panels made from esterified bamboo particles. J Wood Sci 56(3):216–221
Hu X, Li D, Luo B, Li L (2020) Weathering characteristics of wood-plastic composites compatibilized with ethylene vinyl acetate. BioResources 15(2):3930–3944

Ibach RE, Clemons CM (2017) Long term durability of wood-plastic composites made with chemically modified wood. In: Advancements in fiber-polymer composites conference proceedings, Madison, WI, pp 1–17

Ichazo MN, Albano C, Gonzalez J, Perera R, Candal AM (2001) Polypropylene/wood flour composites: treatments and properties. Compos Struct 54(2–3):207–214

Isa A, Toyoda T, Suzuki S, Kojima Y, Ito H, Makise R, Okamoto M (2014) The effects of wet-milled wood flour on the mechanical properties of wood flour/polypropylene composites. J Wood Chem Technol 34(1):20–30

Jebrane M, Sebe G (2007) A novel simple route to wood acetylation by transesterification with vinyl acetate. Holzforschung 61(2):143–147

Kaboorani A (2009) Thermal properties of composites made of heat-treated wood and polypropylene. J Compos Mater 43(22):2599–2607

Kaboorani A, Faezipour M, Ebrahimi G (2008) Feasibility of using heat treated wood in wood/thermoplastic composites. J Reinf Plast Compos 27(16–17):1689–1699

Kajaks J, Zagorska A, Mezinskis A (2015) Some exploitation properties of wood plastic composites (WPC) based on high density polyethylene and timber industry waste. J Mater. Sci 21(3):396–399

Kallakas H, Shamim MA, Olutubo T, Poltimäe T, Süld TM, Krumme A, Kers J (2015) Effect of chemical modification of wood flour on the mechanical properties of wood-plastic composites. Agron Res 13(3):639–653

Källbom S, Lillqvist K, Spoljaric S, Seppälä J, Segerholm K, Rautkari L, Wålinder M (2020) Effects of water soaking–drying cycles on thermally modified spruce wood–plastic composites. Wood Fiber Sci 52(1):2–12

Kamdem DP, Pizzi A, Jermannaud A (2002) Durability of heat-treated wood. Holz Als Roh-Und Werkstoff 60(1):1–6

Kaymakci A, Birinci E, Ayrilmis N (2019) Surface characteristics of wood polypropylene nanocomposites reinforced with multi-walled carbon nanotubes. Compos B Eng 157:43–46

Kaymakci A (2019) Effect of titanium dioxide on some mechanical, thermal, and surface properties of wood-plastic nanocomposites. BioResources 14(1):1969–1979

Khonsari A, Taghiyari HR, Karimi A, Tajvidi M (2015) Study on the effects of wood flour geometry on physical and mechanical properties of wood-plastic composites. Maderas Ciencia Y Tecnología 17(3):545–558

Kim HS, Lee BH, Choi SW, Kim S, Kim HJ (2007) The effect of types of maleic anhydride-grafted polypropylene (MAPP) on the interfacial adhesion properties of bio-flour-filled polypropylene composites. Compos A Appl Sci Manuf 38(6):1473–1482

Kim TW, Lee SY, Chun SJ, Doh GH, Paik KH (2011) Effect of silane coupling on the fundamental properties of wood flour reinforced polypropylene composites. J Compos Mater 45(15):1595–1605

Kocaefe D, Poncsak S, Boluk Y (2008) Effect of thermal treatment on the chemical composition and mechanical properties of birch and aspen. BioResources 3(2):517–537

Kokta BV, Raj RG, Daneault C (1989) Use of wood flour as filler in polypropylene: Studies on mechanical properties. Polym-Plast Technol Eng 28(3):247–259

Koohestani B, Ganetri I, Yilmaz E (2017) Effects of silane modified minerals on mechanical, microstructural, thermal, and rheological properties of wood plastic composites. Compos B Eng 111:103–111

Kordkheili HY, Farsi M, Rezazadeh Z (2013) Physical, mechanical and morphological properties of polymer composites manufactured from carbon nanotubes and wood flour. Compos B Eng 44(1):750–755

Kurimoto Y, Sasaki S (2013) Preparation of acetylated wood meal and polypropylene composites I: acetylation of wood meal by mechanochemical processing and its characteristics. J Wood Sci 59(3):209–215

Kurimoto Y, Sasaki S (2013a) Preparation of acetylated wood meal and polypropylene composites II: mechanical properties and dimensional stability of the composites. J Wood Sci 59(3):216–220

Lafia-Araga RA, Hassan A, Yahya R, Rahman NA, Hornsby PR, Heidarian J (2012) Thermal and mechanical properties of treated and untreated Red Balau (Shorea dipterocarpaceae)/LDPE composites. J Reinf Plast Compos 31(4):215–224

Lee H, Kim DS (2009) Preparation and physical properties of wood/polypropylene/clay nanocomposites. J Appl Polym Sci 111(6):2769–2776

Lee SY, Yang HS, Kim HJ, Jeong CS, Lim BS, Lee JN (2004) Creep behavior and manufacturing parameters of wood flour filled polypropylene composites. Compos Struct 65(3–4):459–469

Lei Y, Wu Q (2010) Wood plastic composites based on microfibrillar blends of high density polyethylene/poly (ethylene terephthalate). Biores Technol 101(10):3665–3671

Lisperguer J, Droguett C, Ruf B, Nunez M (2007) The effect of wood acetylation on thermal behavior of wood-polystyrene composites. J Chil Chem Soc 52(1):1073–1075

Luo S, Cao J, Peng Y (2014) Properties of glycerin-thermally modified wood flour/polypropylene composites. Polym Compos 35(2):201–207

Ma Y, He H, Huang B, Jing H, Zhao Z (2020) In situ fabrication of wood flour/nano silica hybrid and its application in polypropylene-based wood-plastic composites. Polym Compos 41 (2):573–584

Malakani M, Bazyar B, Talaiepour M, Hemmasi AH, Ghasemi I (2015) Effect of acetylation of wood flour and MAPP content during compounding on physical properties, decay resistance, contact angle, and morphology of polypropylene/wood flour composites. BioResources 10 (2):2113–2129

Maldas D, Kokta BV, Raj RG, Daneault C (1988) Improvement of the mechanical properties of sawdust wood fibre—polystyrene composites by chemical treatment. Polymer 29(7):1255–1265

Matuana LM, Woodhams RT, Balatinecz JJ, Park CB (1998) Influence of interfacial interactions on the properties of PVC/cellulosic fiber composites. Polym Compos 19(4):446–455

Mbarek TB, Robert L, Sammouda H, Charrier B, Orteu JJ, Hugot F (2013) Effect of acetylation and additive on the tensile properties of wood fiber–high-density polyethylene composite. J Reinf Plast Compos 32(21):1646–1655

Mengeloğlu F, Karakuş K (2008) Some properties of eucalyptus wood flour filled recycled high density polyethylene polymer-composites. Turk J Agric For 32(6):537–546

Müller M, Radovanovic I, Grüneberg T, Militz H, Krause A (2012) Influence of various wood modifications on the properties of polyvinyl chloride/wood flour composites. J Appl Polym Sci 125(1):308–312

Murayama K, Ueno T, Kobori H, Kojima Y, Suzuki S, Aoki K, Okamoto M (2019) Mechanical properties of wood/plastic composites formed using wood flour produced by wet ball-milling under various milling times and drying methods. J Wood Sci 65(1):1–10

Myers GE, Chahyadi IS, Coberly CA, Ermer DS (1991) Wood flour/polypropylene composites: influence of maleated polypropylene and process and composition variables on mechanical properties. Int J Polym Mater 15(1):21–44

Niemz P, Hofmann T, Rétfalvi T (2010) Investigation of chemical changes in the structure of thermally modified wood. Maderas. Ciencia Y Tecnología 12(2):69–78

Niu Z, Chen Y, Feng J (2016) Preparation, structure, and property of wood flour incorporated polypropylene composites prepared by a solid-state mechanochemical method. J Appl Polym Sci **133**(10)

Oksman K, Clemons C (1998) Mechanical properties and morphology of impact modified polypropylene–wood flour composites. J Appl Polym Sci 67(9):1503–1513

Özmen N, Çetin NS, Mengeloğlu F, Birinci E, Karakuş K (2013) Effect of wood acetylation with vinyl acetate and acetic anhydride on the properties of wood-plastic composites. BioResources 8(1):753–767

Peng Y, Liu R, Cao J (2015) Characterization of surface chemistry and crystallization behavior of polypropylene composites reinforced with wood flour, cellulose, and lignin during accelerated weathering. Appl Surf Sci 332:253–259

Peng Y, Li X, Wang W, Cao J (2020) Photodegradation of wood flour/polypropylene composites incorporated with carbon materials with different morphologies. Wood Mat Sci Eng 15(2):104–113

Reddy MM, Vivekanandhan S, Misra M, Bhatia SK, Mohanty AK (2013) Biobased plastics and bionanocomposites: current status and future opportunities. Prog Polym Sci 38(10–11):1653–1689

Seo YW, Kim DS (2014) Effects of wood flour size on the physical properties of polypropylene/wood flour composites. Polymer Korea 38(3):327–332

Sewda K, Maiti SN (2013) Dynamic mechanical properties of high density polyethylene and teak wood flour composites. Polym Bull 70(10):2657–2674

Shah BL, Matuana LM, Heiden PA (2005) Novel coupling agents for PVC/wood-flour composites. J Vinyl Add Tech 11(4):160–165

Soccalingame L, Bourmaud A, Perrin D, Bénézet JC, Bergeret A (2015) Reprocessing of wood flour reinforced polypropylene composites: impact of particle size and coupling agent on composite and particle properties. Polym Degrad Stab 113:72–85

Stark NM, Berger MJ (1997) Effect of particle size on properties of wood-flour reinforced polypropylene composites. In: Proceedings of the fourth international conference on woodfibre–plastic composites, pp 12–14, May 1997

Tabar MM, Tabarsa T, Mashkour M, Khazaeian A (2015) Using silicon dioxide (SiO_2) nano-powder as reinforcement for walnut shell flour/HDPE composite materials. J Indian Acad Wood Sci 12(1):15–21

Tisserat B, Reifschneider L, Joshee N, Finkenstadt VL (2013) Properties of high density polyethylene–Paulownia wood flour composites via injection molding. BioResources 8 (3):4440–4458

Tuong VM, Li J (2010) Effect of heat treatment on the change in color. BioResources 5(2):1257–1267

Turku I, Kärki T (2014) Research progress in wood-plastic nanocomposites: a review. J Thermoplast Compos Mater 27(2):180–204

Xie Y, Hill CA, Xiao Z, Militz H, Mai C (2010) Silane coupling agents used for natural fiber/polymer composites: a review. Compos A Appl Sci Manuf 41(7):806–819

Yadav SM, Yusoh KB (2019) Sub-surface mechanical properties and sub-surface creep behavior of wood-plastic composites reinforced by organoclay. Sci Eng Compos Mater 26(1):114–121

Yeh SK, Agarwal S, Gupta RK (2009) Wood–plastic composites formulated with virgin and recycled ABS. Compos Sci Technol 69(13):2225–2230

Ye X, Wang H, Wu Z, Zhou H, Tian X (2018) The functional features and interface design of wood/polypropylene composites based on microencapsulated wood particles via adopting in situ emulsion polymerization. Polym Compos 39(2):427–436

Yim H, Kim DS (2012) Physical properties of PVC/aminosilane-treated wood flour/organoclay composites. Polym Adv Technol 23(11):1441–1445

Zhao Y, Wang K, Zhu F, Xue P, Jia M (2006) Properties of poly (vinyl chloride)/wood flour/montmorillonite composites: effects of coupling agents and layered silicate. Polym Degrad Stab 91(12):2874–2883

Zhu L, Cao J, Wang Y, Liu R, Zhao G (2014) Effect of MAPP on interfacial compatibility of wood flour/polypropylene composite evaluated with dielectric approach. Polym Compos 35(3):489–494

Morphological Characterization of the Wood Polymer Composites

M. Chandrasekar, K. Senthilkumar, T. Senthil Muthu Kumar, R. Sabarish, and Suchart Siengchin

Abstract This work addresses the role of various imaging techniques in investigating the morphology of the wood and wood polymer composites (WPC). Micrographs from the scanning electron microscope (SEM) and field emission scanning electron microscope (FESEM) were helpful in assessing the surface changes in the fibre and matrix due to the fibre treatment and coupling agent, interfacial adhesion between the fibre-matrix, failure behavior under various loads and degradation effects due to the aging. Images from the transmission electron microscope (TEM) were effective in determining the dispersion characteristics of additives in the polymer matrix. In addition to the microstructural images, quantitative assessment of the surface changes in the fibre and matrix can be obtained from the laser scanning microscope (LSM) and atomic force microscope (AFM) images.

Keywords Wood polymer composites · Micrographs · Interfacial adhesion · Aging effects · Surface modification

M. Chandrasekar (✉)
School of Aeronautical Sciences, Hindustan Institute of Technology and Science, Padur, Kelambakkam, Chennai 603103, Tamilnadu, India

K. Senthilkumar
Center of Innovation in Design and Engineering for Manufacturing (CoI-DEM), King Mongkut's University of Technology North Bangkok, 1518 Wongsawang Road, Bangsue, Bangkok 10800, Thailand

T. Senthil Muthu Kumar
Department of Mechanical Engineering, Kalasalingam Academy of Research and Education, Anand Nagar, Krishnankoil 626126, Tamilnadu, India

R. Sabarish · S. Siengchin
Department of Materials and Production Engineering, The Sirindhorn International Thai-German Graduate School of Engineering (TGGS), King Mongkut's University of Technology North Bangkok, 1518 Wongsawang Road, Bangsue, Bangkok 10800, Thailand

S. Mavinkere Rangappa et al. (eds.), *Wood Polymer Composites*, Composites Science and Technology, https://doi.org/10.1007/978-981-16-1606-8_5

1 Introduction

Generally, the wood polymer composites (WPC) are fabricated from a variety of wood products: (i) fibre or cellulose (ii) logs (iii) strands (iv) sawn timber (v) chips (vi) wafers. Initially, the leaves and bark of the tree are removed followed by slicing into pieces based on the wood geometry. The wood pieces are subjected to the drying process for removing moistures, followed by selecting the right quality wooden pieces. About 90% of the stem portions are being utilized to produce the

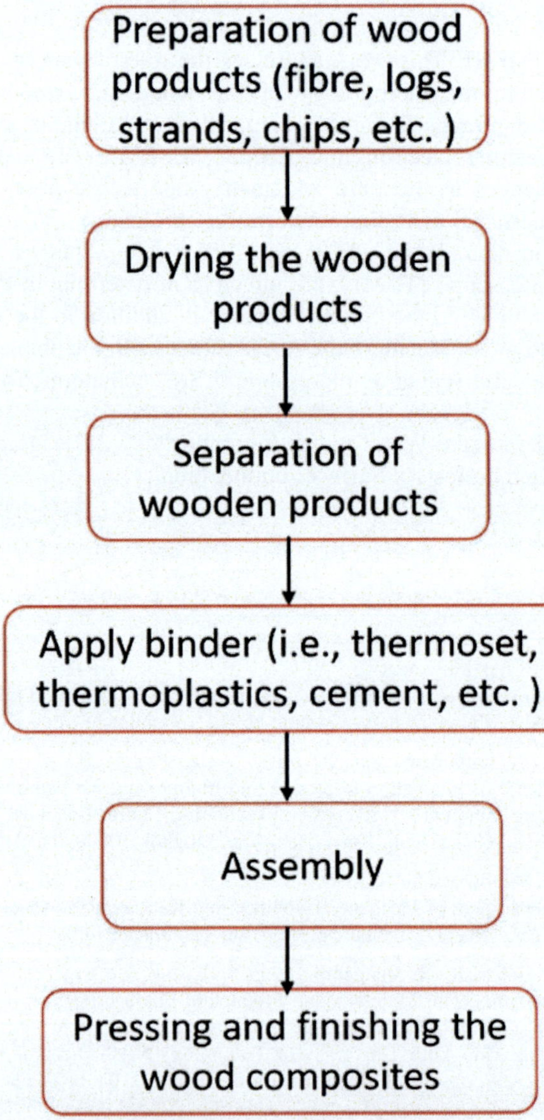

Fig. 1 Step-by-step procedure of fabricating wood polymer composites

WPC rather than other parts. Finally, the composite is obtained by holding wood products together using an adhesive binder. The WPC can be produced by pressing, extrusion, impregnation, etc. Figure 1 shows the step-by-step procedure of fabricating the WPC (Rowell 2012; Ansell 2015b; He et al. 2020; Sun et al. 2020).

1.1 Factors Influencing Properties of the WPC

The properties of the WPC depend upon on their constituents such as wood and polymer. In case of the wood, factors such as (i) the arrangement of the wood (ii) size, and (iii) shape influence the properties. For instance, Fig. 2 shows the microstructures of coniferous Douglas fir, and deciduous English oak respectively (Ansell 2015a). The strength and stiffness of WPC can also be tailored by altering the fibre orientation (Senthilkumar et al. 2018, 2019). On the other hand, use of polymers such as thermoset, thermoplastics and elastomers plays a vital role in the terms of strength and stiffness. Polymers can be further classified based on whether they are synthesized from organic and inorganic sources.

Curing of wood with the polymers should not exceed 200 °C, as further increase in processing temperature during the curing would lead to thermal degradation of the wood. Thermoset polymers such as epoxy, polyester, vinyl ester, etc. have curing temperatures well below 200 °C. Thermoplastic polymers are preferred over the thermoset polymers in their use as adhesive or binder for wood based materials. Thermoplastic has higher curing temperatures in the range between 150 °C–250 °C. A number of thermoplastic polymers such as phenol formaldehyde (PF), high-density

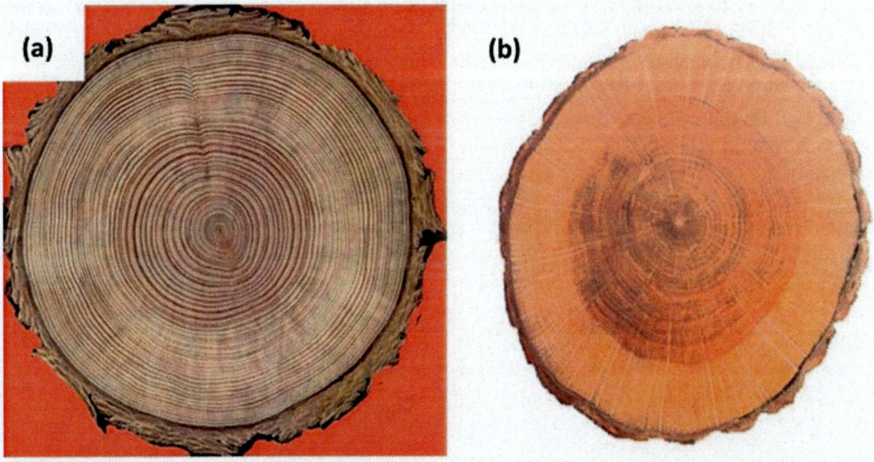

Fig. 2 The microstructure of wood (a) Coniferous Douglas Fir and (b) Deciduous English Oak (reused with permission, License number 4874720409441)

polyethylene (HDPE), low-density polyethylene (LDPE), polyvinyl chloride (PVC), polypropylene (PP), polystyrene (PS) has been used as binders over the thermosets and elastomers (Klyosov 2007). Table 1 shows a wide range of binders obtained from the renewable sources (Mark 2013; Biron 2018; Antov et al. 2020; Pizzi et al. 2020). Some reported work on thermoplastics and thermoset based wood-polymer composites are reported in Table 2.

Table 1 Types of binders produced from the renewable and organic sources

Thermoset binders	Natural binders	Other binders
Urea–Formaldehyde (UF)	Lignin	Elastomeric adhesives (based on natural or synthetic rubbers)
Melamine–Formaldehyde (MF)	Tannin	Hot melt adhesives (HMA)
Phenol–Formaldehyde (PF)	Soybean	Pressure-sensitive adhesives (PSA)
Resorcinol–Formaldehyde (RF)	Proteins	Poly (vinyl acetate) (PVA)
Isocyanates		

Table 2 Reported work on wood-based composites using thermoplastic and thermoset matrix

Type of composites	References
Lacquered spruce or oak solid wood, oak-veneered particle board/ thermoplastic composites	(Lipovac et al. 2020)
Pinus elliottii wood (sawdust)/recycled thermoplastics; Polyethylene terephthalate, HDPE, PP, PVC	(Lopez et al. 2020)
Foamed wood/liner low-density polyethylene	(Guo 2020)
Hardwood Kraft pulp/ PVA based composites	(Ko et al. 2020)
Wood fibre/recycled high-density polyethylene	(Grammatikos et al. 2020)
Silicon carbide ceramics thermoset matrix composites	(Fürst et al. 2018)
Recycled acrylonitrile butadiene styrene/waste bakelite powder (thermosetting) and wood dust composites	(Chawla et al. 2020)
Wood floor/chemically modified soybean oil composites	(Rosu et al. 2019)
Wood/PP composites	(Krause et al. 2018)

Mechanical properties such as strength, stiffness, impact, etc. and physical properties such as density, dimensional stability under curing, etc. are the deciding factors involved in the selection of the WPC in a particular application. The functional attributes of wood composites are defined by using their structural morphology. The wood polymer composites become more durable and attractive, whereas the composites are fabricated using the moisture resisting polymers, i.e., the polymers should have higher moisture resisting property than the wood.

Though the wood fibers have many advantages, their single drawback would be the hydrophilic nature. Thus, the wood fibers are responsible for dimensional changes (i.e., shrinkage and swelling); also, they are susceptible to insects and fungal attacks. Therefore, the wood fibers are mixed with both thermoset and thermoplastics for enhancing their functional properties; also, they tend to reduce the moisture absorption behavior. However, the moistures can be penetrated to the wood composites through the exposed sections. In most of the cases, the wood composites are subjected to coating for avoiding the moisture absorption.

1.2 Applications of the WPC

The demand for WPC is increasing day-by-day. It is a common building material in the construction industry. The applications include (i) structural beams (ii) flooring (iii) lumber (iv) fencing (v) paneling (vi) furniture (vii) palettes (viii) stair treads (ix) millwork (x) roof construction. Since WPC is widely used in many applications, their usage is expected to further increase in the future (Niska and Sain 2008; Mertens et al. 2018; Askadskii et al. 2019; Deviatkin et al. 2019; Yu et al. 2020). Table 3 lists a few of the essential applications of WPC.

2 Morphological Characterization of the WPC

Before the advent of scanning electron microscope (SEM), morphological features of the WPC were examined by images from the optical microscope. With SEM, micrographs of the composite in the magnification range from 50× to as high as 1000× is possible. Micrographs of the WPC specimens obtained from the SEM have been used by research community to investigate the following: (a) Failure behavior under the mechanical loads, (b) interfacial adhesion or wetting characteristic of the fiber and matrix and (c) degradation effects due to the moisture absorption and aging.

Table 3 Types of wood polymer composites, features, and their illustrations (Harris 2015; Hughes 2015; Ormondroyd and Stefanowski 2015) (Figures re-used with permission, License numbers 4874850310530; 4,874,850,591,309; 4,874,850,788,683)

Type of application	Features of wood-composites	Illustration
Medium-density fiberboards	• It was developed for furniture fittings. • It has a uniform fiber distribution. • They are used for garden furniture, ceiling tiles, construction panels, advertising boards, etc.	
laminated veneer lumber	• They are excellent in specific properties. • They are used from airscrew blades to sailing boats. • Also, these composites can be used in packaging, furniture, and construction. • It has good strength to weight ratio. Thus, it can be used in transporting related applications—for instance, flooring for railway wagons.	
Cross-laminated timber	• The building can be constructed in a rapid manner using the cross-laminated timber; primarily, these are used for making wall elements. • It is possible to make zero carbon emissions as the constructors approach these materials.	

A typical wood fiber is made up of the wood rays, fibers, lumens and vessels (Bao et al. 2019). In the engineering applications, wood plies are usually bonded together to form the laminate. For instance, in case of the veneer-plywood laminate used in Table, false partitions, chairs, etc., the veneer ply is bonded to the plywood with the phenol formaldehynde resin (PF). Penetration of the adhesive or PF resin into the wood structure can be assessed from the advancing imaging techniques such as Transimission Electron Microscopy (TEM) and X-ray Computed Topography (XCT) (Bao et al. 2019).

In a recent study, Yu et al. used an ultra-depth 3-dimensional optical microscopy to indicate the presence of the PF resin in the bamboo fiber bundle after their impregnation (Fig. 3(a) and (b)). PF resin impregnated into the wood cures and helps the wood to perform better under mechanical loads as well as to resist moisture. Further, it also helps in bonding the adjacent wood layers in case of the laminates with veneer and plywood or between the plywood's (Yu et al. 2017).

In addition to the micrographs, constituents of the specimen under investigation can also be indicated by energy dispersive x-ray spectroscopy (EDS) integrated along with the SEM. This is particularly useful for characterizing the nanocomposites where the presence of nanoparticles can be identified through the presence of elements in the EDS spectra.

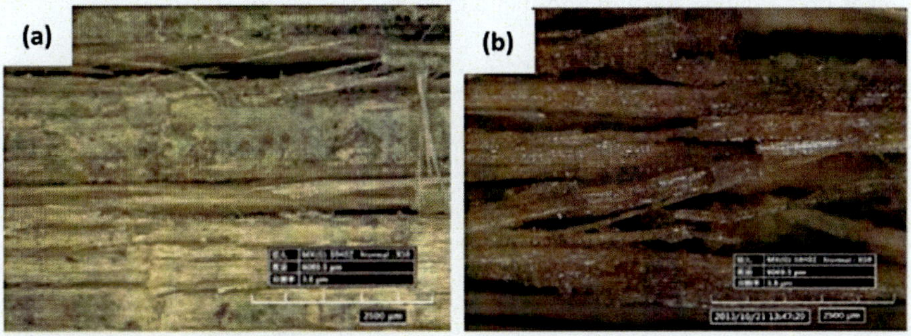

Fig. 3 Bamboo fiber bundle (a) before impregnation with the PF resin and (b) after impregnation with the PF resin (Yu et al. 2017) [Re-used with permission from Elsevier, License Number-4891351168318]

3 Interfacial Properties in Terms of Morphology of the WPC

The interfacial bonding in WPC in terms of the morphology is discussed in this section. WPC show weaker interfacial characteristics between the fiber and matrix. The hydrophilic nature of wood fiber is responsible for its weak adhesion with the polymer matrix. Studies indicate that the wood fiber-matrix adhesion could be improved by the addition of coupling agents, surface treatment method etc. (Bledzki et al. 2002; Gwon et al. 2010; Kim and Pal 2010; Renner et al. 2011). Figure 4 shows the methods that can be used to improve the performance of the WPC.

Kocsis and Czigany studied the failure mechanism of PP/WF composites. The morphological analysis showed that de-bonding and fiber pull-out were the main failure mechanism of the composite. Weak adhesion between fiber and polymer was also confirmed by Acoustic emission measurements (AE) (Kocsis and Czigány 2007). Gulitah and co-workers fabricated WPC using different recycled plastic (PP, HDPE, LDPE) and studied the performance of the resultant composite under the tensile load. Fiber pull-out, fiber breakage and voids were seen in the SEM micrographs (Fig. 5(a)–(f)). At lower fiber volume fraction, voids and fiber pull-out was more prevalent than the composite with higher fiber volume fraction. This behavior confirms the poor interfacial bonding between the fiber and polymer at lower fiber volume fraction (Gulitah and Liew 2019).

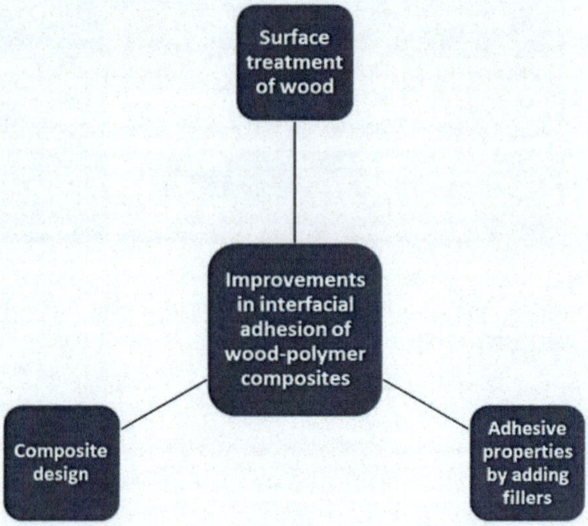

Fig. 4 Methods to improve the performance of the WPC

Yuan et al. fabricated a WPC/HDPE. The morphology of WPC in the presence and absence of coupling agent such as Maleic Anhydride grafting (MA) was examined using the SEM micrographs. They found that in the absence of coupling agent, the surface of WF/HDPE composite is smooth and wood fiber had weak adhesion with the polymer matrix, hence, fiber pull-out and fiber-matrix de-bonding was predominant. However, after incorporating the coupling agent, the surface of the composite turned rough; implying an improvement in the fiber-matrix compatibility. Instead of fiber pull-out, fiber fracture was observed in the composite treated with coupling agent. They also showed that increasing the fiber length was also favorable for their interaction with the polymer matrix (Yuan et al. 2008). According to Poletto (2017), use of MA as coupling agent with the PS resin resulted in significant improvement in the wetting between WF and polymer matrix. Better mechanical interlocking between the fiber and matrix due to the use of coupling led to superior strength and modulus under the tension, bending and impact. Zhang and co-workers investigated the influence of fiber weight of the glass yarn, aramid yarn, aramid roving and carbon yarn on the mechanical properties of the WPC. Based on SEM analysis, the authors suggested that the fiber type employed and the fiber weight plays a crucial role in determining the interfacial compatibility of the WPC. In the case of glass/WPC and carbon/WPC, the fibers were coated with polymer resins indicating a good interfacial bonding between fiber and matrix. Several cracks were visible in the interphase of aramid fibers and polymer while aramid fiber was found to attached to the WF rather than the polymer. Hence aramid yarn/roving reinforced WPC displayed less interfacial compatibility (Zhang et al. 2019). Maadeed et al. fabricated recycled PP/glass fiber/wood flour (WGFRPP) and compared its thermo-mechanical properties to glass/PP

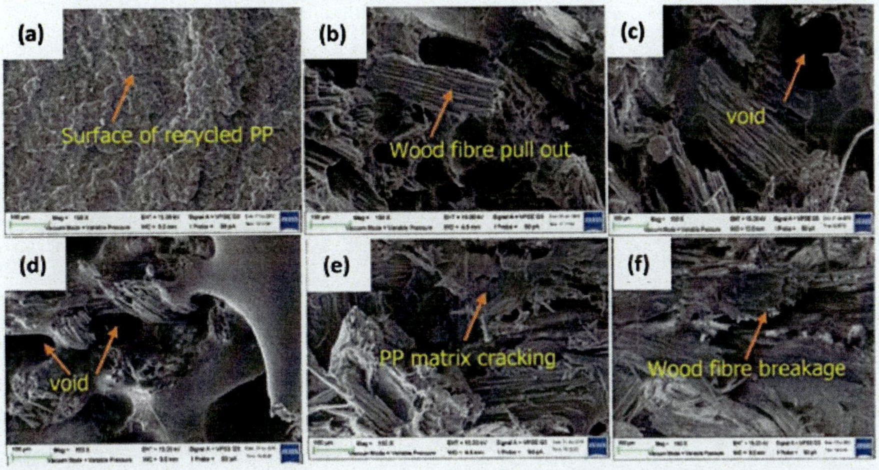

Fig. 5 SEM image of PP-WF (wood fiber) (a) PP-WF (100:0), (b) PP-WF (90:10), (c) PPWF (80:20), (d) PP-WF (70:30), (e) PP-WF (60:40) and (f) PP-WF (50:50) (Gulitah and Liew 2019) (Open access content)

composites (GFRPP). Composite with the glass fiber displayed fewer fiber pull-outs, shorter pull-out depth and less fiber-matrix de-bonding (Fig. 6(b)) than the hybrid composite with wood flour and glass fiber (Fig. 6(a)) (Al-Maadeed et al. 2014).

Silwa et al. employed SEM to study the interaction of WF and pebax matrix, a type of thermoplastic elastomer. Figure 7(a) and (b) shows the SEM images in which the composite were devoid of voids and fiber pull-out. Moreover, the WFs were distributed in the polymer matrix with no gap at their interface. This observation implies a good fiber-matrix interaction (Sliwa et al. 2012). Panaitescu et al. studied the interfacial bonding of cellulose fiber from wood waste and (poly (3-hydroxybutyrate) PHB by SEM analysis. The result showed that fibers were

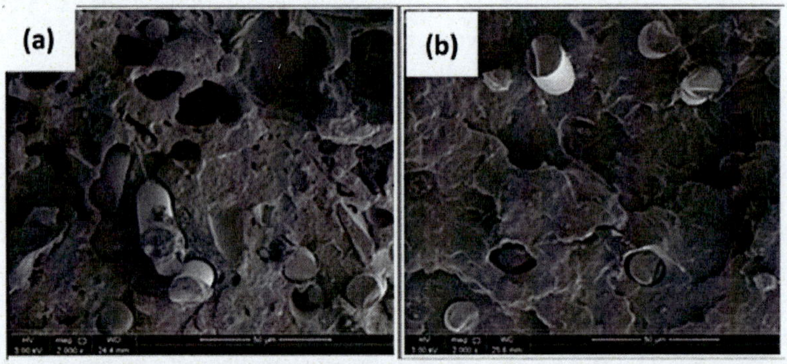

Fig. 6 SEM micrograph (a) WGFRPP and (b) GFRPP (Al-Maadeed et al. 2014) [Re-used with permission from Elsevier, License Number-4820690120435]

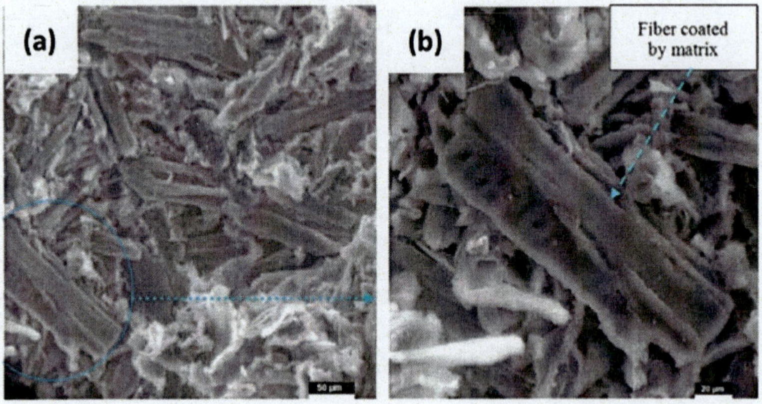

Fig. 7 SEM images (a) WF/pebax composite and (b) better wettability between WF and pebax matrix in the composite (Sliwa et al. 2012) [Re-used with permission from Elsevier, License Number-4823860808962]

uniformly distributed in the PHB matrix with no gap in their interface. Broken fibers and voids were also observed in the SEM image of the composite (Fig. 8(a)–(d)). However, the fiber pull out is small and thus implies a strong adhesion between the polymer and fibers. They further suggested that the hydrophobic lignin present in the WF is responsible for its greater adhesion with the matrix (Panaitescu et al. 2020). Similar failure mode i.e. fiber pull-out and fiber breakage was reported for WF reinforced PHBV composite by Singh et al. (Singh and Mohanty 2007). However, according to their observation on increasing the WF content, the fiber pull-out length increases. This could be due to greater fiber–fiber interactions which reduces the fiber-matrix adhesion. As a result, the fibers are easily pulled out from the matrix and the composite displayed poor mechanical strength.

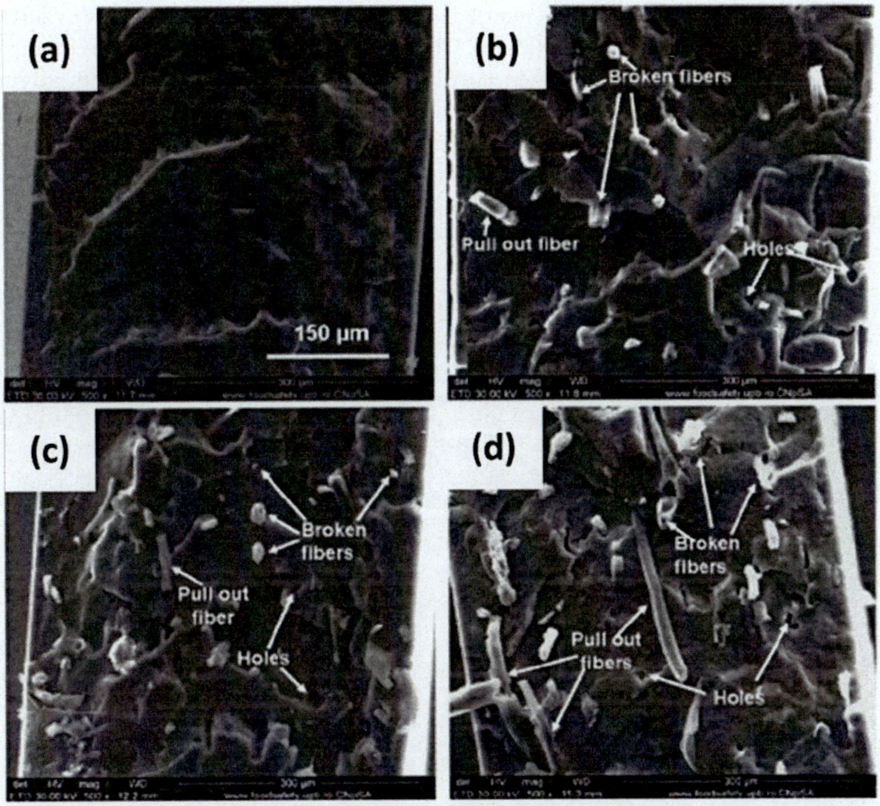

Fig. 8 SEM images (×500): (a) PHB, (b) PHB/CF5, (c) PHB/CF10, and (d) PHB/CF15 (Panaitescu et al. 2020) [Re-used with permission from Elsevier, License Number-4823870277868]

Shah et al. fabricated PLA/wood flour composite and studied its mechanical properties. They observed that with addition of wood flour, the mechanical properties of the composite improved significantly. This could be due to better adhesion between fiber and matrix as visible from the uniform dispersion of wood flour in the PLA matrix from the SEM images. The fracture surface morphology of the composite showed more fiber breakage rather than fiber the pull-out providing a strong evidence for the good compatibility between PLA and WF (Shah et al. 2008).

4 Surface Modification of WF

Islam et al. employed alkali and diazonium treatment to improve the adhesion of wood flour and polymers. The interfacial adhesion of the composite was evaluated using SEM as shown in Fig. 9(a)–(c). The morphological analysis of the composite with the untreated fiber indicated a visible gap between wood flour and polymer thus implying a poor compatibility in their interface. However, in case of the composite with treated fiber, the gap between wood flour and matrix was absent suggesting better adhesion between the reinforcement and matrix (Islam et al. 2012).

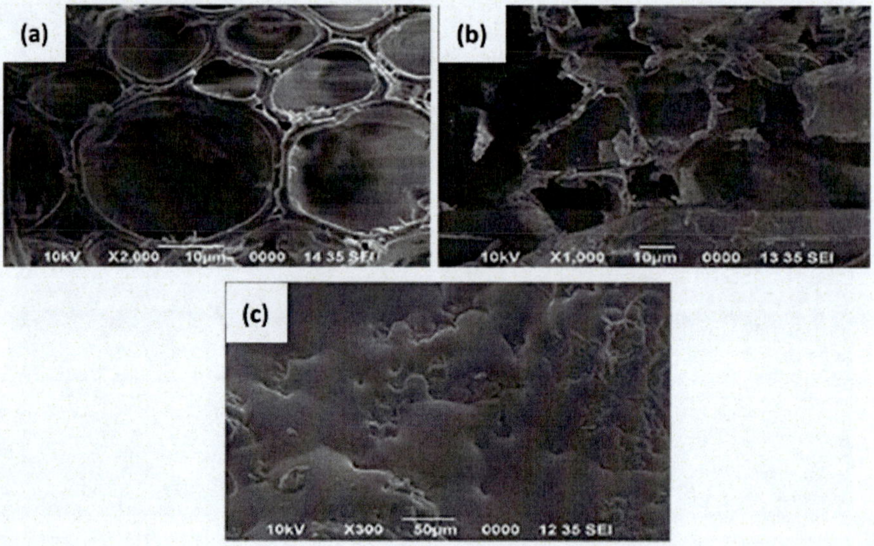

Fig. 9 SEM micrographs of (i) untreated wood, (ii) Untreated wood polymer composites and (iii) Pretreated wood polymer composites (Islam et al. 2012) [Re-used with permission from Elsevier, License Number- 4824260015434]

Hunnekens et al. performed plasma treatment for 60 s on the wood flour/PP composite. WPC Surface became rougher due to the plasma treatment as visible from the atomic force microscope (AFM) and laser scanning microscope (LSM) images in the Fig. 10(a) and (b) and Fig. 11(a) and (b) respectively. AFM and LSM also provide quantitative assessment of the morphological changes due to the plasma treatment in terms of the surface roughness (Hünnekens et al. 2018).

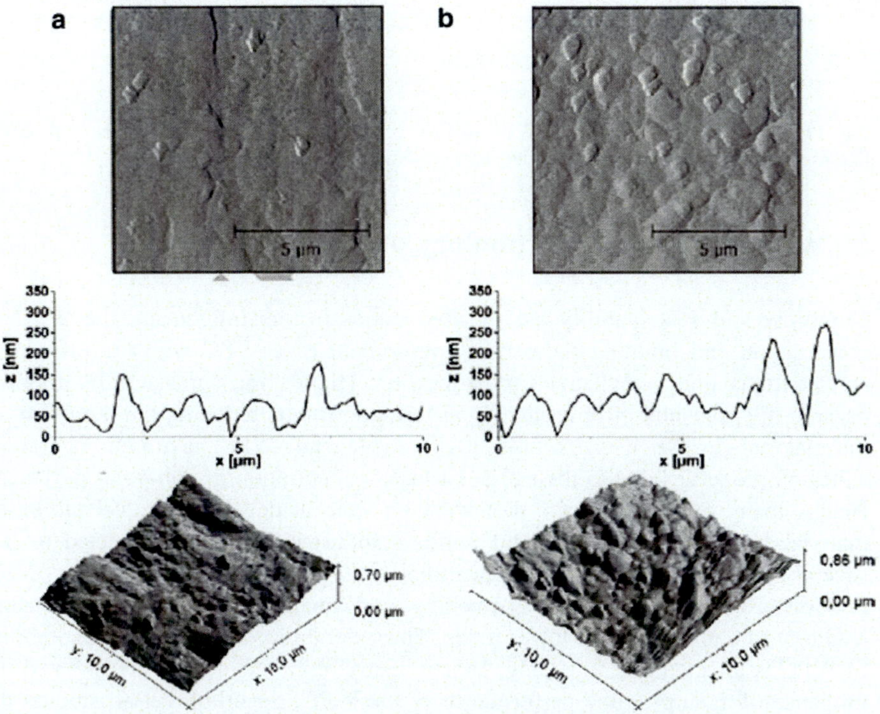

Fig. 10 AFM images (a) untreated WPC and (b) plasma treated WPC (Hünnekens et al. 2018) [Re-used with permission from Elsevier, License Number-4891381183598]

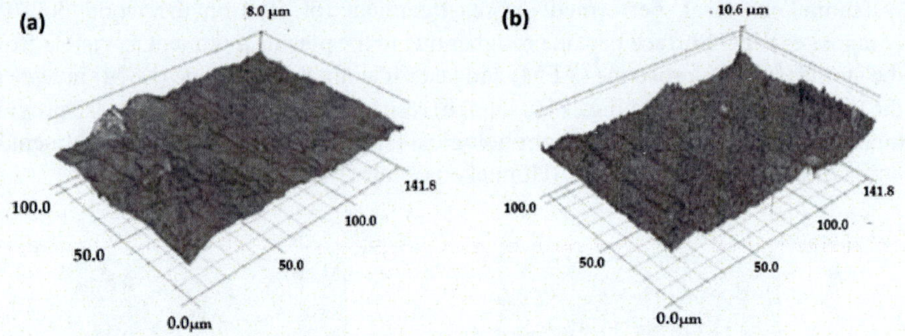

Fig. 11 LSM images (a) untreated WPC and (b) plasma treated WPC (Hünnekens et al. 2018) [Re-used with permission from Elsevier, License Number- 4891381183598]

5 Aging Effects on Morphology of the WPC

Durability and sustainability are the two major factors influencing the WPC in construction and building industry. Development of a WPC with the prolonged service life is a huge challenge for the industry. Higher water uptake of hydrophilic organic fillers results in fungi and frost formation, thus reducing the durability of the material (Friedrich and Luible 2016). In general, WPC have poor interfacial adhesion properties due to their high surface concentration in non-polar polymers. Hence, many researchers have devoted their time in developing novel bio-based materials for the construction and building industry which are expected to last several decades in the outdoor applications.

Moisture absorption induces swelling and micro-cracks in the composites leading to failure under the lower loads. Thus, durability and dimensional stability of WPC is affected due to ageing. Hence, many researchers have investigated the influence of ageing on the performance of the WPC. Segerholm et al. studied the effect of artificial ageing on the unmodified and acetylated wood/recycled high density polyethylene composites. In their study, the composites were artificially aged by water soaking, artificial weathering and white or brown-rot decay methods. Their results revealed that the artificial weathering of WPC induced cracks on the matrix surface and also resulted in delamination (Fig. 12(a) and (b)). Furthermore, it was also found that, acetylation of wood and the corresponding composites exhibited reduced moisture sorption rate when compared to its unmodified counterpart (Segerholm et al. 2012).

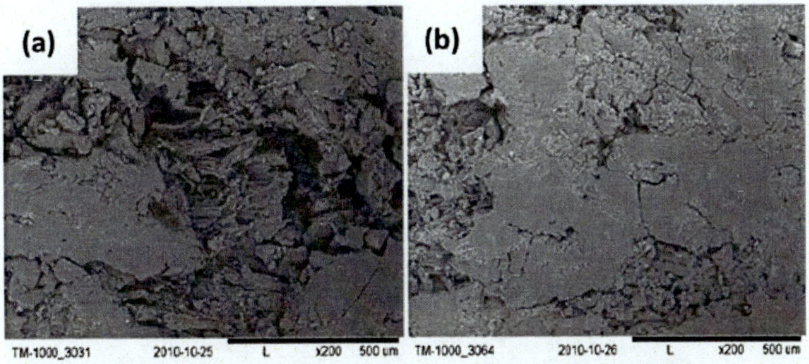

Fig. 12 Formation of cracks due to ageing (a) unmodified WPC and (b) acetylated WPC (Segerholm et al. 2012) (Open Access Content)

Cantero et al. analyzed the effects of ageing on the mechanical properties MA grafted PP copolymer/wood flour composite. The addition of MAPP increased both stiffness and strength marginally. It was also found that hygrothermal ageing had significant impact on the mechanical properties (Cantero et al. 2003). The poor hydrogen bonding between water molecules and the hydroxyl groups present in cellulose component of the wood floor was responsible for the degradation effects (Avella et al. 1998). Poor interfacial adhesion between the fiber and matrix can promote water penetration and intensifies the hygroscopic effects of the composites.

Wood flake reinforced HDPE composites were subjected to different ageing conditions viz., ultraviolet, oxidation and heat to analyze their effect on the mechanical properties. Under ageing the color of the wood composites faded and many visible cracks were found on the surface to the depth commonly known as the environment stress cracks (ESC) as shown in Fig. 13. The tensile properties such as strength, modulus and energy absorption of the weathered composites were reduced to at least 50 to 70% of the unweathered composites. The surface layer was degraded initially due to weathering which lowers the stiffness of the composites. Similarly, the fracture toughness of the composites was reduced to about 40% and the impact strength was reduced about 20%. The reduced mechanical properties for the composites subjected to UV radiation were due to propagation of the ESC in the composites under various loads. ESC can also promote higher water uptake in the composites (Li 2000).

Darie et al. studied the effects of reinforcement of different weight percentage of oak wood floor/LDPE composite on the tensile and impact properties with and without artificial ageing. The results showed that after accelerated weathering, the composites had higher retention rate of the mechanical properties. This could be attributed to the stabilizing effect of wood flour and the filler has acted as a screen restricting the photo-degradation to the outer surface layers of the composites (Darie et al. 2011).

Fig. 13 UV radiation induced environmental stress cracks on the surface of the composite (Li 2000) [Re-used with permission from Elsevier, License Number- 4891381389755]

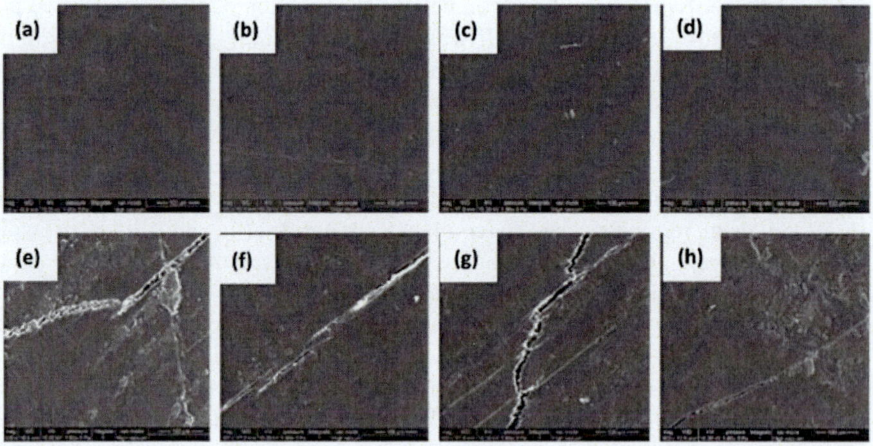

Fig. 14 Surface morphologies of WPCs with different light stabilizers; Samples (a, e): control; (b, f): 2%UV-327; (c, g): 2% HS-944; (d, h): 2% nano-SiO_2 after 0 h and 2500 h of UV exposure, respectively (Liu et al. 2018) (Open Access Content)

In an investigation performed by Chen et al., an attempt was made to enhance the resistance of HDPE/wood flour composites against the UV aging. This was achieved by incorporating light stabilizers such as UV-327, HS-944 and nano-SiO_2 to the shell of the WPC. From the results of the study, it was found that adding UV-327 and nano-SiO_2 resulted in better retention rate of the flexural strength and modulus of the WPC than HS-944 filler. Through the morphological images shown in Fig. 14(a)–(f) the effect of light stabilizers was analyzed. The surface of the WPC with 2% UV-327 was the found to be the smoothest among the investigated configurations. Composites with lesser number of cracks on the surface yielded better retention in the properties. Hence, the light stabilizers protected the WPCs against damage from the UV radiation (Liu et al. 2018).

6 Conclusion

A typical wood polymer composite (WPC) that consists of a polymer matrix and wood as the reinforcement can be fabricated in the following ways: (a) Wood can be used in particulate form and blended with thermoplastic resin to form a WPC, (b) thermoset resin or liquid thermoplastic binders can be impregnated into the WF and (c) adhesive bonding of plywood to form a laminate. Following are the conclusions from the review on the morphological properties of the WPC:

- Morphological features of the wood, interfacial adhesion between the WF and matrix as well as failure behavior of the WPC under various loads is characterized through the microstructural images from the scanning electron microscope (SEM), field emission scanning electron microscope (FESEM), transmission electron microscope (TEM), Atomic force microscope (AFM), x-ray computed tomography (XCT), etc.
- Resin impregnated into the wood fills the lumens in the wood structure and strengthens the material and serves as binder as well.
- Weak interfacial adhesion between the hydrophilic WF and hydrophobic matrix was identified from the fiber-matrix de-bonding, larger extent of the fiber pull-out and fiber pull-out of larger depth.
- Use of the fiber treatment and coupling agent enhances the wettability between the fiber-matrix by providing better mechanical interlocking. Fiber treatment modifies the fiber surface to rough while the addition of coupling agents induces surface roughness of the resin as identified from the microstructural images from the TEM and AFM.
- Micrographs from the SEM and FESEM were helpful in identifying the morphological degradation of the WPC subjected to moisture absorption, hygrothermal conditions and accelerated weathering.

Acknowledgements We hereby acknowledge and sincerely appreciate unalloyed supports from the managements of the following institutions: Hindustan Institute of Technology & Science, Tamilnadu, India and King Mongkut's University of Technology North Bangkok (KMUTNB), Thailand.

References

Al-Maadeed MA, Shabana YM, Khanam PN (2014) Processing, characterization and modeling of recycled polypropylene/glass fibre/wood flour composites. Mater Des 58:374–380

Ansell MP (2015a) Wood microstructure-A cellular composite. Elsevier Ltd., Amsterdam

Ansell MP (2015b) Wood composites. Woodhead Publishing, Cambridge

Antov P, Savov V, Neykov N (2020) Sustainable bio-based adhesives for eco-friendly wood composites a review. Wood Res 65:51–62

Askadskii A, Matseevich T, Askadskii A (2019) Structure and properties of wood-polymer composites (WPC). Cambridge Scholars Publishing, Newcastle upon Tyne

Avella M, Casale L, Dell'erba R et al (1998) Broom fibers as reinforcing materials for polypropylene-based composites. J Appl Polym Sci 68:1077–1089

Bao M, Li N, Huang C et al (2019) Fabrication, physical–mechanical properties and morphological characterizations of novel scrimber composite. Eur J Wood Wood Prod 77:741–747

Biron M (2018) Thermoplastics and thermoplastic composites. William Andrew, Cambridge

Bledzki AK, Sperber VE, Faruk O (2002) Natural and wood fibre reinforcement in polymers. iSmithers Rapra Publishing

Cantero G, Arbelaiz A, Mugika F et al (2003) Mechanical behavior of wood/polypropylene composites: effects of fibre treatments and ageing processes. J Reinf Plast Compos 22:37–50

Chawla K, Singh J, Singh R (2020) On recyclability of thermosetting polymer and wood dust as reinforcement in secondary recycled ABS for nonstructural engineering applications. J Thermoplast Compos Mater 0892705720925135

Darie R, Bercea M, Kozlowski M, Spiridon I (2011) Evaluation of properties of LDPE/oak wood composites exposed to artificial ageing. Cellul Chem Technol 45:127–135

Deviatkin I, Khan M, Ernst E, Horttanainen M (2019) Wooden and plastic pallets: a review of life cycle assessment (LCA) studies. Sustainability 11:5750

Friedrich D, Luible A (2016) Investigations on ageing of wood-plastic composites for outdoor applications: a meta-analysis using empiric data derived from diverse weathering trials. Constr Build Mater 124:1142–1152. https://doi.org/10.1016/j.conbuildmat.2016.08.123

Fürst C, Plank B, Senck S, et al (2018) Bio based silicon carbide ceramics from extruded thermoset based wood polymer composites. In: Proceedings of the ECCM18—18th European conference of composite materials, Athens, Greece, pp 24–28

Grammatikos S, Jøsendal HB, Papatzani S, et al (2020) Recycled wood plastic composites as floor panels for ship containers. In: IOP conference series: materials science and engineering. IOP Publishing, p 12006

Gulitah V, Liew KC (2019) Morpho-mechanical properties of wood fiber plastic composite (WFPC) based on three different recycled plastic codes. Int J Biobased Plast 1:22–30

Guo G (2020) Density reduction behaviors and cell morphology in extrusion of LLDPE/wood fiber composites with physical and chemical blowing agents. J Appl Polym Sci 137:48829

Gwon JG, Lee SY, Chun SJ et al (2010) Effects of chemical treatments of hybrid fillers on the physical and thermal properties of wood plastic composites. Compos Part A Appl Sci Manuf 41:1491–1497

Harris R (2015) Cross laminated timber. Elsevier Ltd., Amsterdam

He M, Sun X, Li Z, Feng W (2020) Bending, shear, and compressive properties of three-and five-layer cross-laminated timber fabricated with black spruce. J Wood Sci 66:1–17

Hughes M (2015) Plywood and other veneer-based products. Elsevier Ltd., Amsterdam

Hünnekens B, Avramidis G, Ohms G et al (2018) Impact of plasma treatment under atmospheric pressure on surface chemistry and surface morphology of extruded and injection-molded wood-polymer composites (WPC). Appl Surf Sci 441:564–574

Islam MS, Hamdan S, Hasan M et al (2012) Effect of coupling reactions on the mechanical and biological properties of tropical wood polymer composites (WPC). Int Biodeterior Biodegradation 72:108–113

Kim JK, Pal K (2010) Recent advances in the processing of wood-plastic composites. Springer, Heidelberg

Klyosov AA (2007) Wood-plastic composites. Wiley, Hoboken

Ko H, Kim JW, Kim HC et al (2020) Esterified PVA-lignin resin by maleic acid applicable for natural fiber reinforced composites. J Appl Polym Sci 137:48836

Kocsis Z, Czigány T (2007) Investigation of the debonding process in wood fiber reinforced polymer composites by acoustic emission. In: Materials science forum. Trans Tech Publ, pp 199–206

Krause KC, Sauerbier P, Koddenberg T, Krause A (2018) Utilization of recycled material sources for wood-polypropylene composites: effect on internal composite structure, particle characteristics and physico-mechanical properties. Fibers 6:86

Li R (2000) Environmental degradation of wood-HDPE composite. Polym Degrad Stab 70:135–145. https://doi.org/10.1016/S0141-3910(00)00099-9

Lipovac D, Podrekar N, Burnard MD, Šarabon N (2020) Effect of desk materials on affective states and cognitive performance. J Wood Sci 66:1–12

Liu C, Mei C, Xu B et al (2018) Light stabilizers added to the shell of co-extruded wood/high-density polyethylene composites to improve mechanical and anti-UV ageing properties. R Soc Open Sci 5:180074

Lopez YM, Gonçalves FG, Paes JB, et al (2020) Resistance of wood plastic composite produced by compression to termites Nasutitermes corniger (Motsch.) and Cryptotermes brevis (Walker). Int Biodeterior Biodegradation 152:104998

Mark HF (2013) Encyclopedia of polymer science and technology, concise. Wiley, Hoboken

Mertens O, Krause KC, Weber M, Krause A (2018) Performance of thermomechanical wood fibers in polypropylene composites. Wood Mater Sci Eng

Niska KO, Sain M (2008) Wood-polymer composites. Elsevier, Amsterdam

Ormondroyd GA, Stefanowski B (2015) Fibreboards and their applications. Elsevier Ltd., Amsterdam

Panaitescu DM, Nicolae CA, Gabor AR, Trusca R (2020) Thermal and mechanical properties of poly (3-hydroxybutyrate) reinforced with cellulose fibers from wood waste. Ind Crops Prod 145:112071

Pizzi A, Papadopoulos AN, Policardi F (2020) Wood composites and their polymer binders. Polymers (Basel) 12:1115

Poletto M (2017) Mechanical, dynamic mechanical and morphological properties of composites based on recycled polystyrene filled with wood flour wastes. Maderas Cienc Y Tecnol 19:433–442

Renner K, Móczó J, Pukánszky B (2011) Polymer/wood composites. Wiley Encycl Compos 1–19

Rosu L, Mustata F, Varganici C et al (2019) Thermal behaviour and fungi resistance of composites based on wood and natural and synthetic epoxy resins cured with maleopimaric acid. Polym Degrad Stab 160:148–161

Rowell RM (2012) Handbook of wood chemistry and wood composites. CRC Press, Boca Raton

Segerholm BK, Ibach RE, Wålinder MEP (2012) Moisture sorption in artificially aged wood-plastic composites. BioResources 7:1283–1293. https://doi.org/10.15376/biores.7.1.1283-1293

Senthilkumar K, Kumar TSM, Chandrasekar M, et al (2019) Recent advances in thermal properties of hybrid cellulosic fiber reinforced polymer composites. Int J Biol Macromol

Senthilkumar K, Saba N, Rajini N et al (2018) Mechanical properties evaluation of sisal fibre reinforced polymer composites: a review. Constr. Build, Mater

Shah BL, Selke SE, Walters MB, Heiden PA (2008) Effects of wood flour and chitosan on mechanical, chemical, and thermal properties of polylactide. Polym Compos 29:655–663

Singh S, Mohanty AK (2007) Wood fiber reinforced bacterial bioplastic composites: fabrication and performance evaluation. Compos Sci Technol 67:1753–1763

Sliwa F, Charrier F, Marin G, Malet F (2012) Mechanical and interfacial properties of wood and bio-based thermoplastic composite. Compos Sci Technol 72:1733–1740

Sun X, He M, Li Z (2020) Novel engineered wood and bamboo composites for structural applications: state-of-art of manufacturing technology and mechanical performance evaluation. Constr Build Mater 249:118751

Yu M, Zhang G, Saunders T (2020) Wood-derived ultra-high temperature carbides and their composites: a review. Ceram Int 46:5536–5547

Yu Y, Liu R, Huang Y et al (2017) Preparation, physical, mechanical, and interfacial morphological properties of engineered bamboo scrimber. Constr Build Mater 157:1032–1039

Yuan Q, Wu D, Gotama J, Bateman S (2008) Wood fiber reinforced polyethylene and polypropylene composites with high modulus and impact strength. J Thermoplast Compos Mater 21:195–208

Zhang J, Li Y, Xing D et al (2019) Reinforcement of continuous fibers for extruded wood-flour/HDPE composites: effects of fiber type and amount. Constr Build Mater 228:116718

Mechanical Properties of Wood Polymer Composites

Rupam Gogoi and Gaurav Manik

Abstract The world has been facing a huge environmental impact due to global warming, which sometimes lead to unprecedented natural disasters creating a crisis for the present and future generations to come. Sustainable development, which was a far cry in the past, has now been a burning topic and desire in all aspects of human endeavour towards development. In polymer composites, reinforcing materials from natural resources such as wood have been attracting a lot of interest from researchers worldwide due to their low cost, and abundance in making an environment-friendly polymer composite. Waste or by-products from wood milling industries in the form of fibre, sawdust or wood flour (WF) have been potentially used as a filler material in various wood polymer composites (WPCs). Although lower-cost attracts their use as a reinforcer and/or filler, WPCs do have serious issues in low-performance aspect and higher moisture absorption. These disadvantages are scopes for researchers and organisation to address. Nonetheless, various WPCs as commercial products are available in the market and are being successfully used. To leverage the benefits of WPCs, a critical criterion for consideration is their mechanical performance. Bulk mechanical properties such as tensile, flexural, impact etc., are crucial to find the limitations and also to evaluate their load carrying capacity which can potentially define their area of application. In the present study, an in-depth analysis of different mechanical properties reported for various WPCs along with different theoretical modeling techniques, used previously have been summarized.

Keywords Wood flour · Wood polymer composites · Environment friendly · Mechanical properties · Micromechanical model

---

R. Gogoi · G. Manik (✉)
Department of Polymer and Process Engineering, Indian Institute of Technology Roorkee, Saharanpur Campus, Paper Mill Road, Saharanpur 247001, UP, India
e-mail: gaurav.manik@pe.iitr.ac.in; manikfpt@iitr.ac.in

R. Gogoi
e-mail: rgogoi@pe.iitr.ac.in

S. Mavinkere Rangappa et al. (eds.), *Wood Polymer Composites*,
Composites Science and Technology, https://doi.org/10.1007/978-981-16-1606-8_6

1 Introduction

Wood polymer composites (WPCs) are essentially polymer composites which are filled or reinforced with various naturally available wood materials in dust, powder or flour and fibre form. Typically, wood fillers are blended with desired thermoplastic polymer melt to form pallets, which is then extruded, injection or compression molded to obtain desired products. Figure 1 depicts a typical process flow chart involved in WPC manufacturing.

Fillers are dried well before the mixing process so as to avoid any entrapped moisture which can otherwise deteriorate the product quality and reduce its performance. Polymers such as Polypropylene (PP), Polyethylene (PE), high density PE (HDPE), polyvinyl chloride (PVC) etc., are majorly used in WPCs along with various coupling agents and additives. Coupling agents, which help in enhancing the filler-matrix interfacial adhesion, improve the bulk mechanical properties of WPCs, such as tensile, flexural, and impact strength. Coupling agents for instance maleic anhydride-grafted-PP (MA-g-PP), and maleic anhydride-grafted-PE (MA-g-PE) are commonly used in the manufacturing of WPCs. In addition to coupling agents, various mineral additives like talc, calcium carbonate ($CaCO_3$), silica, clay, wollastonite ($CaSiO_3$), etc. (Klyosov 2007) and impact modifiers like ethylene propylene diene monomer (EPDM) rubber, ethylene-propylene rubber (EPR), styrene-ethylene-butylene-styrene (SEBS), etc. (Oksman and Clemons 1998; Clemons 2010) are also used during fabrication of WPCs to enhance different properties like modulus and impact resistance. Extrusion, injection, and compression molding are the main manufacturing processes involved in producing the final products as illustrated in Fig. 1. Recently, WPCs gained commercial focus and benefits in the construction, building, and furniture applications mainly due to their various advantages over raw wood products. Few fields of application are illustrated in Fig. 2. Resistance against fire, termites, water, etc. along with good aesthetics have promoted its penetration in different markets (Niska and Sain 2008). It was estimated that the market size of WPCs in the year 2019 was around 5.3 billion USD with a future growth rate of 11.4% (Grand View Research 2020). The automotive sector has also shown great interest in using WPCs based products in

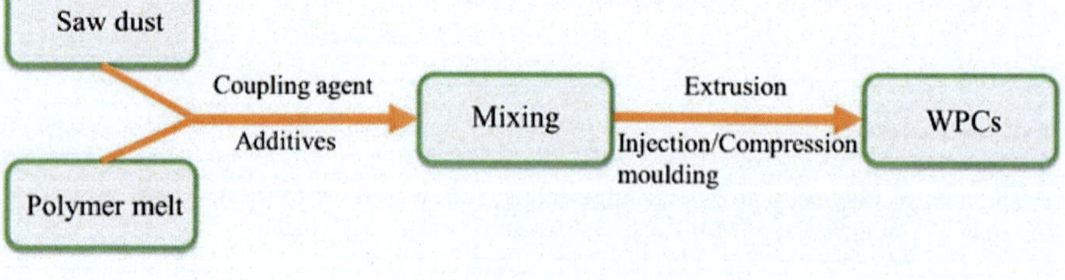

Fig. 1 Schematic diagram of a typical WPC manufacturing process involving various raw materials

car interiors due to their low density and enhanced acoustic performance (Ashori 2008; Markets and Markets 2020).

Use of low density materials not only reduces the overall weight of vehicles but also increases fuel efficiency and cut down CO_2 emissions (Ashori 2008). Furthermore, hybrid WPC-fibre reinforced polymer (FRP) composites find wide applications in civil infrastructures, building and bridges, railroad ties etc. due to their additional combined features aided by the FRP composites (Naghipour et al. 2011). WPCs have potential for far wider application in near future based on their performance characteristics. Depending upon various factors of the reinforcing fillers, the mechanical properties of WPCs can vary accordingly. Main factors such as filler type, volume, size and shape, inherent tensile strength, stiffness etc. play key roles (Chan et al. 2018). On the other hand, interfacial interactions, type of base polymer matrix and coupling agents, types of filler surface treatments also equally affect the mechanical properties of WPCs. The role and effect of all the aforementioned parameters have been discussed extensively by many researches previously (Stark and Rowlands 2003; Lee et al. 2004; Bengtsson et al. 2005; Clemons 2010; Valente et al. 2011; Kumar et al. 2017; Gogoi et al. 2019). Le et al. studied the variation of WF content from 0–40 wt.% and the effect of MA-g-PP on tensile properties of WF/PP composites (Lee et al. 2004). It was observed that in absence of MA-g-PP, the tensile strength decreases with increasing WF content, while the presence of MA-g-PP improves the strength by imparting better filler-matrix interfacial adhesion. Effect of vinyltrimethoxy silane grafting of HDPE on thermal and mechanical properties of spruce and pine WF reinforced WPCs were investigated by

Fig. 2 Commercial products made from wood plastic composites: (a) decking (Changxing Hanming Technology Co. Ltd. 2020), (b) flower boxes (Shanghai Seven Trust Industry Co. Ltd. 2020), (c) car interior (Jeluplast 2020) (d) park furniture (Integrated Recycling 2020a), and (e) architectural screening (Integrated Recycling 2020b)

Bengtsson et al. (Bengtsson et al. 2005). The crosslinking formed between the WFs and the polymer chains of HDPE not only improved the tensile properties of the WPCs but also reduced creep deformation and water uptake. To study the effect of different filler sizes on mechanical properties of WPCs, Stark et al. fabricated Ponderosa pine WF reinforced PP composites with filler size of 35, 70, 120, and 235 mesh (Stark and Rowlands 2003). Keeping the WF content fixed at 40 wt.% it was observed that with increasing filler aspect ratio the tensile and flexural strength of the composites also increased. While the unnotched impact energy decreased, the notched impact energy increased with increasing filler size. Hybrid incorporation of glass fibre (both virgin and recycled) in WPCs based on LDPE and PP matrix was studied by Valente et al. (Valente et al. 2011). With increasing WF and glass fibre loading, it was observed that the flexural modulus of the composites improved but the flexural strength decreased. Apart from the role of fillers, the mechanical properties of WPCs are also equally affected by the type of base polymer matrix as mentioned earlier. In this regard, Clemons (Clemons 2010) studied the effect of polymer blends of PP and HDPE modified with elastomers (EPDM, MA-g-EPDM) on the mechanical properties of WPCs. Coupling agents MA-g-PP and MA-g-PE were also separately used for the neat and blends of PP and HDPE respectively. In case of the blends, the observed tensile properties were proportional to the relative quantity of PP and HDPE in the blend composition. The addition of EPDM to the blend reduced both the strength and modulus but increased the strain at break along with impact energy absorption. This was more pronounced for the HDPE rich blends due to its higher toughness than PP. Clemson also observed that the incorporation of WF makes the composites brittle and weak but the reduction could be compensated reasonably by incorporating few combinations of elastomer and coupling agents (Clemons 2010).

In addition to aforementioned experimental works on the fabrication of various WPCs based on different combination of polymers and fillers, researchers have also given emphasis on the theoretical modelling of such WPCs (Połeć et al. 2010; Migneault et al. 2011; Hančič et al. 2012). Theoretical modelling of polymer composite materials involves the prediction of their bulk properties based on the correlation of intrinsic material properties of the constituent materials and microarchitecture of the inclusions (Neagu and Gamstedt 2008). The most common and simple is the rule of mixture shortly known as ROM, which expresses, the composite property as the summation of volume-weighted properties of the base matrix and the inclusions. But often debate over ROM appeared as the micromechanical behavior of composite is not only dependent on the relative volume fraction of the constituents but also on the geometrical parameters of the inclusion such as shape, aspect ratio, orientation, distribution and dispersion (Neagu and Gamstedt 2008). Halpin-Pagano and Halpin–Tsai have considered in their models the aspect ratio of filler and its random orientation within a continuous matrix phase (Halpin and Pagano 1969; Affdl and Kardos 1976). These models are based on the classical laminate theory with the assumption that the composite is a stack of lamina where each lamina is essentially a sheet of discontinuous fibres reinforced in a continuous phase and aligned in a particular angle (Halpin and Pagano 1969).

Similarly, the Kelly–Tyson model which is a modification of ROM, not only considers the aspect ratio of fillers but also includes the filler-matrix interfacial shear strength into the model for an efficient properties prediction (Migneault et al. 2011). Apart from the aforementioned empirical relations, some widely used models are Mori–Tanaka (Mori and Tanaka 1973; Hančič et al. 2012) and Cox-Krenchel models (Połeć et al. 2010). In addition, researchers have also investigated the mechanical properties of WPCs through direct numerical simulations (Gusev 1997; Hine et al. 2002) and generalized method of cells (Hančič et al. 2012). Most of the micromechanical models developed so far have considered inclusions to be of simpler geometry with less anisotropy. On the other hand, WFs used in WPCs are anisotropic in nature having irregular shape, rough surface with pits and dislocations. Hence, to exploit the micromechanical models in case of WPCs, fillers like WFs are assumed to be (a) transversely isotropic and (b) homogeneous and cylindrical in nature (Neagu and Gamstedt 2008). In the present review, different mechanical tests used for WPCs are discussed in detail considering the aforementioned parameters under the subtopics; tensile, flexural and impact test. In addition, the general micromechanics involved in WPCs which defines their bulk properties by theoretical modelling have also been discussed in the subsequent sections.

2 Tensile Test

In a typical tensile test, load vs. elongation curve is obtained which is converted into a stress–strain plot to measure the material strength and stiffness. Tensile test of WPCs is measured generally under the ASTM D638 and ISO527 standards. The tensile behaviour varies from composites to composites which depends majorly upon filler loading, filler size including their aspect ratio (L/D ratio), filler polymer interfacial interactions influenced by presence of coupling agents, surface treatments of filler, and also grafting of polymers.

The effect of WF loading on tensile strength was studied by Le et al. by varying its content from 0–40 wt.% in a PP matrix with and without addition of MA-g-PP (Lee et al. 2004). As seen in Fig. 3, with increasing filler loading the composites without MA-g-PP showed a linear decrease in tensile strength, whereas composites having 3 wt.% MA-g-PP showed an increment until 40 wt.% of WF reinforcement. The decrease in the tensile strength for the former was due to the inefficient transfer of load from the polymer matrix to the WF which was aided by the inferior filler matrix adhesion. In comparison to neat PP and for a 40 wt.% WF loading, a 9.2% decrease in tensile strength was observed for composite without MA-g-PP, whereas an increase of 51.7% was observed in presence of MA-g-PP.

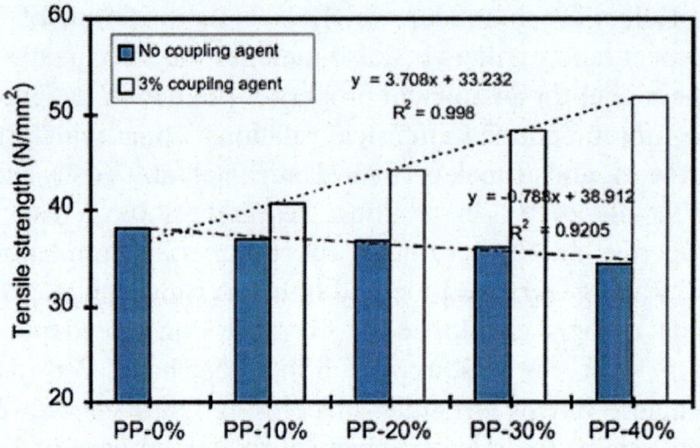

Fig. 3 Illustration of the variation of tensile strength with increasing WF content with and without coupling agent (Lee et al. 2004) "Reprinted from publication tittle, Vol. 65, Sun-Young Lee, Han-Seung Yang, Hyun-Joong Kim, Chan-Seo Jeong, Byeong-Soo Lim, Jong-Nam Lee, Creep behavior and manufacturing parameters of wood flour filled polypropylene composites, 459–469, Copyright (2004), with permission from Elsevier."

Fig. 4 Illustration of the stress–strain curve of XLPE and its composites with different wt.% of WF (Bengtsson et al. 2005) "Reprinted from publication tittle, Vol. 65, Magnus Bengtsson, Paul Gatenholm, Kristiina Oksman, The effect of crosslinking on the properties of polyethylene/wood flour composites, 1468–1479, Copyright (2005), with permission from Elsevier."

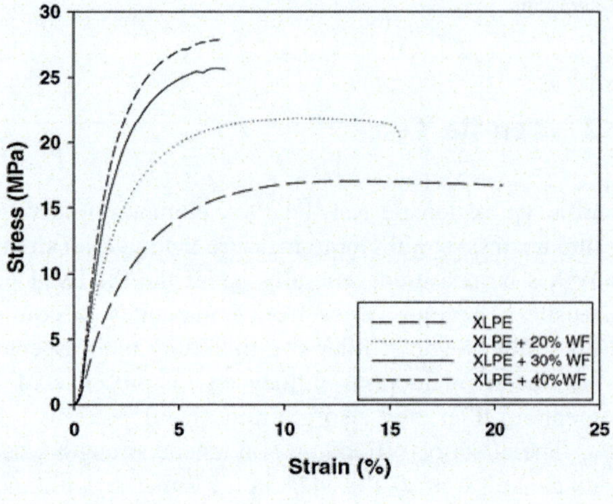

Bengtsson et al. investigated the effect of spruce and pine WF content in a vinyltrimethoxy silane grafted HDPE (XLPE) matrix for various thermal and mechanical properties (Bengtsson et al. 2005). Interestingly, it was observed that the tensile strength and modulus improved upon increasing filler content as seen in Fig. 4.

In absence of any coupling agent it is unusual for WF filled polymer composites to have an increase in tensile strength upon increasing filler loading as have been discussed by many researchers (Lee et al. 2004; Bengtsson et al. 2005). Here, the silane grafting on the HDPE backbone had a possible crosslinking with the hydroxyl groups present on the surface of WF creating covalent or hydrogen bonding as shown

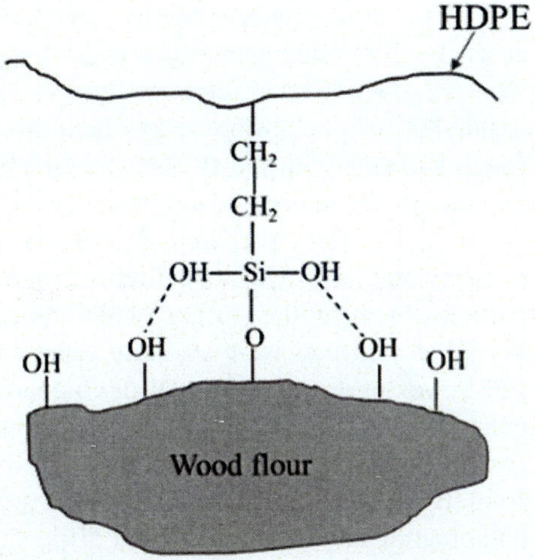

Fig. 5 Illustration of probable crosslinking between WF and vinyltrimethoxy silane grafted HDPE (Bengtsson et al. 2005) "Reprinted from publication tittle, Vol. 65, Magnus Bengtsson, Paul Gatenholm, Kristiina Oksman, The effect of crosslinking on the properties of polyethylene/ wood flour composites, 1468–1479, Copyright (2005), with permission from Elsevier."

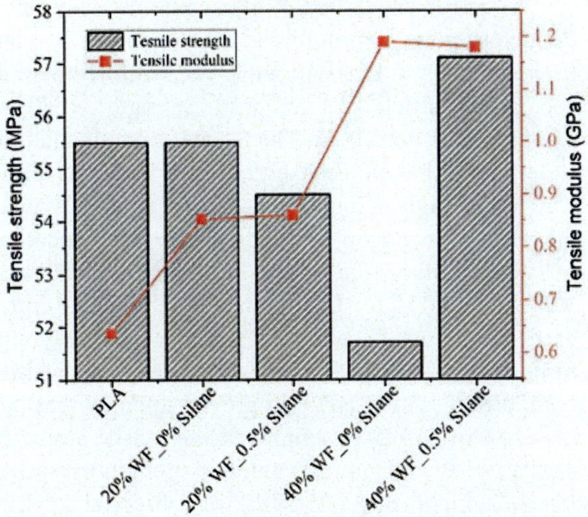

Fig. 6 Variation of tensile strength and modulus of Pine WF reinforced PLA composites with and without silane treatment of the WFs (Pilla et al. 2008)

in the schematic in Fig. 5. Nevertheless, XLPE reinforced with 40% WF showed the highest tensile strength and modulus of 27.3 ± 1.4 and 1481.5 ± 155.1 MPa which was ~61% and ~165% above neat XLPE respectively.

Pilla et al. (Pilla et al. 2008) have analysed the effect of filler content and silane treatment on the tensile properties of Pine WF reinforced Polylactide (PLA) composites. A ductile to brittle failure behavior was observed for the Pine WF/ PLA composite with filler loading. Interestingly, it was observed that the ultimate tensile strength was almost consistent irrespective of the WF loading, but the modulus improved significantly from ~638 MPa to ~1180 MPa, as seen in Fig. 6.

Unlike, in the study made by Lee et al. (Lee et al. 2004) (see Fig. 3) where in absence of coupling agent the tensile strength of WF/PP composite decreased upon increasing WF loading, here the Pine WF/PLA composites did not display the same trend. This can be attributed to a better interfacial adhesion of the Pine WF having rough surfaces with the PLA matrix. Hybrid incorporation of WF and nano silica together in PP matrix, on the mechanical properties was studied by Ma et al. (Ma et al. 2020). They prepared in situ hybrid WF/nano silica filler by reacting the hydroxylated WF with tetraethyl orthosilicate (TEOS), which lead the nano silica particles to sit on the surface of WF. For a comparative study, hybrid nano silica/WF/PP composites were also fabricated by simple blending of nano silica, WF and PP. It was observed from that in comparison to blending, the in situ modification enhanced the tensile properties of hybrid composites. Tensile strength was highest for 20 wt.% in situ modification, while it gradually decreased for blended system from 0–40 wt.% filler loading. Nano particles as such have high tendency to agglomerate and hence inefficient dispersion and distribution of the particles occur. Thus, the poor performance is caused by the stress concentration generated around the discontinuous phases within the polymer composite.

The effect of different filler particle size on the mechanical properties of Ponderosa pine WF reinforced PP composites was studied by Stark et al. (Stark and Rowlands 2003). For a 40 wt.% WF reinforcement in PP, it was observed that with decreasing particle size the tensile strength initially increased, reached a maximum and thereafter decreased. The tensile strength and modulus of the WF/PP composite were 25.5 MPa and 3.61 GPa respectively for an average WF size of 0.215 mm, which was highest among all. Cavus et al. (Cavus and Mengeloglu 2020) studied the effect of mahogany WF size on the mechanical properties of PP and recycled PP (rPP) based composites using 3 wt.% MA-g-PP as a compatibilizer. Three range of WF sizes; 0.074 to 0.149 mm, 0.177 to 0.250 mm and 0.400 to 0.841 mm were used for the investigation with a 40 wt.% WF content fixed for all the composite formulation. It was observed that in absence of MA-g-PP the tensile strength for both WF/PP and WF/rPP improved upon increasing filler size. In comparison, the presence of MA-g-PP improved the tensile strength significantly for both WF/PP and WF/rPP with similar values of strength irrespective of the filler size. Maximum tensile strength of 24.25 MPa was observed for filler size of 0.177 to 0.250 mm, probably due to the improved compatibility of WF with PP matrix via MA-g-PP. Furthermore, upon increasing filler size the tensile modulus increased in a range of 754 to 876 MPa. Interestingly, WF/rPP had shown higher modulus than WF/PP which was presumably due to the shorter PP chains obtained after recycling which create more crystallinity within the composite making it more stiffer (Cavus and Mengeloglu 2020). Chaudemanche et al. (Chaudemanche et al. 2018) used image analysis on a mixture of WF (Pinus pinaster and Picea sitchensis) to study their size distribution and also fabricated WF/HDPE composites having WF of three different size range. The filler size was categorized by sieve sizes of 800 μm, 500 μm and 200 μm. It was observed that with increasing filler size the composite density decreased, which was due to the presence of inherent porosity within and at the interfaces of WFs. To understand the effect of filler size, its distribution and orientation, tensile test specimens were cut in extrusion or machine direction (MD) and transverse direction (TD) of the fabricated sample. Both the tensile strength and modulus improved with increasing flour size for the MD samples, whereas it decreased for the TD samples. MD tensile samples with WF 800 μm showed the highest tensile strength of 15 MPa and tensile modulus of ~3.24 GPa. It was observed from the morphological analysis of fractured samples that the WFs with higher L/D ratio aligned in the extrusion direction whereas finer particles were less oriented making the composite stronger and stiffer in the extruded direction (Chaudemanche et al. 2018).

Wang et al. investigated the effect of maleated polybutadiene-grafted-polypropylene (MAPB-g-PP) on the mechanical properties of WF reinforced PP composites (Wang et al. 2020). For a fixed wt.% of WF among 40, 50 and 60%, they varied the MAPB-g-PP content from 0–20 wt.% in the composites. It was observed that out of all the prepared formulations, a combination of 50 wt.% WF and 12 wt.% of MAPB-g-PP in PP displayed the highest tensile strength among all. An in-depth study on the effect of elastomer modified polymer blends on the mechanical properties of WPCs was conducted by Clemons (Clemons 2010). HDPE and PP in various proportion along with MA-g-PP, MA-g-PE, EPDM and MA-EPDM were used in different quantity for this study. Table 1 shows the composite formulations along with different mechanical properties. All the composites are reinforced with a fixed loading of 30 wt.% WF. The tensile strength and modulus of neat HDPE and PP were 20.9 MPa and 0.76 GPa, and 38.2 MPa and 2.06 GPa respectively. With only 30 wt.% WF reinforcement the tensile modulus increased significantly (HDPE: 2.82 GPa and PP: 4.08 GPa), due to the restriction in the polymer chain mobility. But the tensile strength only slightly increased for HDPE up to 21.4 MPa and it reduced in case of PP down to 29.3 MPa. Whereas, addition of only 3 wt.% MA-g-PE/MA-g-PP to the WPCs improved the strength very significantly with moderate increase in modulus as observed for 100/0 and 0/100 in Table 1.

In case of the WPCs with different blends proportion, the one with higher portion of PP showed better strength and modulus in comparison to the similar proportional blends where HDPE is major matrix phase. HDPE being major phase, the tensile properties improved upon blending it with PP while the properties deteriorate when PP was a major phase and HDPE was incorporated for blending. Since the properties of neat PP are superior than HDPE. HDPE/PP ratio of 75/25 and 25/75 showed the highest tensile strength and modulus of 26.1 MPa and 3.33 GPa, and 36.2 MPa and 4.02 GPa respectively when 3 wt.% MA-g-PE and MA-g-PP were added. The addition of only 10 wt.% EPDM reduced the tensile strength of the WPCs to the lowest among the blends, 18.5 MPa for 75/25 ratio and 22.9 MPa for 25/75 ratio. While replacing EPDM with same quantity of MA-EPDM increased the tensile strength but reduced the modulus to lowest among the respective blends, 2.26 GPa for 75/25 ratio and 2.71 GPa for 25/75 ratio. Csikós et al. (Csikós et al. 2015) prepared maleic anhydride-grafted-poly(lactic acid) (MA-PLA) by reactive extrusion in different grafting percentage of maleic anhydride. The efficiency of the prepared MA-PLA was checked by preparing composite samples of WF/PLA with varying MA-PLA content and grafting degree and also by varying WF content, as observed in Fig. 7.

Table 1. Composite formulations along with their mechanical properties (Clemons 2010) “Reprinted from publication tittle, Vol. 41, Craig Clemons, Elastomer modified polypropylene–polyethylene blends as matrices for wood flour–plastic composites, 1559–1569, Copyright (2010), with permission from Elsevier.”

HDPE/ PP ratio (%/%)	EPDM	MA-g-PE	MA-g-PP	MA-EPDM	Tensile modulus (GPa)	Tensile strength (MPa)	Flexural modulus (GPa)	Flexural strength (MPa)	Notched impact energy (J/m)
100/0					2.82	21.4	1.80	31.0	48.4
100/0		3			2.92	26.6	1.76	36.9	46.4
75/25					3.34	23.1	2.21	35.9	44.1
75/25	10				2.49	18.5	1.47	26.5	64.2
75/25		3			3.33	26.1	2.10	37.0	40.5
75/25	10	3			2.63	20.7	1.44	25.3	61.1
75/25				10	2.26	25.0	1.36	30.6	71.9
50/50					3.77	25.0	2.67	41.7	28.6
50/50	10				2.77	20.7	2.03	31.5	45.9
25/75					3.90	28.3	3.04	47.1	26.5
25/75	10				3.11	22.9	2.45	37.1	40.9
25/75			3		4.02	36.2	3.05	58.6	24.4
25/75	10		3		3.12	30.5	2.28	46.1	49.7
25/75				10	2.71	26.9	1.95	37.2	52.7
0/100					4.08	29.3	3.16	45.8	26.0
0/100			3		4.16	39.0	3.13	60.8	24.7

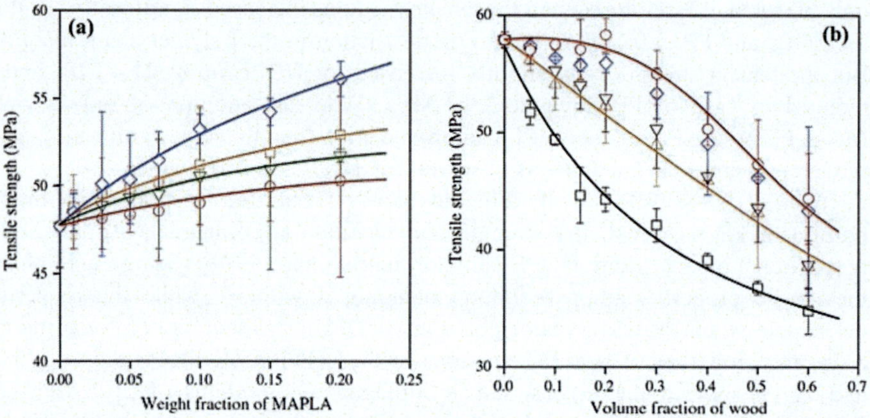

Fig. 7 Tensile strength of WF/PLA composite as a function of (a) MAPLA wt.%; symbols: (○) 0.2, (▽) 0.7, (□) 1.6, (◇) 2.5 MA group per PLA chain and (b) volume fraction of WF; symbols: (□) 0, (△) 5, (▽) 10, (◇) 15, (○) 20 wt.% MAPLA (Csikós et al. 2015) “Reprinted from publication tittle, Vol. 68, Áron Csikós, Gábor Faludi, Attila Domján, Károly Renner, János Móczó, Béla Pukánszky, Modification of interfacial adhesion with a functionalized polymer in PLA/wood composites, 592–600, Copyright (2015), with permission from Elsevier.”

The WF/PLA composite samples used for the tensile test, were fabricated keeping the WF content fixed at 30 wt.%. As can observed in Fig. 7(a), the tensile strength increased with increasing of MA content per PLA chain and also with the wt.% of MA-PLA within the composite. The highest tensile strength was observed for composites with 20 wt.% of MA-PLA having 2.5 MA group/PLA chain. To study the reinforcement efficiency of the prepared MA-PLA, Csikós et al. also analysed the variation of WF content on the tensile strength as shown in Fig. 7(b). For 0 wt.% MA-PLA content the tensile strength sharply decreases with increasing WF loading, which is typical for most of the polymers. With increasing MA-PLA content the decrease in tensile strength with increasing WF content gradually decreases and also saturation is shown by composites with 20 wt.% MAPLA up till 20 vol.% WF.

The drastic effect of moisture content in a WPC on its mechanical properties was studied in detail by Stark (Stark 2001). The moisture uptake of WF/PP composites were found to be directly proportional to the exposure time and relative humidity (RH). The equilibrium or saturation in moisture absorption was attained after an exposure time of 1500 and 1200 h for 20 and 40 wt.% WF reinforced PP composites. It was observed that while there was a very negligible effect of different RH on tensile strength and modulus, the water soak test showed significant reduction in both the properties. The degradation in the interfacial adhesion of WF and PP in composites and absence of any compatibilizer may have combinedly affected this deterioration (Stark 2001).

3 Flexural Behavior

The flexural test determines the flexural strength and flexural stiffness/modulus of a material. This test is significant for materials which are subjected to frequent bending forces, for instance structural beams. ISO 178, ASTM D198-94, ASTM D790 and ASTM D 7264 are the standards generally used for measuring flexural properties which involves 3-point or 4-point bending procedures.

Valente et al. prepared hybrid composites of used glass fibre and WF of hardwood beech with LDPE and PP matrixes in which they studied the effect of different WF and glass fibre (both virgin and recycled) content on the flexural properties of the composites (Valente et al. 2011). It was observed for both LDPE and PP composites (containing only WF) that with increasing WF content the flexural strength decreased while the flexural modulus increased. This phenomenon is quite common in thermoplastic composites filled with particulates or short fibres. Since the only micromechanical deformation process is the debonding of the filler material from the continuous polymer phase, which is more pronounced in composites containing no coupling agents. On the other hand, the flexural modulus depends very moderately on the interfacial adhesion. Increasing filler addition ceases mobility of polymer chains which ultimately leads to a higher stiffness (Valente et al. 2011). Nevertheless, with 20 wt.% WF, PP composites showed a flexural strength and flexural modulus of ~22 MPa and ~2.75 MPa, while LDPE composites displayed ~13 MPa and ~1.25 MPa respectively. In case of the hybrid composites containing both WF and glass fibre, it was observed that recycled glass fibres provided similar strength to the composites as the virgin. Ma et al. have studied the effect of hybrid filler system of nano silica and WF on the flexural properties (Ma et al. 2020). It was observed that flexural modulus increased for both the blend modified and in situ modified hybrid fillers with their increasing content. The flexural strength on the other hand, increased initially, reached a maximum for 20 wt.% filler loading and thereafter decreased for the hybrid filler system. Zong et al. investigated the effect of increasing content of WF in the shell layer on the flexural strength of co-extruded wood plastic/lumber composite (Zong et al. 2020). WPCs as such suffer significantly from brittle fracture and creep deformation which restricts their use as structural components. Laminated veneer lumber (LVL) co-extruded with WF reinforced polyvinyl chloride composite (WPVC) as shell layers showed improved flexural properties in comparison to WPVC alone with increasing WF content from 20 to 60 wt.%. The pure LVL showed a flexural strength and modulus of 79.8 MPa and 10.3 GPa respectively, which decreased by 13% and 30% upon coating with polyvinyl chloride alone due to its brittle nature. While the flexural strength increased to 87.8 MPa and then dropped to 86.1 MPa for a WF content of 40 wt.% and 60 wt.% in WPVC respectively. The further increase in filler content may have led to agglomeration of the WF within the polymer matrix, thereby, making it weaker (Zong et al. 2020).

The effect of filler particle size on flexural properties was studied by Stark et al. for WF/PP composites (Stark and Rowlands 2003). It was observed that Ponderosa pine WF with an average particle size of 0.128 mm in a PP matrix showed the highest flexural strength of 42.9 MPa. While the highest flexural modulus was observed for an average particle size of 0.215 mm. Cavus et al. (Cavus and Mengeloglu 2020) studied the effect of Mahogany WF size on various mechanical properties. In the flexural test it was observed that for both WF/PP and WF/rPP flexural strength and modulus improved upon increasing filler size in absence of MA-g-PP, with highest being 37.32 and 2178.12 MPa respectively in the filler size of 0.400 to 0.841 mm. Upon the addition of 3 wt.% MA-g-PP, these properties improved further but showed similar values irrespective of the filler size range. In filler size range of 0.400 to 0.841 mm, WF/PP showed the highest flexural strength and modulus of 41.86 and 2343.93 MPa respectively (Cavus and Mengeloglu 2020).

In a study made by Wang et al. it was observed that MAPB-g-PP is an effective coupling agent for WF/PP composite systems (Wang et al. 2020). With 12 wt.% content of MAPB-g-PP, the flexural strength improved by 55.4% for a WF/PP composite having 50 wt.% of WF. Along with coupling agent polymer blends also have significant contribution to the flexural properties of WPCs. In a study made by Clemons (Clemons 2010), it was concluded that WPCs having polymer blends like HDPE/PP as base matrix can have flexural strength higher than WPCs with single polymer as matrix as seen in Table 1. Initially, with only 30 wt.% WF addition the flexural strength and modulus of neat HDPE increased to 31.0 MPa and 1.80 GPa from 20.2 MPa and 0.56 GPa respectively. Furthermore, the addition of 3 wt.% MA-g-PE to the WF/HDPE composite improved the flexural strength to 36.9 MPa but reduced the modulus to 1.76 GPa, as seen for 100/0 in Table 1. MA-g-PE helps in improving interfacial adhesion, thereby, effective load transfer from matrix to WF increases making the composite stronger. On the other hand, addition of 30 wt.% WF in neat PP reduced its flexural strength from 50.1 MPa to 45.8 MPa but improved the modulus from 1.74 GPa to 3.16 GPa. However, addition of 3 wt.% MA-g-PP brought the flexural strength to 60.8 MPa, which is higher than the neat PP, as seen for 0/100 in Table 1. In case of polymer blends of HDPE/PP in proportions of 75/25, 50/50 and 25/75, it was observed that without the addition of any additives like EPDM or maleic anhydride-based coupling agent the flexural property behaved based on the rule of mixture. The property of the major phase dominated the mechanical behaviour of the polymer blend based WPCs (Clemons 2010).

Absorption of moisture is a critical concern, as it deteriorates the mechanical performance of WPCs. Care must be taken while fabricating such composites by proper drying of WFs and use of appropriate compatibilizer. For a 40 wt.% WF reinforced PP composite, it was observed by Stark (Stark 2001) that the flexural strength and modulus were affected significantly by water soak test. A decrease of ~22 and ~39% in flexural strength and modulus was observed after a 3000-h water soak test of the composites.

4 Impact Strength

Impact strength is defined as the ability of a material to withstand a particular amount of energy when subjected to an impact load. Two types of impact tests, Izod and Charpy, are conducted for evaluating impact strength. For plastics and polymer composites, Izod impact test is mostly considered. ASTM D256, is the standard normally followed to conduct such test.

Stark et al. have studied the effect of particle size on the notched and unnotched Izod impact strength of WF/PP composites (Stark and Rowlands 2003). They observed that the impact energy of composites increased with increasing particle size for the notched samples while it decreased for the unnotched samples. The notched impact energy explains the energy required for crack propagation and the unnotched impact energy explains the energy required for crack initiation. Since larger particles having higher surface area will require more energy for the already initiated crack to propagate, and hence, the highest notched impact energy of 22.3 J/m was shown by WF/PP composites with largest particle size of 0.513 mm. Contrarily, minimum energy (54 J/m) was required for initiating a crack on unnotched sample for the largest particle size 0.513 mm. Since stress concentration at the interphase of hydrophilic WF and the hydrophobic polymer matrix was highest for the largest considered particle size.

Wang et al. fabricated WF/PP composites with 40, 50 and 60 wt.% of WF and varied the MAPB-g-PP content from 0–20% (Wang et al. 2020). A significant improvement in the Izod impact strength was observed for both 50 and 60 wt.% WF reinforced PP composites at a MAPB-g-PP content of 12%. Composite with 50 wt.% WF showed 60.1% increase in the impact strength, which was highest among all the composite formulations. Wang et al. also compared efficiency of MAPB-g-PP with MA-g-PP and found that the former is a better choice for improving both the impact strength and impact toughness. It was also concluded in the study that MAPB-g-PP serves both as a compatibilizer and also as a toughener (Wang et al. 2020). Clemons (Clemons 2010) studied the effect of polymer blend of HDPE and PP as a matrix for WPCs and investigated various mechanical properties. It was observed that as such virgin HDPE is significantly tougher than PP, with the former exhibiting notched impact strength of 125.8 J/m which is ~346% higher than that of PP. Addition of 30 wt.% WF reduced the impact strength of both the WPCs with HDPE and PP as base matrix, as seen in Table 1 for 100/0 and 0/100. The stress concentration generated in the interphase of filler particles along with weak filler–polymer interfacial adhesion supports the propagation of cracks easily, making the composites less impact resistant. In case of composites fabricated using polymer blends, the behaviour of impact strength followed the rule of mixture with proportional wt.% of HDPE and PP. The addition of only 10 wt.% EPDM tremendously increased the impact strength of composites with blend 75/25 by ~46% and that of 25/75 by ~54%. In addition, use of 10 wt.% EPDM in combination with 3 wt.% coupling agents for the composites with blends further improved the impact strength. But the highest value of impact strength of 71.9 J/m and 52.7 J/m was observed when 10 wt.% MA-EPDM was used alone for the composites with blend proportion of HDPE/PP in 75/25 and 25/75 respectively, as seen in Table 1 (Clemons 2010). The effect of different impact modifiers; maleated SEBS (SEBS-MA), EPDM and maleated EPDM (EPDM-MA) on the impact properties of WF/PP composites was studied by Oksman et al. (Oksman and Clemons 1998). All the prepared composites had a fixed WF content of 40 wt.% and impact modifier content of 10 wt.%. Composites composed of only MA-g-PP had also a fixed content of 2 wt.%. It was observed that impact modifier, SEBS-MA had a significant positive role in improving the impact strength in comparison to the EPDM and EPDM-MA alone, which further improves upon the addition of MA-g-PP. The notched and unnotched impact energies of the WF/PP composites increased by ~107% and ~95% over the virgin PP. Maleated EPDM and SEBS helped to serve both as a compatibilizer and impact modifier which resulted in improved interfacial adhesion, and thereby, better mechanical strength and impact resistance was displayed by the composites. The apparently lower storage modulus confirmed that, while the interphase formed by EPDM-MA was soft, the interphase formed by SEBS-MA was hard making the latter to transfer load more effectively (Oksman and Clemons 1998). Similar to the aforementioned study, Guo et al. (Guo and Wang 2007) also studied the effect of varying SEBS-MA content from 0 to 10 wt.% with an increment of 2 wt.% on the impact fracture behavior of WF/PP composites. It was observed that 8 wt.% of SEBS-MA was optimum for the 40 wt.% WF reinforced PP composites to demonstrate the highest impact resistance.

5 Micromechanical Modelling of WPCs

Various empirical relations defining the micromechanics of composite materials have been developed by many researchers over the years which can predict different bulk mechanical properties with initial input of constituent material properties. Based on the type of composite, the different models developed so far can be categorized under continuous, short fibre or discontinuous and particulate composites. In case of short fibre and particulate composites micromechanical models such as; rule of mixture (ROM), Halpin–Tsai, Tsai-Pagano, shear-lag theory, laminate analogy approach, Nielsen model, Mori–Tanaka model, self-consistent model etc., have been famously applied to estimate composite properties. All these models mostly consider the filler inclusion to be geometrically regular and uniform with little to no self-interactions among the fillers or reinforcing members. WFs, on the other hand, are rough, have a low L/D ratio, irregular in shape and size, possess pits and dislocations and are anisotropic in nature. Nevertheless, the aforementioned models should suffice for the closest property estimations of WPCs considering WF to be a short fibre and/or particle inclusion. Micromechanical models have been abundantly used for natural fibre reinforced polymer composites, but very few of them have been reported for WPCs consisting of WFs or sawdust (Mansor et al. 2013). The simplest and most widely used empirical relation is ROM given by Eq. 1, which is used to calculate the elasticity of composite material based on the proportional property contribution of the filler and base matrix.

$$E_c = V_f E_f + (1 - V_f) E_m \tag{1}$$

where, E_c, E_f and E_m are the elastic modulus of composite, WF and polymer matrix respectively. V_f is the volume fraction of WF in the WPC. It is worth noticing here that the ROM neither consider the filler geometrical parameters, orientation nor the filler–polymer interfacial adhesion. The random orientation of fillers was addressed in the empirical relation coined for E_c by Halpin-Pagano (Halpin and Pagano 1969) for short fibre reinforcing a continuous matrix and is given by the Eq. (2).

$$E_c = \frac{3}{8} E_L + \frac{5}{8} E_T \tag{2}$$

where, E_L and E_T are the longitudinal and transverse elastic moduli of an oriented short fibre composite and can be estimated by the Halpin–Tsai equations (Affdl and Kardos 1976) as given in Eq. (3) and (4).

$$E_L = \frac{1 + 2\left(\frac{l_f}{d_f}\right)\eta_L V_f}{1 - \eta_L V_f} E_m \tag{3}$$

$$E_T = \frac{1 + 2\eta_T V_f}{1 - \eta_T V_f} E_m \tag{4}$$

where,

$$\eta_L = \frac{\frac{E_f}{E_m} - 1}{\frac{E_f}{E_m} + 2\left(\frac{l_f}{d_f}\right)} \tag{5}$$

$$\eta_T = \frac{\frac{E_f}{E_m} - 1}{\frac{E_f}{E_m} + 2} \tag{6}$$

In Eq. (3) and (5), l_f and d_f are the length and diameter of the fibre inclusion. In case of WF the L/D ratio (or l_f/d_f) is not very large, but generally greater than 1 (Chaudemanche et al. 2018). Nevertheless, the Halpin-Pagano relation has been used by researchers to predict the elasticity of WPCs (Pilla et al. 2008; Huang et al. 2013). It was observed by Pilla et al. (Pilla et al. 2008) and Huang et al. (Huang et al. 2013) that although both the experimental and theoretical data of elastic modulus followed similar trend but the predictions were less accurate. Hence, arguments on the empirical relation were made based on disregard against filler agglomeration, perfect filler dispersion, interfacial adhesion etc. Mirbagheri et al. have used Halpin–Tsai and ROM for hybrid polymer composites reinforced with WF and short kenaf fibres (Mirbagheri et al. 2007). It was observed that the Halpin–Tsai model followed a non-linear trend and it underestimated the modulus of hybrid composite at lower filler loading while overestimated the same at higher loading. Mori–Tanaka model (MTM) considers elliptical inclusions with their different orientations in a continuous matrix (Mori and Tanaka 1973). The model considers Eshelby's tensor (Eshelby 1957) S, to relate the perturbed strain of a single inclusion to the eigenstrain. The Mori–Tanaka empirical relation presents stress generated by a composite for a finite inclusion and with an eigen strain ε^*, as shown in Eq. (7).

$$\langle\sigma\rangle_m = C\langle\varepsilon\rangle_m = -v_1 C(S\varepsilon^* - \varepsilon^*) \tag{7}$$

where, $\langle\sigma\rangle_m$ is the stress in the matrix or composite, C is the effective stiffens and v_1 is the volume fraction of inclusion and S is the Eshelby's tensor. The average effective stiffness, can be calculated by Eq. (8) (Qin and Yang 2008) and applied in Eq. (7) to predict the average stress.

$$\bar{C} = C^0 + v_1(C^1 - C^0)A'(v_0 I + v_1 A')^{-1} \quad (8)$$

where, A' is the strain concentration tensor, I is the unit tensor and the subscript and superscript 0 and 1 denote matrix and inclusion respectively. Hančič et al. (Hančič et al. 2012) applied the MTM to predict the elasto-plastic responses of PP and PS based WPCs. Out of many variations for applying average stress concept of MTM to composites, Hančič et al. used the Benveniste's solution (Benveniste 1987). Hančič et al. also applied a generalised method of cells (GMC) model to predict the composite properties and did a comparative study with MTM and experimental values. The GMC is essentially an analytical micromechanical model which assumes the composite to be an arrangement of a number of tiny cells in form of building blocks. Each cell, consisting of fibre surrounded by the matrix, is representative of the bulk composite. Figure 8(a) and (b) shows the periodically arranged fibre inclusion in a matrix and the representative cell. In the detailed explanation of GMC, it was derived that the average stress of the composite is given by Eq. (9).

$$\bar{\sigma}_{ij} = \frac{1}{V}\sum_{\alpha=1}^{2}\sum_{\beta=1}^{2}\sum_{\gamma=1}^{2} v_{\alpha\beta\gamma}\bar{\sigma}_{ij}^{(\alpha\beta\gamma)} \quad (9)$$

where, $\bar{\sigma}_{ij}$ is the average stress in the subcell, $V = (d_1 + d_2)(h_1 + h_2)(l_1 + l_2)$ and $v_{\alpha\beta\gamma} = d_\alpha h_\beta l_\gamma$. Similarly, the average strain in the composite is given by

$$\bar{\varepsilon}_{ij} = \frac{1}{V}\sum_{\alpha=1}^{2}\sum_{\beta=1}^{2}\sum_{\gamma=1}^{2} v_{\alpha\beta\gamma}\bar{\varepsilon}_{ij}^{(\alpha\beta\gamma)} \quad (10)$$

The results predicted by both MTM and GMC models are shown in Fig. 9 and 10 (Hančič et al. 2012). It was observed that both the models made a very close predictions of the experimental values for tensile stress and elastic modulus as a function of fibre volume fraction.

At lower fibre volume MTM was effective while GMC at higher loading, as seen in Fig. 9. It was concluded that in the elastic region the predictions made by models were accurate while near the failure region they were less accurate (Hančič et al. 2012).

Cox-Krenchel model (Cox 1952; Krenchel 1964) is a micromechanical model similar to ROM with some modifications. The empirical relation is given by Eq. (11).

$$E_c = \eta_o \eta_L E_f V_f + E_m(1 - V_f) \quad (11)$$

where, η_o and η_L are the orientation efficiency factor and shear lag factor respectively.

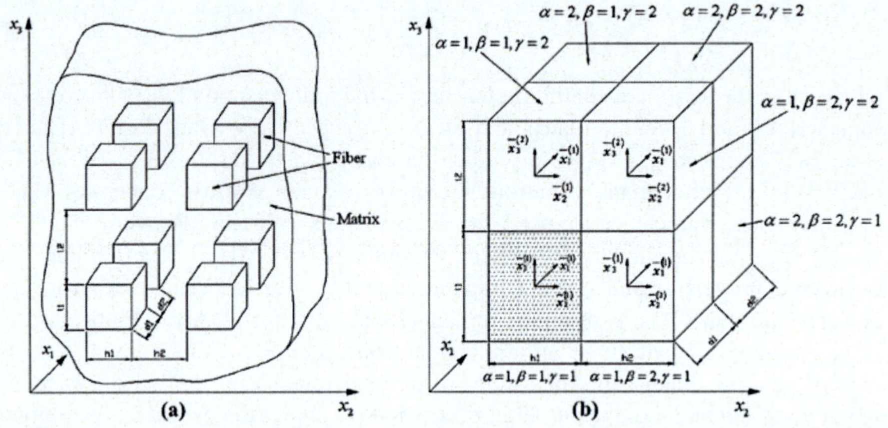

Fig. 8 (a) Composite with fibre inclusions, arranged in periodic array, (b) representative cell with eight subcells α, β, γ = 1,2 (Hančič et al. 2012) "Reprinted from publication tittle, Vol. 43, Aleš Hančič, Franc Kosel, Karl Kuzman, Janez Marko Slabe, Comparison of two micromechanical models for predicting elasto-plastic response of wood–plastic composites, 1500–1507, Copyright (2012), with permission from Elsevier."

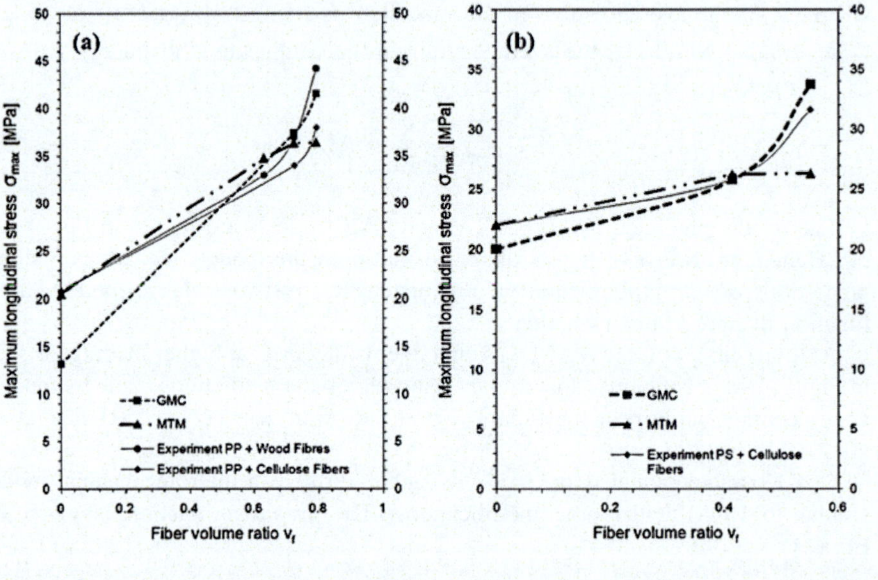

Fig. 9 Illustration of the variation of maximum longitudinal stress of WPCs as a function of fibre volume fraction (Hančič et al. 2012) "Reprinted from publication tittle, Vol. 43, Aleš Hančič, Franc Kosel, Karl Kuzman, Janez Marko Slabe, Comparison of two micromechanical models for predicting elasto-plastic response of wood–plastic composites, 1500–1507, Copyright (2012), with permission from Elsevier."

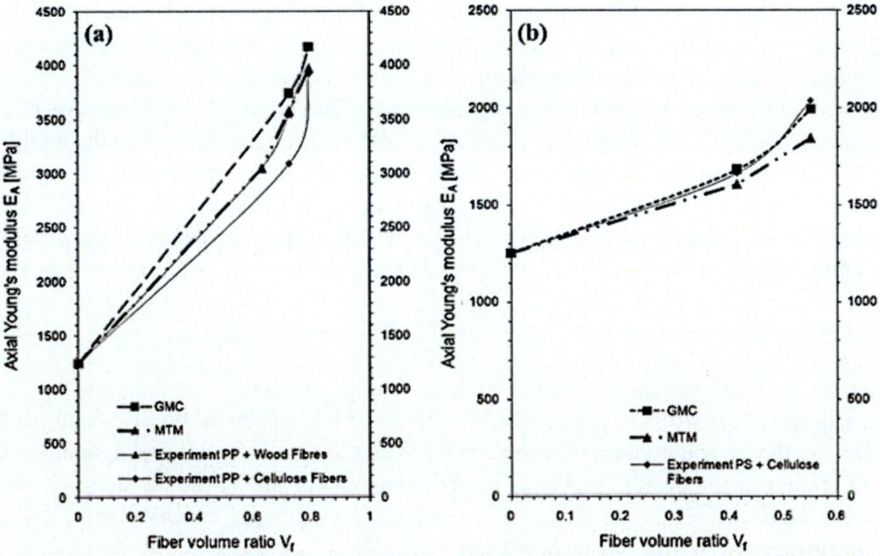

Fig. 10 Illustration of the variation of axial young modulus of WPCs as a function of fibre volume fraction (Hančič et al. 2012) "Reprinted from publication tittle, Vol. 43, Aleš Hančič, Franc Kosel, Karl Kuzman, Janez Marko Slabe, Comparison of two micromechanical models for predicting elasto-plastic response of wood–plastic composites, 1500–1507, Copyright (2012), with permission from Elsevier."

$$\eta_o = \sum_n a_n \cos^4 \text{ where, } \sum_n a_n = 1 \tag{12}$$

η_o = 3/8 for a 2D random fibre orientation.
η_o = 1/5 for a 3D random fibre orientation

$$\eta_o = \left(1 - \frac{\tanh(\beta A_r)}{\beta A_r}\right) \tag{13}$$

For a hexagonal filler packing system

$$\beta^2 = 2\left(\frac{G_m}{E_f ln\left(2\pi/V_f\sqrt{3}\right)}\right)^{1/2} \tag{14}$$

where, G_m is the shear modulus and A_r is the aspect ratio of filler.

Połec' et al. (Połeć et al. 2010) applied the Cox-Krenchel model for die drawn WPCs with two different WF namely softwood (90% spruce and 10% fir) and hardwood (100% beech) at two different nominal draw ratios $R_N = 3$ and $R_N = 10$. It was observed that a good match of the experimental and theoretical values of composite modulus was obtained in case of the 2D random orientation of fibre when void fraction was taken into consideration (Połeć et al. 2010). Migneault et al. applied Kelly-Tyson (Migneault et al. 2011) model to evaluate the tensile strength of WPC reinforced with birch pulp fibre with varying fibre aspect ratio from 8.3 to 21.3. The Kelly-Tyson model is given by the Eq. (15). Unlike the general ROM model, it considers both the aspect ratio and the fibre-matrix interfacial shear stress component.

$$\sigma_c = \alpha_o 2\tau(L/D)v_f + \sigma_m v_m \tag{15}$$

where, α_o is the orientation factor, σ_c is the composite tensile strength and τ is the interfacial shear stress of the incorporated fibre. It was observed by Migneault et al. that the theoretical predictions made by the Kelly-Tyson were in good agreement to the experimental results for both injection moulded and extruded samples with different fibre orientation factor. For injection molded and extruded samples, α_o considered was 1 and 3/8 respectively.

The prediction of flexural behaviour of WPCs is very important as they are often subjected to bending loads during application. Naghipour et al. (Naghipour et al. 2011) investigated the flexural behaviour of hybrid WPC composite beams retro-fitted separately with 1–3 layers of glass and carbon FRP using both experimental and numerical analysis. The non-linear analysis of the *I* section of WPC beam considered in the study was done using classical beam theory to evaluate the load versus displacement curve from the theoretically determined bending moment.

In order to calculate the bending moment, Naghipour et al. proposed an exponential function for stress–strain behavior of WPC in both tension and compression sides parallel to the fibres, which was given by Eq. (16).

$$\sigma = A\left(1 - e^{-B\epsilon}\right) \tag{16}$$

where, σ and ϵ are the stress and strain, while A and B are material parameters. With an incremental strain of $\Delta\epsilon$, the bending moment was calculated until the maximum strain at failure for WPC in compression and FRP in tension was reached. It was observed that in comparison to pure WPC beam, the flexural load carrying capacity of the hybrid WPC-glass and carbon FRP improved by 80.3% and 76.6% with an average error of prediction by the theoretical model being 4% and 5.25% respectively. The credible results of the theoretical model suggest its use in predicting and accessing the flexural response of hybrid WPC beams (Naghipour et al. 2011).

6 Conclusion

This study covers an in-depth analysis of mechanical properties of WPCs along with different empirical models used to theoretically predict their bulk properties. The variation of different parameters such as filler loading and size, interfacial adhesion, effect of coupling agents etc. on the tensile, flexural and impact behaviour were reviewed and analysed. The general trend of strength and modulus shown by WPCs with increasing WF loading is that of reducing and increasing in nature respectively, if no coupling agent is added or no modification of filler and matrix is done (Lee et al. 2004). Interestingly, the addition of coupling agent or grafting on the base polymer chain may potentially improve the interfacial adhesion between WF and polymer matrix leading to an efficient stress transfer from matrix to filler resulting an in improved stiffness and strength of the WPCs (Lee et al. 2004; Bengtsson et al. 2005). The unnotched impact resistance also decreases with increasing filler size and its loading and depends on effective interfacial adhesion along with impact modifiers for improving the resistance (Stark and Rowlands 2003). Comparatively, with a larger surface area WPCs with smaller filler sizes can potentially have more surface interaction which can result in improved mechanical properties. But to exploit the same the interfacial adhesion has to be improved via addition of coupling agents or compatibilizers. Apart from concentrating on WF role, the mechanical properties of WPCs can also be optimized by a number of different combinations of base polymer blending, coupling agent, modifiers etc. depending upon the property requirements for desired application (Clemons 2010). Furthermore, different theoretical models describing the micromechanical behaviour of WPCs have been also discussed. It was observed that the accuracy of predicting modulus or strength increases for the models which extends the inclusion of parameters such as filler orientation, aspect ratio, interfacial adhesion etc. Nonetheless, among several models in literature Kelly-Tyson and Mori–Tanaka models discussed here have comparatively shown a closer proximity with the experimental results. The kind of knowledge created in the present study by the combined analysis of experimental tests and theoretical models is not only crucial but also a pre-requisite for fabricating WPCs with desired mechanical properties for various general and specialized applications.

Acknowledgements The first author wishes to thank the Ministry of Human Resource and Development (MHRD), Govt. of India for the financial support provided in the form of research fellowship.

References

Affdl JCH, Kardos JL (1976) The Halpin-Tsai equations: a review. Polym Eng Sci 16:344–352. https://doi.org/10.1002/pen.760160512

Ashori A (2008) Wood-plastic composites as promising green-composites for automotive industries! Bioresour. Technol. 99:4661–4667

Bengtsson M, Gatenholm P, Oksman K (2005) The effect of crosslinking on the properties of polyethylene/wood flour composites. Compos Sci Technol 65:1468–1479. https://doi.org/10.1016/j.compscitech.2004.12.050

Benveniste Y (1987) A new approach to the application of Mori-Tanaka's theory in composite materials

Cavus V, Mengeloglu F (2020) Effect of wood particle size on selected properties of neat and recycled wood polypropylene composites. In: BioResources. https://ojs.cnr.ncsu.edu/index.php/BioRes/article/view/BioRes_15_2_3427_Cavus_Wood_Particle_Size_Recycled_Wood. Accessed 27 Apr 2020

Chan CM, Vandi L-J, Pratt S et al (2018) Polymer reviews composites of wood and biodegradable thermoplastics: a review composites of wood and biodegradable thermoplastics: a review. Polym Rev 58:444–494. https://doi.org/10.1080/15583724.2017.1380039

Changxing Hanming Technology Co. Ltd. (2020) Swimming pool wpc decking or Anti-UV swimming pool deck flooring. https://www.cxhanming.com/product/swimming-pool-wpc-decking/. Accessed 28 May 2020

Chaudemanche S, Perrot A, Pimbert S et al (2018) Properties of an industrial extruded HDPE-WPC: the effect of the size distribution of wood flour particles. Constr Build Mater 162:543–552. https://doi.org/10.1016/j.conbuildmat.2017.12.061

Clemons C (2010) Elastomer modified polypropylene-polyethylene blends as matrices for wood flour-plastic composites. Compos Part a Appl Sci Manuf 41:1559–1569. https://doi.org/10.1016/j.compositesa.2010.07.002

Cox HL (1952) The elasticity and strength of paper and other fibrous materials. Br J Appl Phys 3:72. https://doi.org/10.1088/0508-3443/3/3/302

Csikós Á, Faludi G, Domján A et al (2015) Modification of interfacial adhesion with a functionalized polymer in PLA/wood composites. Eur Polym J 68:592–600. https://doi.org/10.1016/j.eurpolymj.2015.03.032

Eshelby JD (1957) The determination of the elastic field of an ellipsoidal inclusion, and related problems. Proc R Soc London Ser a Math Phys Sci 241:376–396. https://doi.org/10.1098/rspa.1957.0133

Gogoi R, Kumar N, Mireja S et al (2019) Effect of hollow glass microspheres on the morphology, rheology and crystallinity of short bamboo fiber-reinforced hybrid polypropylene composite. JOM 71:548–558. https://doi.org/10.1007/s11837-018-3268-3

Grand View Research (2020) Wood Plastic Composite Market Size | Industry Report, 2027. https://www.grandviewresearch.com/industry-analysis/wood-plastic-composites-market. Accessed 30 Apr 2020

Guo CG, Wang QW (2007) Effect of maleic anhydride grafted styrene-ethylene-butylene-styrene (MA-SEBS) on impact fracture behavior of polypropylene / wood fiber composites. J For Res 18:203–207. https://doi.org/10.1007/s11676-007-0041-9

Gusev AA (1997) Representative volume element size for elastic composites: a numerical study. J Mech Phys Solids 45:1449–1459. https://doi.org/10.1016/S0022-5096(97)00016-1

Halpin JC, Pagano NJ (1969) The laminate approximation for randomly oriented fibrous composites. J Compos Mater 3:720–724. https://doi.org/10.1177/002199836900300416

Hančič A, Kosel F, Kuzman K, Slabe JM (2012) Comparison of two micromechanical models for predicting elasto-plastic response of wood-plastic composites. Compos Part B Eng 43:1500–1507. https://doi.org/10.1016/j.compositesb.2011.11.024

Hine PJ, Rudolf Lusti H, Gusev AA (2002) Numerical simulation of the effects of volume fraction, aspect ratio and fibre length distribution on the elastic and thermoelastic properties of short fibre composites. Compos Sci Technol 62:1445–1453. https://doi.org/10.1016/S0266-3538(02)00089-1

Huang R, Kim B-J, Lee S, et al (2013) co-Extruded Wood-Plastic Composites with Talc-Filled Shells: Morphology, Mechanical, and Thermal Expansion Performance. BioResources 8 (2):2283–2299. https://ojs.cnr.ncsu.edu/index.php/BioRes/article/view/BioRes_08_2_2283_Huang_Wood_Plastic_Composites

Integrated Recycling (2020a) Outdoor Furniture – Integrated Recycling. https://www.integratedrecycling.com.au/outdoor-furniture/. Accessed 28 May 2020

Integrated Recycling (2020b) Architectual Screens – Integrated Recycling. https://www.integratedrecycling.com.au/screens/. Accessed 28 May 2020

Jeluplast (2020) WPC granulate for injection-mould in. https://www.jeluplast.com/en/wpc-granulate-in-injection-moulding/. Accessed 28 May 2020

Klyosov AA (2007) Wood-plastic composites. Wiley. https://books.google.co.in/books?hl=en&lr=&id=KmuK4w_D7UUC&oi=fnd&pg=PR7&ots=CB9T-_2ysw&sig=MBBIwOlc2yXlutuwxwEyDFwag48&redir_esc=y#v=onepage&q&f=false. Accessed 30 Apr 2020

Krenchel H (1964) Fibre reinforcement; theoretical and practical investigations of the elasticity and strength of fibre-reinforced materials. agris.fao.org

Kumar N, Mireja S, Khandelwal V et al (2017) Light-weight high-strength hollow glass microspheres and bamboo fiber based hybrid polypropylene composite: a strength analysis and morphological study. Compos Part B Eng 109:277–285. https://doi.org/10.1016/j.compositesb.2016.10.052

Lee SY, Yang HS, Kim HJ et al (2004) Creep behavior and manufacturing parameters of wood flour filled polypropylene composites. Compos Struct 65:459–469. https://doi.org/10.1016/j.compstruct.2003.12.007

Ma Y, He H, Huang B et al (2020) *In situ* fabrication of wood flour/nano silica hybrid and its application in polypropylene-based wood-plastic composites. Polym Compos 41:573–584. https://doi.org/10.1002/pc.25389

Mansor MR, Sapuan SM, Zainudin ES et al (2013) Stiffness prediction of hybrid kenaf/glass fiber reinforced polypropylene composites using rule of mixtures (ROM) and Rule of Hybrid Mixtures (RoHM). J Polym Mater 30(3):321–334

Markets and Markets (2020) Wood Plastic Composite Market by Type & Application - Global Forecast 2021 | MarketsandMarkets | Last Updated on April-2020. https://www.marketsandmarkets.com/Market-Reports/wood-plastic-composite-market-170450806.html. Accessed 23 Apr 2020

Migneault S, Koubaa A, Erchiqui F et al (2011) Application of micromechanical models to tensile properties of wood-plastic composites. Wood Sci Technol 45:521–532. https://doi.org/10.1007/s00226-010-0351-5

Mirbagheri J, Tajvidi M, Hermanson JC, Ghasemi I (2007) Tensile properties of wood flour/kenaf fiber polypropylene hybrid composites. J Appl Polym Sci 105:3054–3059. https://doi.org/10.1002/app.26363

Mori T, Tanaka K (1973) Average stress in matrix and average elastic energy of materials with misfitting inclusions. Acta Metall 21:571–574. https://doi.org/10.1016/0001-6160(73)90064-3

Naghipour M, Nematzadeh M, Yahyazadeh Q (2011) Analytical and experimental study on flexural performance of WPC-FRP beams. Constr Build Mater 25:829–837. https://doi.org/10.1016/j.conbuildmat.2010.06.104

Neagu RC, Gamstedt EK (2008) Micromechanical modelling of wood-polymer composites. In: Wood-polymer composites. Elsevier Ltd., pp 118–141

Niska KO, Sain M (2008) Wood-Polymer Composites. Elsevier Ltd, Amsterdam

Oksman K, Clemons C (1998) Mechanical properties and morphology of impact modified polypropylene–wood flour composites. J Appl Polym Sci 67:1503–1513. https://doi.org/10.1002/(SICI)1097-4628(19980228)67:9%3c1503::AID-APP1%3e3.0.CO;2-H

Pilla S, Gong S, O'Neill E et al (2008) Polylactide-pine wood flour composites. Polym Eng Sci 48:578–587. https://doi.org/10.1002/pen.20971

Połeć I, Hine PJ, Bonner MJ et al (2010) Die drawn wood polymer composites. II. Micromechanical modelling of tensile modulus. Compos Sci Technol 70:53–60. https://doi.org/10.1016/j.compscitech.2009.09.004

Qin Q-H, Yang Q-S (2008). Macro-Micro Theory on Multi-Field Coupling Behavior of Heterogeneous Materials. https://doi.org/10.1007/978-3-540-78259-9

Shanghai Seven Trust Industry Co. Ltd. (2020) WPC Flower Boxes | Composite Wood Planter Boxes. https://shanghaiwpc.com/products/composite-flower-boxes. Accessed 28 May 2020

Stark N (2001) Influence of moisture absorption on mechanical properties of wood flour-polypropylene composites. J Thermoplast Compos Mater 14:421–432. https://doi.org/10.1106/UDKY-0403-626E-1H4P

Stark NM, Rowlands RE (2003) Effects of wood fiber characteristics on mechanical properties of wood/polypropylene composites

Valente M, Sarasini F, Marra F et al (2011) Hybrid recycled glass fiber/wood flour thermoplastic composites: manufacturing and mechanical characterization. Compos Part a Appl Sci Manuf 42:649–657. https://doi.org/10.1016/j.compositesa.2011.02.004

Wang X, Guo C, Song K (2020) The effects of maleated polybutadiene-grafted polypropylene (MAPB-g-PP) content on the properties of wood flour/polypropylene composites. J Vinyl Addit Technol 26:17–23. https://doi.org/10.1002/vnl.21711

Zong G, Hao X, Hao J et al (2020) High-strength, lightweight, co-extruded wood flour-polyvinyl chloride/lumber composites: effects of wood content in shell layer on mechanical properties, creep resistance, and dimensional stability. J Clean Prod 244:118860. https://doi.org/10.1016/j.jclepro.2019.118860

Structure of Wood Fiber and Factors Affecting Mechanical Properties of Wood Polymer Composites

Neeraj Dubey, Rajesh Purohit, and H. Mohit

Abstract Wood polymer composites (WPCs) are continuously extending the field of applications of composites and already they have replaced several conventional materials from exterior and interior building components. The suitability of WPC for an application primarily depends upon their mechanical properties such as strength, stiffness and toughness. These properties of WPC depend upon the constituent materials, such as fiber & matrix, and bonding between them. However, some other parameters such as moisture content and manufacturing techniques also play an important role. Although, there are several methods successfully established to improve the strength and performance of WPCs. This chapter discusses about, various parameters on which mechanical properties of cellulosic fibers and polymeric matrix material depends. It also includes study of mechanical properties of natural fiber composites (NFCs) and WPCs. The experimental and theoretical method for evaluating mechanical properties of WPCs have also been presented. It also discusses various techniques for improving the mechanical properties of WPCs.

Keywords Wood polymer composites (WPC) · Natural fiber composites (NFC) · Wood fiber · Mechanical properties · Chemical treatment

1 Introduction

The polymer matrix composites (PMCs) replaced conventional materials such as metal, wood, plastic and glass in different essential fields such as aerospace, automotive, defence and house-hold applications. The reinforcement of synthetic fibers such as glass, carbon, and Kevlarin the polymeric matrix led to production of

N. Dubey (✉)
Department of Mechanical Engineering, LNCT, Bhopal, M.P., India

R. Purohit
Department of Mechanical Engineering, MANIT, Bhopal, M.P., India

H. Mohit
Centre for Nanoscience, Composite Research Center, Chennai, Tamil Nadu, India

S. Mavinkere Rangappa et al. (eds.), *Wood Polymer Composites*,
Composites Science and Technology, https://doi.org/10.1007/978-981-16-1606-8_7

versatile composite materials. PMCs' output grew at a high-speed rate due to the various advantages such as lightweight, chemically inactive, water resistance, easy to process, low cost, and excellent specific strength. Consequently, the PMCs as waste material after their end-use have become a great threat to the environment. The main cause is attributed to non-biodegradability of the plastics. It turned the focus of the material researchers towards the synthesis of eco-friendly and sustainable materials. This search led to a renaissance of the plant fibers for reinforcement. However, there is no match of strength of synthetic fibers with that of the natural fibers, but for the moderate strength applications the natural fiber composites (NFCs) have proved as a noble material. In addition to biodegradability, the other advantages of NFCs are less wear of processing tools, no production cost of fibers, good specific properties, low cost, abundant availability, human health-friendly and easy waste management.In the past three decades, many researchers have developed several NFCs by reinforcing polymers with strong natural fibers such as flax, jute, hemp, sisal, coir etc. Nowadays,the NFCs have gained an appreciable commercial value. But to compete in a market, the ease of manufacturing process and high rate of production is a critical factor. It gave an impetus to the production of another class of NFCs as wood polymer composites (WPCs).

WPCs are the class of composites in which short fibers of wood (wood flour or saw dust) are mixed with thermoplastic resins such as Polypropelene (PP), Polyvinyl chloride (PVC), Low density polyethylene (LDPE), High density polyethylene (HDPE) etc. In a compounding extruder system the plastic and wood fibers are processed to form pellets of WPC. These pellets are used as raw material in different manufacturing processes such as injection moulding, compression moulding, and extrusion. However, the WPCs with theromoset resins and wood floor have also developed but 95% WPCs are manufactured as wood flour/PP (Marcovich et al. 2009; Du et al. 2013). WPCs' mechanical properties lie intermediate between both constituent, i.e., plastic and wood, and the WPCs have mainly emerged as an alternate material for wood. The inherent drawbacks of wood as an engineering material are a tendency to absorb moisture and vulnerability to fungus and insects. Due to moisture absorption, the wood swells and then contracts on drying this make it dimensionally unstable.

Moreover, due to the government policies for preservation of trees and forests the wood has become a costly and scarcely available material. In case of WPCs, the wood particles are coated by the polymer, which protects them from moisture and fungal attack. It also makes WPCs more stable and durable in comparison to wood. All these advantages promoted WPCs' use in various applications such as Decking and boardwalks, Flooring, Cladding (siding), Fencing and Railing, subcomponents of building, and furniture.

This chapter mainly focuses on the mechanical properties and various factors affecting the strength of WPCs. WPC is a category of NFCs, so the factors associated with the strength issue are the same. In which the compatibility between the materials of hydrophilic fibers and a hydrophobic matrix is the critical factor. This factor influences the interfacial strength between the fiber and matrix, as well as the ability to transfer load from matrix to fiber. Without a strong interfacial strength advantage of reinforcement cannot be taken. However, several research works have been published on various methods to improve interfacial strength. This chapter also includes the discussions over some of these methods.

2 Natural Fiber

The cellulosic natural fibers can be subdivided into annual plant fiber and wood fiber (Madsen and Kristofer 2013), and they can be classified as presented in Fig. 1. According to their utilization, the plants from which natural fibers are obtained can be categorized as primary and secondary. Plants which are grown particularly for fibers are called primary such as jute, cotton, sisal, hemp, kenaf, whereas plants in which fibers are obtained as by-product are called secondary plants such as pineapple, banana, flex, oil palm, sugar palm, coconut (Jawaid and Khalil 2011; Santulli 2007).

It has been estimated that at least 1000 types of plant produce usable fibers or fiber bundles. However, the numbers that are currently grown, specifically for fiber production, are relatively small. The domain for wood fiber composites (WPCs) is rather wider because it includes all the natural fibers (chopped), Agro-waste (Rice husk, wheat straw etc.), and wood flour (Saw dust) of all soft and hard wood timbers such as Teak, pine, white cedar, black spruce etc.

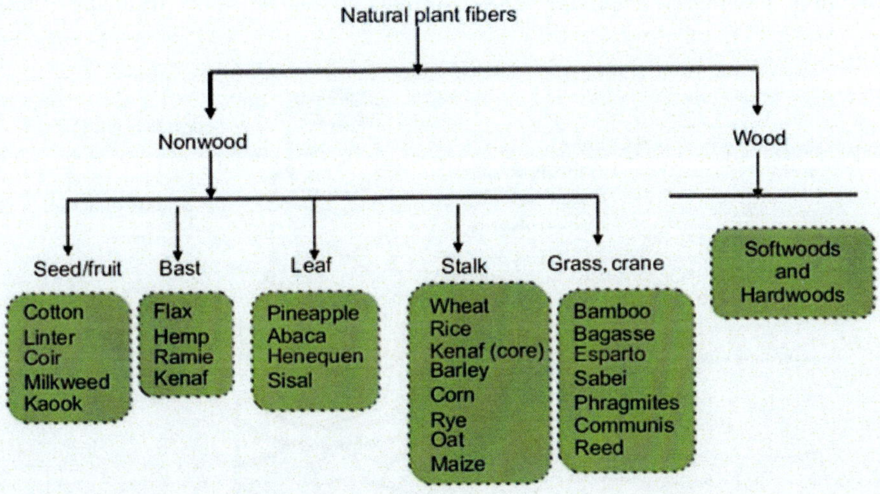

Fig. 1 Classification of Cellulosic Natural fibers

2.1 Structure of Wood or Natural Fiber

Natural fibers derived from plants are basically itself exists as fiber reinforced composite in which cellulose fibrils are reinforced in a hemicellulosic matrix. The structure of a biofiber is shown in Fig. 2. It can be observed that each fiber consists of a complex layered structure which includes primary cell wall and three secondary cell walls (S1, S2, and S3). The thick middle layer of secondary cell wall (S2) consists of a series of helically wound cellular micro-fibrils formed from long chain of crystalline cellulose molecules which is mainly responsible for mechanical strength of fiber. In a fiber, each cell wall is formed from three main components, which are cellulose, lignin, and hemicelluloses. In each cell wall, lignin and hemicelluloses act as a matrix in which cellulose fibrils are reinforced in helical form. Other components include pectin, oil, and waxes. Unlike synthetic fibers each natural fiber has a hollow structure, and that central hollow part is called the lumen. The structural representation of cellulose, hemicelluloses, pectin, and lignin, are shown in Fig. 3.

(D) lignin (Westman et al. 2010) (Open Access Content).

2.1.1 Cellulose

Cellulose is the fundamental structural element of all plant fibers. In 1999, it was reported that several plants' cell walls consist of a common substance, which was called cellulose (Bledzki and Gassan 1999). The cellulose exists as a straight buildup polymer comprising D-anhydroglucopyranose units (glucose units)

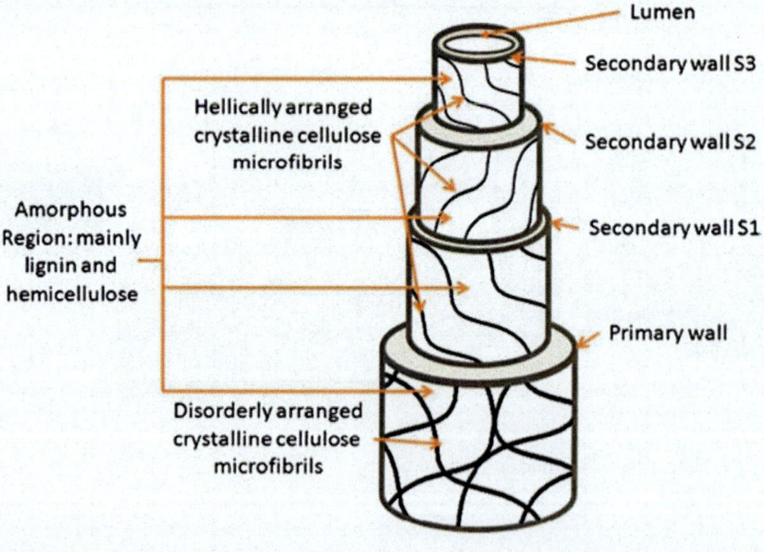

Fig. 2 Structure of a biofiber (Rong et al. 2001) [reused with permission from Elsevier, License number 4978080511062]

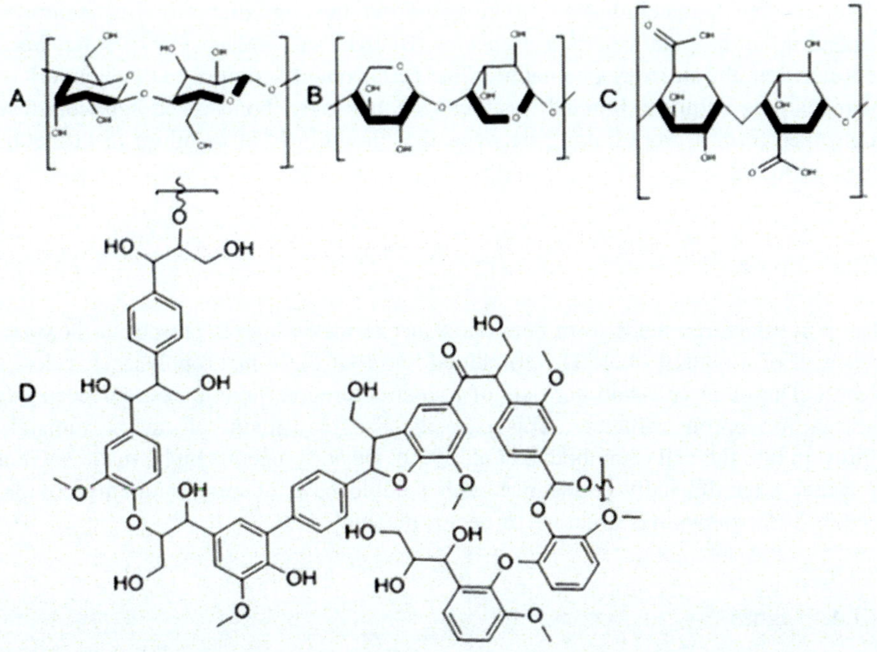

Fig. 3 Structural representations of (A) cellulose, (B) hemicellulose, (C) pectin and

consolidated by β-1, 4-glycosidic bonds. These units are reinforced together through 1 and 4 carbons (Fig. 3 A) to frame celloboise. One cellobiose unit's length is roughly 1 nm long, and the normal length of a cellulose chain is 5000 cellobiose units (5 μm). The crystalline region is formed by these tightly bonded flat chains (Moshibudi 2005).

The chemical and physical properties of the cellulose are attributed to its supra-molecular structure. The mechanical properties of a cellulosic fiber depend on the cell geometry its cellulose type (Bledzki and Gassan 1999).

2.1.2 Hemicellulose

Hemicellulose exists in the form of different types of sugar chain. After removing lignin from the cell wall of natural fiber, the hemicelluloses remain attached with cellulose. It exists as branched and short chains of polysaccharides bonded together (Fig. 3 B). There are three basic differences between cellulose and hemicellulose. The first is, cellulose contains only 1, 4-β-D-glucopyranose units, but hemicellulose contains several different sugar units. Secondly, cellulose is exactly a linear polymer, whereas hemicellulose shows a significant degree of chain branching. Thirdly, cellulose possesses a higher degree of polymerization (10–100 times) in comparison to hemicellulose. It has been proved that the cellulose also has more strongly packed structure than hemicelluloses. Unlike cellulose, the ingredients of hemicellulose depend upon the plant's parts and species (Bledzki and Gassan 1999). It has been reported that, the structure of a natural fiber cell is mainly comprised of a network of hemicellulose-cellulose bonded together with hydrogen bonds. The hemicellulose molecules form a cross-linking between the microfibrils of cellulose (Stamboulis et al. 2000).

2.1.3 Pectin

The polygalacturon acids form heteropolysaccharides which is present in the outer cell wall of a natural fiber. The combined name of these heteropolysaccharides is Pectin. The outer cell wall consists of non-structured carbohydrates and pectin as well as has porous structure. This porosity also play main role in the moisture diffusion into the cell (Stamboulis et al. 2000). Besides, pectin can be made solvent in water after the halfway balance with soluble base or ammonium hydroxide. Figure 3 (C) shows the structure of pectin molecule.

2.1.4 Lignin

In a natural fiber cell wall, both aromatic and aliphatic constituents form a complex polymer called lignin. The monomer units of lignin consist of complex structured various ring substituted phenyl-propane. Lignin is the main compound, which is responsible for the rigidity of natural fiber. The structure of lignin differs from plant to plant as well as with different parts of the plant.

Lignin acts as a natural polymer matrix which binds the cellulose fibers in plant fiber. The monomers of wood lignin are appeared in Fig. 3 (D). In the structure of lignin, the cross connecting happens through the C4 oxygen and at the empty aromatic ring carbons. The amount of potential cross-linking sites on lignin differentiates it from cellulose and hemicelluloses. It provides a non-crystalline (amorphous) structure. Lignin exists as an extract material in a plant fiber, which can be removed by the pulping process. The removed lignin can be utilized as a phenol substitute in lattice materials, even though lignin's mechanical properties are sub-par compared to the cellulose (Bledzki and Gassan 1999). Moreover, lignin is a thermoplastic material which gets soften at temperatures around 90 °C and start to flow at temperatures around 170 °C. Generally, lignin's low content is desirable for WPCs' production because its removal process requires polluting bleaching techniques (Stamboulis et al. 2000).

2.1.5 Extractives

Besides the constituents mentioned above the plant fibers also constitute waxes, resin, triglycerides, fatty acids, steryl esters and sterols (Bledzki and Gassan 1999).

These components collectively are called as extractives. The extractives act as a coating on the cell wall of the plant fibers and protect them from moisture (Bayer et al. 2017).

Table 1 shows chemical composition of various wood fibers, agro-waste fiber, and natural fibers reported in a journal. Also, in Table 2, the major constituents of wood fiber, which affects most to the mechanical properties, and their associated properties, have been presented as distinctively.

Table 1 Chemical composition of various cellulosic fibers

Fiber	Lignin%	Cellulose%	Extractive%	References
Wood fiber				
Hard wood	21.0–34.0	31.0–64.0	0.1–7.7	Bayer et al. 2017
Soft wood	21.0–37.0	30.0–60	0.2–8.5	
Chinese fir	37.1	45.1	4.0	Hung et al. 2017
Taiwan red pine	28.3	37.7	11.7	
Indian charcoal trema	25.4	42.6	4.4	
Makino Bamboo	20.9	43.7	4.0	
Aspen	21.9	50	5	Migneault et al. 2015
Birch	22.1	47.3	3.5	
Spruce	27.0	43.0	5	
Agro-waste				
Oil seed stalk	23.5	55.7	3.5	Nourbakhsh and Ashori 0.2010
Corn stalk	14.2	46.5	3.6	
Reed stalk	22.7	48.2	3.2	
Rice husk	20.0	35.45	14–17	Bayer et al. 2017
Rice straw	8–19	41–57	8–38	
Bagasse	25.3	55.2	–	
Long natural fibers				
Jute	11.8	64.4	0.7	Azwa and Yousif 2013
Flax	2.0	64.1	1.5–3.3	
Sisal	9.9	65.8	0.8–1.1	
Hemp	3.7	74.4	0.9–1.7	

Table 2 Chemical composition of wood fibers (Mohanty et al. 2005)

Constituents	Composition%	Nature	Degradation Temperature in °C	Affinity for moisture	Role
Lignin	20–30	Amorphous	110	Hydrophobic	Stiffness
Cellulose	44–50	Crystalline	200	Hydrophilic	Strength
Extract ive	0–10	Polymeric	100–200	Hydrophobic	Encrusting

3 Strength of Natural Fibers

Fibers are the main load bearing components in polymer matrix composites. Fibers carry a load along the direction of their length and provide strength and stiffness in that direction. Thus fibers can be oriented in a specific direction to provide desired properties in the direction of primary load. The properties and geometry of natural fibers are depending highly upon its natural habitat. Important factors effecting properties of natural fibers are as follows: (Dittenber and GangaRao 2012; John and Thomas 2008; Wong et al. 2010; Lefeuvre et al. 2015).

- Species
- Growing condition
- Whether condition
- Cambium age
- Chemical composition
- Internal fiber structure
- Microfibrillar angle
- Separating process
- Defects like pits and voids

The dependency of above mentioned factors on nature make difficult to analyze natural fibers and the mechanism of interface in natural fiber composites. The mechanical properties of natural fibers mainly depend upon microfibrillar angle as smaller angle provides higher strength and larger angle leads to higher ductility. The favorable factors for higher mechanical strength of a natural fiber are longer cell length, higher cellulose content and lower microfibrillar angle (Methacanon et al. 2010; Azwa and Yousif 2013). The fibril angle in a normal wood ranges 14–19°. Due to this low value of fibril angle in the thick S2 layer (Fig. 2), wood fiber does not swell or shrink in the longitudinal direction (Rowell 2005). The presence of voids in natural fibers leads to higher porosity in the natural fibers grown in the wet habitat. The higher porosity results into the higher moisture absorption which is undesirable for natural fiber composites (Wong et al. 2010; Yousif and El-Tayeb 2009; Yousif et al. 2012). Moreover, the highly oriented crystalline structure of cellulose improves the stiffness and tensile properties, but also increases the tendency to form kink bands under compressive loading. The formation of kink bands adversely affects the compressive as well as tensile strength (Al-Bahadly 2013). In a study on wood fibers, the presence of extractible (waxes) material interacts with the hydroxyl group of lignin and cellulose and makes the fiber protected from moisture absorption (Ashori and Nourbakhsh 2010).

Moreover, the mechanical properties of the wood fiber critically depend upon species, part of the tree and pulping process. Hard wood fibers are stiffer and have higher impact strength than soft wood fibers (Bledzki et al. 1998; Bledzki and Faruk 2003). Mechanical properties of WPC also depend upon l/d ratio of the fibers, due to this fact enhanced mechanical performance is achieved with wood fiber (pulp) in comparison to wood flour (Caulfield et al. 2010).

Table 3 Mechanical properties of Natural fibers (Yan et al. 2014; Wambua et al. 2003) and Wood fiber (Bledzki et al. 1998)

Fibers	Diameter (μm)	Relative density	Tensile strength (MPa)	Young's modulus (GPa)	Specific modulus (GPa x cm^3/g)	Elongation at failure (%)
E-glass	< 15	2.4–2.6	2100–3500	71–76	29	1.7–4.8
Soft wood	30	1.5	500–1500	10–50	40	3–4
Hard wood	30	1.5	1000–2100	10–80	45	1–2
Flax	12–600	1.4–1.5	343–2000	27.6–103	45	1.2–3.3
Jute	20–200	1.3–1.49	320–800	30	30	1–1.8
Hemp		1.4–1.5	270–900	23.5–90	40	1–3.5
Kenaf	-	1.4	223–930	14.5–53	24	1.5–2.7
Sisal	8–200	1.33–1.5	363–700	9.0–38	17	2.0–7.0

Mechanical properties of some commercial natural fibers and wood fibers are shown in Table 3 in comparison to the glass fiber which clearly shows that the strength of glass fiber has no match with strength of natural fibers but many natural fibers have better specific strength than that of glass fiber which is mainly due to lighter weight or low density.

4 Strength of Wood Plastic Composites

WPCs are the most appropriate composite materials for replacing wood as an engineering material. WPCs take advantage of both primary ingredients (wood fiber and polymer) and possess the mechanical properties such as stiffness and strength that lie in between them.Wood as engineering material has drawback of moisture absorption and vulnerability for the attack of fungus and insect. But, in a WPC the polymer forms a protective coating for the wood fibers which protect them from moisture and microbial attack. In return, wood fibers provide the advantages such as strength, lightness and biodegradability. Though both kinds of polymers have been used to fabricate WPCs but thermoplastics such as PP, PVC, HDPE and LDPE are the most common. In the production of WPCs the thermoplastic polymers are compounded with wood fibers. During compounding, the centre hollow portion of wood fiber cell, which is called Lumen, is filled with polymer. This filling of lumen is the main key of performance of a product of WPC. The presence of hydroxyl group in the ingredients of wood fibers (cellulose, lignin, hemicellulose) creates the problem of incompatibility between the fiber and matrix. This deteriorates the interfacial strength between fiber and matrix. Many researchers have reported several solutions to this problem. Also, the methods for testing various physical and mechanical properties of WPCs have been standardized.

4.1 Mechanical Testing of WPCs

The mechanical properties and their testing methods and test standards that are generally adopted for ensuring uniform product quality in the view of commercialization of a WPC are as follows.

4.1.1 Chemical Composition of Wood Fiber

The chemical composition of a wood fiber such asα-cellulose, holocellulose, Klason lignin and extractives are determined according to the ASTM D1106-96, ASTM D1104-56, ASTM D1103-60, and ASTM D1107-96 standard respectively (Hung et al. 2017).

4.1.2 Measurement of Physical Properties

For WPC analysis, physical properties such as moisture absorption, density, and thermal conductivity are measured. Density of WPC is measured according to the ASTM D792standard which is basically a water displacement method. The water soaking test measures moisture absorption according to the ASTM D1037 standard and thermal conductivity is evaluated in the Guarded-Hot-Plate Apparatus according to the ASTM C177-92 standard.

4.1.3 Tensile Test

Tensile test of WPC is conducted on computerized UTM in which stress strain curve is generated. The mechanical properties such as tensile strength, young modulus (stiffness), and elongation at break are evaluated from the stress–strain characteristic curve analysis. Dumbbell shaped specimens are fixed between the two jaws of the machine and pulled until failure takes place. The size of the specimen and testing parameters are adopted according to ASTM D-638or ENISO 527–2 standard. Figure 4 shows the tensile testing of a specimen on a UTM (Instron 3382).

4.1.4 Flexural Test

Generally, a three-point bending test is conducted to determine the flexural properties such as flexural strength and flexural modulus. A computerized UTM is used for the flexural test in which stress strain curve is generated. Study of a typical stress strain curve also reveals the properties like ductility and resilience. The specimen size and test parameters are decided according to ASTM D790 or EN ISO 178s-tandard. Figure 5 shows the setup for flexural test.

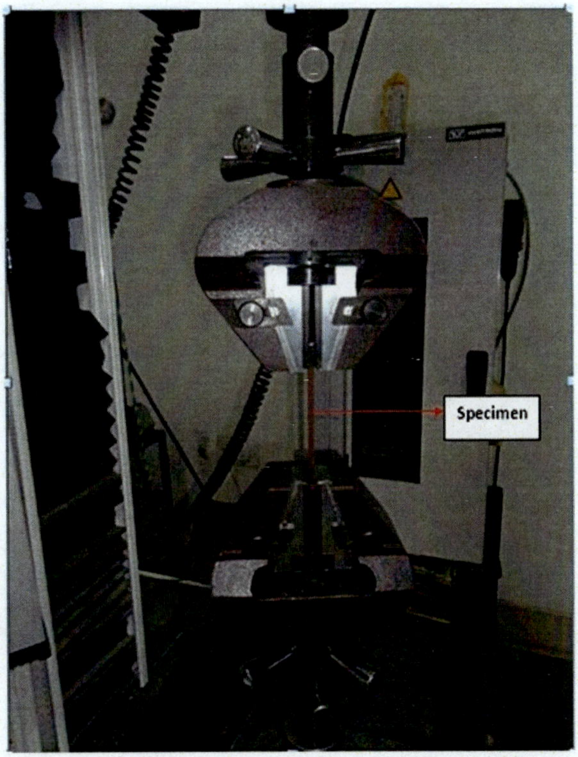

Fig. 4 Tensile test setup

4.1.5 Compression Test

The compression test was conducted on the specimens of the composites on the UTM machine according to ASTM D 695. Figure 6 shows set-up for the compression test. Generally the cubical shaped specimens of size 25 × 25 × 25 mm are used. The specimens are compressed between two cross heads of the UTM fitted with compression attachment.

4.1.6 Impact Test

Two types of impact tests are conducted which are called Charpy and Izod in which specimens are fixed in simply supported and cantilever position respectively. In this test a notched specimen of the WPC is broken in a single blow of hammer of impact tester. Figure 7 shows the setup for impact testing. Impact test determine the toughness (in KJ/m^2) and energy absorbed before fracture (in Joule). The procedure and parameters for the WPC impact test are adopted according to the ASTM D256 or EN ISO 179 standard.

Fig. 5 Setup for flexural test

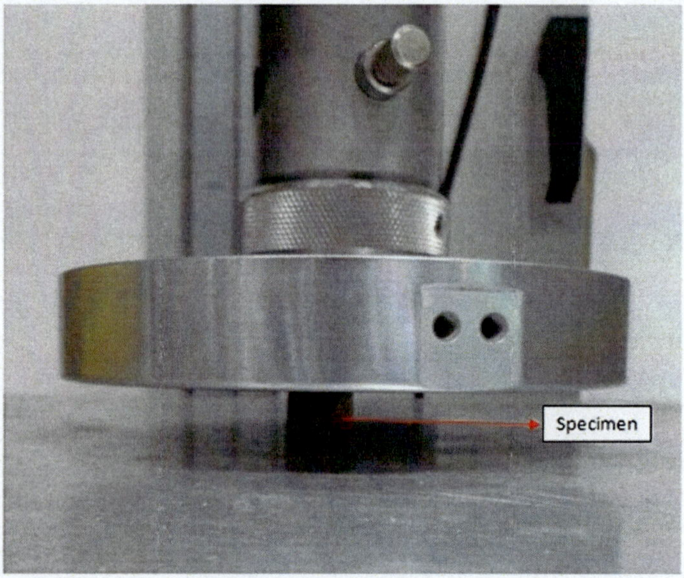

Fig. 6 Set-up for the compression test

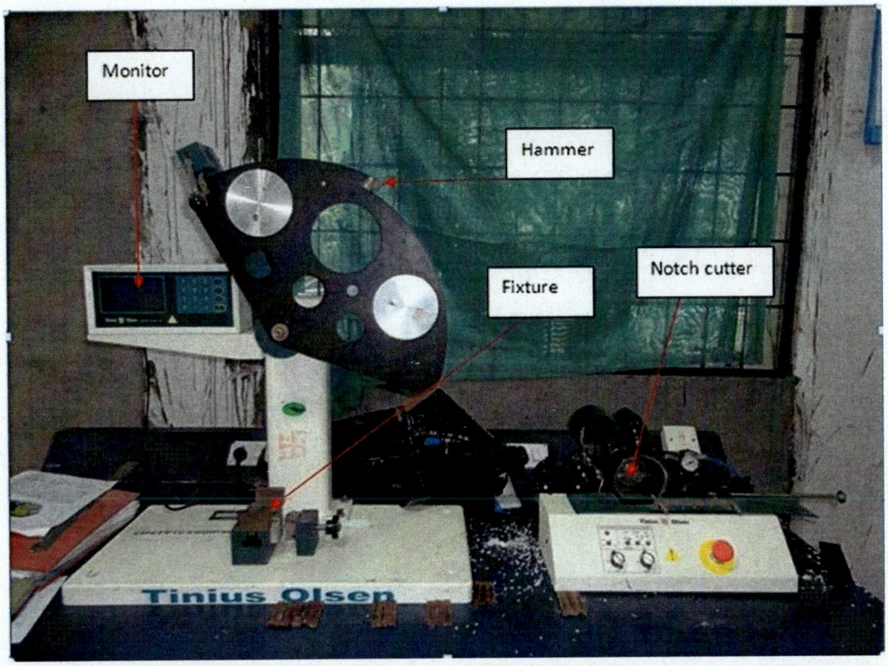

Fig. 7 Impact test setup

Mechanical tests described so far are primary tests for the mechanical characterization of WPC. However, some other tests such as hardness test and fastener holding strength test were also reported by some researchers. In which for hardness test ASTM D143 and ASTM E18 standards were followed whereas, for the fastener holding strength ASTM D1037 standard was adopted (Sain and Pervaiz 2008). Some standard have been established specifically for testing WPC products such as ASTM D7031-11 Standard Guide for Evaluating Mechanical and Physical Properties of WPC Products and ASTM D7032-15 Standard Specification for Establishing Performance Ratings for WPC Deck Boards and Guardrail Systems (Guards or Handrails) (Biron 2018).

Because of the WPCs' field of applications, the flexural strength is primarily investigated by the researchers. Table 5 shows WPCs' mechanical properties, such as Modulus of elasticity (MOE) and Modulus of rupture (MOR), based on different wood fiber species published elsewhere.

4.2 Mathematical models

Various mathematical models have been suggested to determine the natural fiber composites and WPCs' mechanical properties. Table 4 shows a chart of popular

Table 4 Mathematical models for fiber reinforced composites (Machado and Knapic 2017) [reused with permission from Elsevier, License number 4978090185883]

Models	
Parallel model (rule of mixture) $M_c = M_f V_f + M_m V_m;\ T_t = T_f V_f + T_m V_m$ Series model (inverse rule of mixture) $M_c = \frac{M_m M_f}{M_m V_f + M_f V_m};\ T_c = \frac{T_m T_f}{T_m V_f + T_f V_m}$ Hirsch's model $M_c = x(M_m V_m + M_f V_f) + 1(1-x)\frac{M_f M_m}{M_m V_f + M_f V_m}$ $T_c = x(T_m V_m + T_f V_f) + 1(1-x)\frac{T_f T_m}{T_m V_f + T_f V_m}$ Halpin – Tsai model $M_c = M_m\left(\frac{1+A\eta V_f}{1-\eta V_f}\right);\eta = \frac{M_f/M_m - 1}{M_f/M_m + A}$ $T_c = T_m\left(\frac{1+A\eta V_f}{1-\eta V_f}\right);\ \eta = \frac{T_f/T_m - 1}{T_f/T_m + A}$ Modified Halpin – Tsai model $M_c = M_m\left(\frac{1+A\eta V_f}{1-\eta\psi V_f}\right);$ $T_c = T_m\left(\frac{1+A\eta V_f}{1-\eta\psi V_f}\right)\psi = 1 + \left(\frac{1-\phi_{max}}{\phi_{max}}\right);$ $A = K - 1; K = 1 + 2L/d$ Cox model $M_c = M_f V_f\left(1 - \frac{tanh\beta 1/2}{\beta 1/2}\right) + M_m V_m$ $T_c = T_f V_f\left(1 - \frac{tanh\beta 1/2}{\beta 1/2}\right) + T_m V_m$ $\beta = \left[\frac{2\pi G_m}{M_f A_f \ln(R/r)}\right]^{1/2};R = r\left(\frac{\pi}{4V_f}\right)^{1/2}$; square packed fibres $R = \left(\frac{2\pi r^2}{3^{1/2} V_f}\right)^{1/2}$ hexagonal packed fibres	T – Tensile strength M – Young moduli c, f, m – composite, matrix, and fibre, respectively x – parameter, which determines the stress transfer between fibre and matrix A – fibre geometry, distribution and loading conditions ψ – parameter dependent upon particle packing fraction A – determined from the Einstein coefficient, K ϕ_{max} – maximum packing fraction L – average fibre length A_f – fibre cross-sectional area R – radius of the representative volume element R – fibre radius G_m – matrix shear modulus

models for calculating the tensile properties of a fiber reinforced composites. These models depend upon the ingredients' mechanical properties, their fraction of composition, and the fibers' orientation in WPCs. These models consider ideal bonding between fiber and matrix, whereas WPCs' actual strength mainly depends upon interfacial strength between the fiber and matrix, which is also a critical issue in all-natural fiber composites. However, the mechanical properties obtained from these models can be used for analysis purposes. Previous literature shows that the Halpin–Tsai model was the most preferred one for WPCs' analysis (Kalaprasad et al. 1997; Ku et al. 2011; Facca et al. 2006).

Table 5 Mechanical properties of different WPCs

Fiber species	Matrix	MOE (GPa)	MOR (MPa)	Reference
Wood flour based WPCs				
Chinese fir	HDPE	2.80	29.50	Hung et al. 2017
Taiwan red pine	HDPE	1.70	17.50	
Indian charcoal trema	HDPE	1.90	22.50	
Makino Bamboo	HDPE	2.50	21.30	
Rubberwood	HDPE	1.90	18.79	Thanate and Kampanart 2018
Rubberwood	PP	2.60	37.72	
Rubberwood	PVC	0.90	15.60	
Rubberwood	LDPE	1.00	6.39	
Rubberwood	PS	3.50	30.03	
Aspen	HDPE	2.55	52.90	Migneault et al. 2015
Birch	HDPE	2.31	49.20	
Spruce	HDPE	2.45	46.90	
Jelutong	MMA + HMDIC	8.10	56.50	Islam et al. 2012
Terbulan	MMA + HMDIC	9.43	68.10	
Batai	MMA + HMDIC	8.75	63.30	
Rubber	MMA + HMDIC	12.96	111.00	
Pulai	MMA + HMDIC	6.23	46.80	
Hard wood (Lignocell HBS 150–500)	PP	4.10	66.10	Bledzki and Faruk 2003
Soft wood (Lignocell BK 40–90)	PP	5.00	60.00	
White cedar	HDPE	1.71	32.5	Bouafif et al. 2009
Jack pine	HDPE	2.32	40.00	
Black spruce	HDPE	2.30	39.00	
Ponderosa pine wood	PP	3.15	42.60	Stark and Rowlands 2003
Poplar	HDPE	2.60	24.90	Fabiyiand McDonald 2010
Douglas-fir	HDPE	2.60	24.30	
Black locust	HDPE	1.80	22.70	
White oak	HDPE	2.00	21.10	
Ponderosa pine	HDPE	1.50	22.70	
Agro-waste based WPCs				
Oil seed stalk	PP	3.60	33.00	Nourbakhsh and Ashori 2010
Corn stalk	PP	3.30	30.00	
Reed stalk	PP	3.20	31.00	

5 Factors Effecting Strength of WPCs

The Strength of WPC depends upon many factors. Those factors mainly related to constituent materials. However, several investigations are being carried out to improve WPCs' strength and performance, which are mainly focused on improving the water or moisture resistance and interfacial bonding. Researchers have reported many effective additives and treatments to improvise the overall performance of the WPCs. Some critical factors affecting the performance of WPCs are presented in the following sections.

5.1 Effect of Matrix Material

In the manufacturing of WPCs the most common thermoplastics used are Polyolefins (HDPE and LDPE), PP and PVC. Among the polyolefins higher mechanical properties such as stiffness and strength are obtained with HDPE but advantage of ductility is with LDPE. PP is most widely used and most suitable polymeric matrix for the WPCs. The strength and durability of PP is better than polyolefin (Lukkassen and Meidell 2007). The crystallization temperature was increased due to PP's wood fiber content, whereas no significant effect was found for HDPE (Dikobe and Luyt 2017). PVC is scarcely used, due to the presence of the halogen group and generation of hydrogen chloride during the incineration. Moreover, for all the thermoplastic polymers, better mechanical properties were reported in virgin plastics than recycled polymers (Bledzki et al. 1998). Some researchers have investigated the Polymer blends for the matrix material of WPCs. The tensile strength and elongation at break reduces due to wood powder content in the polyblends of PP/HDPE/WP and MAPP/HDPE/WP. Though, elongation at break was higher in MAPP/HDPE/WP. Also, improved mechanical properties were obtained in case of MAPP/HDPE/WP (Dikobe and Luyt 2017; Clemmons 2010). The WPC of wood fibers of five kind of wood species with methyl methacrylate (MMA) and hexamethylene diisocynate (HMDIC) monomer mixture and chemical treatment with benzene diazonium salt provided enhanced mechanical properties as well as bio-degradation resistance (Islam et al. 2012). Five types of thermoplastic matrix (PP, HDPE, LDPE, PVC, and PS) were investigated with rubber wood flour as fiber for their effect on mechanical strength. The highest strength was found with PP matrix whereas lowest strength was found with LDPE and the strength of their WPCs in decreasing order was as PP, PS, HDPE, PVC and LDPE (Thanate and Kampanart 2018).

5.2 *Effect of Wood species*

Different wood species have different characteristics such as density, aspect ratio, mechanical properties and physical structure. In an investigation, the WPCs with two species of wood i.e. oak/PP and Pine/PP were synthesised. The better mechanical properties were found in Pine/PP composite which was mainly due to the higher aspect ratio of the pine wood fiber than oak wood fiber (Ashori and Nourbakhsh 2010). In another study on WPCs based on commercial hard wood fiber (Lignocel HBS 150–500)and soft wood fiber (Lignocel BK 40–90), the mechanical properties obtained was greater in case of hard wood fibers (Bledzki et al. 2002). The mechanical properties of PP based WPCs with wood flour of three different species of wood i.e. aspen, spruce, and birch were evaluated. The young modulus of birch wood flour composite was lowest (1.7 GPa) and aspen wood flour composite was highest (1.9 GPa) (Kokta et al. 1989). In another study, four different species of wood white cedar, jack pine, black spruce and jack pine bark was studied for the HDPE based WPC (Bouafif et al. 2009). It was found that the tensile and flexural properties were higher in case of jack pine and black spruce fibers whereas bark fiber based composite showed lowest strength. These results were correlated with atomic concentration ratio of oxygen to carbon (O/C) on the fiber surface. The presence of lignin indicates low O/C ratio and hinder the formation of bond between fiber and coupling agent (MAPE).

5.3 *Effect of Particle Size*

To investigate the effect of wood particle size, the different WPCs with soft wood fibers, hard wood fibers, wood chips and long wood fiber were fabricated with PP +MAPP. It was found that composites with wood chips showed better tensile and flexural performance whereas composite with hard wood fibers showed better impact resistance (Bledzki and Faruk 2003).Also, in an investigation the WPCs with fiber lengths ranges 0.5–0.9 mm was studied, in which the young modulus was found to be increased with increase in fiber length whereas no significant change in tensile strength was observed (Kokta et al. 1989). The fiber aspect ratio was found as a key parameter for a WPC as the mechanical properties were improved with increase in fiber aspect ratio (Stark and Rowlands 2007). However, the use of tiny particles (0.24–0.35 mm) in WPCs was suggested to provide higher specific surface area and uniform distribution of fibers, in comparison to long fibers. Consequently, interfacial strength between the fiber and matrix was improved, and swelling and tendency to break during processing were also decreased (Yam et al. 1990; Kokta et al. 1983). The mechanical properties were improved with increased particle size in HDPE+MAPE-based WPCs with fibers of four different wood species (Bouafif et al. 2009).

5.4 Effect of Fraction of Wood Fiber

It was observed that, the mechanical properties such as Tensile strength, Flexural strength and Impact strength, decreases with increase in wood fiber content (Kokta et al. 1989). Also, the variation of mechanical properties such as tensile and flexural strength, for both hard wood and soft wood based WPCs was observed to be decreased with increase in wood fiber content. But, when coupling agent was (MAPP) used in these WPCs then the reverse trend was observed. This change in trend was attributed to increase in interfacial strength between the fibers and matrix due to the coupling agent (Bledzki et al. 2002).However, the weight fraction as 50% was found optimum for the wood fiber content (Bledzki and Faruk 2003).

5.5 Effect of Additives

The additives are required with coupling agents as initiators for graft-copolymerization of coupling agents. The organic peroxides are widely used as additives which includes dicumyl peroxide (DCP), tertbutyl peroxy benzonate (TBPB), lauroyl peroxide (LPO), benzoyl peroxide (BPO) and tertbutylperoxy benzonate (TBPB). In which DCP is primarly used for graft copolymerization of MAPP/MAPE and PMPPIC. However, the excessive concentration of peroxide cause share-out of molecular chain between polymer and cellulose fiber and adversely affect the mechanical properties. Thus, the concentration of peroxide is kept limited to 0.5–1% of weight (Lu et al. 2000). Some additives are also used to improve the performance of WPCs such as for flame retardation, boric acid ($H3BO_3$), or borax ($Na_2B_4O_7$), magnesium oxide (MgO), alumina trihydrate ($Al(OH)_3$ are used and for preventing wood fiber from thermal degradation during processing, magnesiumoxide and boron compounds are used (Lu et al. 2000). The mechanical properties such as tensile strength, flexural strength and impact strength was improved by compounding micro crystalline cellulose (MCC)and polypropylene-graft-maleic anhydride (MAPP) with wood flour in a Wood flour/PP WPC (Ashori and Nourbakhsh 2010). Impregnation of wood fibers with Boric acid (BA) and borax (BX) mixture in a WPC fabricated with sapwood of Scots pine and venyl monomer was investigated. Impregnation of BA and BX improved the mechanical properties as well as properties such as moisture resistance, Specific gravity and decay resistance, were also improved (Baysal et al. 2007).

5.6 *Effect of Manufacturing Process*

The primary methods for manufacturing of the WPCs are extrusion and injection molding (IM).Very few researches have been carried out to investigate effect of manufacturing process on the performance of WPCs. In an investigation on HDPE/ wood flour WPC the mechanical properties for Extruded samples and IM samples were compared in which the higher modulus of rupture (MOR) for IM samples was observed whereas modulus of elasticity (MOE) was found similar. In the same research, the density of the IM samples was found higher which confirmed more intimate contact between the fiber and matrix, which also resulted in higher interfacial strength and mechanical properties (Stark et al. 2004). In another investigation, the IM sample's wood fibers were oriented in the flow direction in the outer layer and perpendicular to the flow path in WPCs samples' core layer. The Extruded samples had no defined orientation of the fibers (Michaeli and Menges 1989). The mechanical properties such as specific tensile strength, stiffness, and toughness were found higher in the IM sample than the extruded sample of PE-based WPCs in further research (Migneault et al. 2009; Bledzki and Faruk 2004). In a water absorption test the IM samples were found to be lesser swelled than extruded samples which also confirmed less water absorption by IM samples (Clemons and Ibach 2004). Thus most of the researches proved that the IM process is superior to Extrusion for the manufacturing of WPCs.

6 Improvement of Strength and Water Resistance of WPCs

There are only two critical factors that affect WPCs' performance. The first is the moisture absorption tendency of wood fibers; the second is compatibility between hydrophilic wood fibers and hydrophobic polymer matrix. The tendency of moisture absorption causes other defects such as swelling of fibers, dimensional instability, poor interfacial adhesion and attack of fungi and other microbes. But, in WPCs, if the wood fibers are perfectly coated with the polymer and the lumen of the fibers are completely filled with polymer than moisture cannot penetrate into the fibers. Only the exposed portion of wood is vulnerable for moisture and presence of extractives in the cell wall of wood fibers also protects them from moisture. However, gaps and flaws in the core layer are another cause of water absorption (Hung et al. 2017).

The inherent incompatibility between fiber and matrix in WPCs cause weak interfacial strength, due to which load transfer mechanism from matrix to fiber become partial. In other words, the advantage of reinforcement cannot be taken fully without perfect bonding between matrix and the fibers. The main reason for nature of incompatibility as well as moisture affinity is presence of Hydroxyl (–OH) group on the surface of wood fiber due to presence of Hemicellulose and cellulose.

This makes wood fiber's surface polar in nature whereas almost all thermoplastic are non polar in nature. To increase the compatibility between cellulosic fibers and thermoplastic polymers chemical coupling agents are introduced at the interface. Coupling agents are substances used in small quantities to treat a surface to bond with another surface. Generally, coupling agents may be classified according to their chemical composition as organic, inorganic and organic–inorganic. The coupling agents can further be classified as bonding agents, compatibilizers and dispersing agents. A bonding agent's mechanism in a WPC is to establish a bridge between wood fibers and polymer by forming a covalent bond or secondary interaction (like Hydrogen bond) or polymer chain entanglement. A compatibilizer reduces the interfacial tension and makes two combining surface compatible to each other. In WPCs the compatibilizer reduces the surface energy of fibers and makes it non-polar. A dispersing agent reduces the interfacial energy and facilitates uniform distribution of wood fibers in a polymer matrix such as stearic acid and its metallic salts are used to disperse wood fibers in polymer matrix.

Some compatibilizers which act as bonding agents are maleated polypropelene (MAPP), maleated polyethylene (MAPE) and styrene-maleic anhydride(SMA). MAPP and MAPE are most commonly used compatibilizer in PP and polyolefin based WPCs. The MAPP or MAPE are prepared by modifying PP or PE matrix by maleic anhydride by graft-copolymerization and the bonding mechanism of these copolymers with cellulosic fibers is shown in Fig. 8. Several researcheshave been carried out to investigate the effect of these copolymers on the mechanical

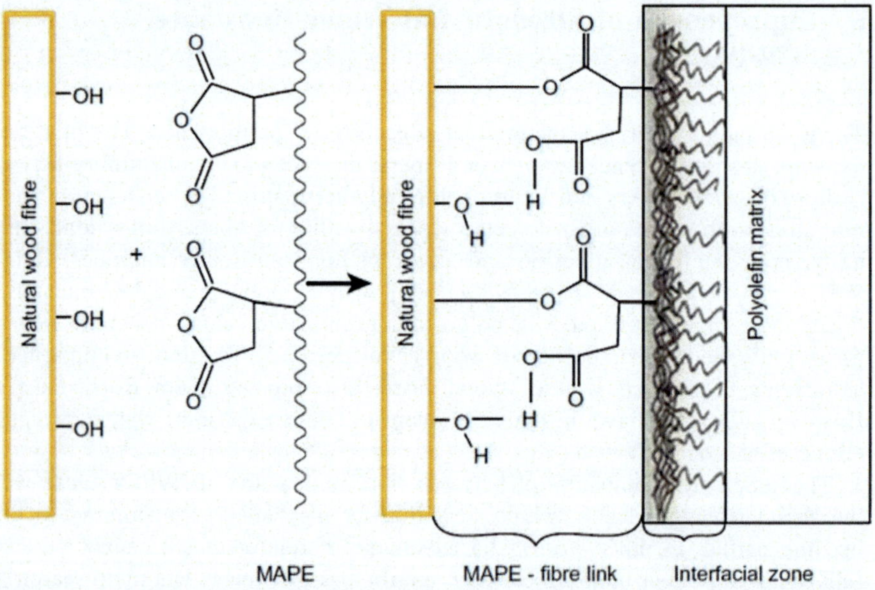

Fig. 8 Bonding mechanism of MAPP in WPC (Bayer et al. 2017) [reused with permission from Elsevier, License number 4977391366524]

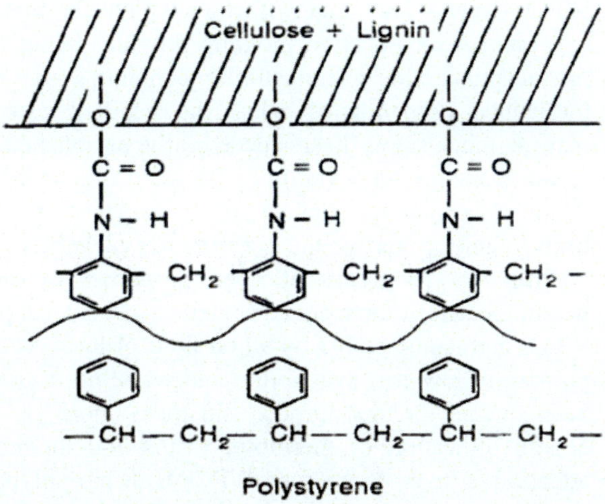

Fig. 9 Bonding of PMPPIC with cellulose and PS (Bledzki and Gassan 1999) [reused with permission from Elsevier, License number 4977391064606]

properties of WPCs, and in all previous researches improvement in mechanical properties was reported (Bledzki et al. 2002; Ashori and Nourbakhsh, 2010; Stamboulis et al. 2000; Beg and Pickering 2008). The optimum amount of coupling agent (MAPP) mainly depends upon fiber content as it increases with increase in fiber content, which ranges 4–8% w/w of fiber However, excessive coupling agents cause deterioration of interfacial properties attributed to the self entanglement of MAPP molecules (Bayer et al. 2017). The coupling agents those are used in polystyrene (PS) and PVC based WPCs are isocyanates.

Polymethylene–polyphenyl–isocyanate (PMPPIC) was found as most effective coupling agent in comparison to other isocyanates such as Toluene 2,4-diisocyanate (TDJC), Hexamethylene diisocyanate (HMDIC), and Ethyl isocyanate (EIC), in PS and PVC based WPCs The reason may be assigned to the PMPPIC and PS/PVC can adhere to each other through strong interaction between the benzene rings' delocalized π electron. Also, PMPPIC makes strong covalent bond with cellulose cell wall. Figure 9 shows the mechanism of bonding of PMPPIC with cellulose and PS matrix (Bledzki and Gassan 1999). Different researchers have investigated many other coupling agents, and a state of the art review of about 40 coupling agents and additives has been reported (Lu et al. 2000).

7 Conclusions and Future Perspectives

From the above studies, it can be concluded that the WPCs are the noble material for the replacement of wood. The wood flour of many plant species has been investigated to manufacture WPCs. However, the mechanical properties of WPCs are inferior to other long natural fiber composites but it can be compromised with the advantages of low cost production at high rate. The mechanical properties of

WPCs depend upon many factors which are mainly related to their ingredient materials. Also, injection molding provides better mechanical properties to WPC products than other manufacturing techniques such as extrusion. For evaluating the mechanical properties of WPCs the standard tests procedures as well as mathematical modelshave been successfully established. The moisture absorption by wood fibers and incompatibility between fiber and matrix material are the primary factors affects the mechanical properties. However, many chemical treatments have been found to improve the performance of WPCs.

Primarily, properties of fiber, properties of matrix material, and interfacial bonding between them are attributed to mechanical properties of a WPC. Generally, WPCs are manufactured by wood flour of local wood species. This also serves the purpose of low cost availability and waste utilization. Many species of wood have been studied for this purpose and many more yet to be explored. After the successful utilization of thermoset resins and thermoplastics as matrix the recent attention is given to produce WPCs with bio-polymers and elastomers. Moreover, to achieve an ideal bonding between fiber and matrix is still waiting for a solution.

References

Al-Bahadly EAO (2013) The mechanical properties of natural fiber composites. PhD thesis, Swinburne University of Technology, Australia

Ashori A, Nourbakhsh A (2010) Reinforced polypropylene composites: effects of chemical compositions and particle size. Biores Technol 101:2515–2519

Ashori A, Nourbakhsh A (2010) Performance properties of microcrystalline cellulose as a reinforcing agent in wood plastic composites. Compos B 41:578–581

Azwa ZN, Yousif BF (2013) Characteristics of kenaf fibre/epoxy composites subjected to thermal degradation. Polym Degrad Stab 98:2752–2759

Bayer J, Granda LA, Méndez JA, Pèlach MA, Vilaseca F, Mutjé P (2017) Cellulose polymer composites (WPC). A book chapter in Advanced High Strength Natural Fibre Composites in Construction. Woodhead Publishing, pp 115–139

Baysal E, Yalinkilic MK, Altinok M, Sonmez A, Peker H, Colak M (2007) Some physical, biological, mechanical and fire properties of wood polymer composite (WPC) pretreated with boric acid and borax mixture. Constr Build Mater 21:1879–1885

Beg MDH, Pickering KL (2008) Mechanical performance of kraft fibre reinforced polypropylene composites: influence of fibre length, fibre beating and hygrothermal ageing. Compos Part A Appl Sci Manuf 39:1748–1755

Biron M (2018) Thermoplastic composites in thermoplastics and thermoplastic composites, 3rd edn. William Andrew Applied Science Publishers, Oxford. ISBN 978-0-08-102501-7

Bismarck A, Mishra S, Lampke T (2005) Plant fibers as reinforcement for greencomposites. Chapter 2. In: Mohanty AK, Misra M, Drzal T (eds.), Natural fibers, biopolymers, biocomposites. CRC Press, New York, USA

Bledzki AK, Gassan J, Theis S (1998) Wood-filled thermoplastic composites. Mech Compos Mater 34:563–568

Bledzki AK, Reihmane S, Gassan J (1998) Thermoplastics reinforced with wood fillers: a literature review. Polym Plast Technol Eng 37(4):451–468

Bledzki AK, Faruk O (2003) Wood fibre reinforced polypropylene composites: effect of fibre geometry and coupling agent on physico-mechanical properties. Appl Compos Mater 10:365–379

Bledzki AK, Faruk O (2004) Extrusion and injection moulded microcellular wood fibre reinforced polypropylene composites. Cell Polym 23(4):211–227

Bledzki AK, Gassan J (1999) Composites reinforced with cellulose based fibers. Progress Pol Sci 24:221–274

Bledzki AK, Omar F, Monimul H (2002) Physico-mechanical studies of wood fiber reinforced composites. Polym-Plast Technol Eng 41(3):435–451

Bouafif H, Koubaa A, Perré P, Cloutier A (2009) Effects of fiber characteristics on the physical and mechanical propertiesof wood plastic composites. Compos Part A 40:1975–1981

Caulfield D, Clemons C, Rowell RM (2010) Wood thermoplastic composites. CRC Press, Boca Raton

Clemmons C (2010) Elastomer modified polypropylene–polyethylene blends as matrices for woodflour–plastic composites. Compos Appl Sci Manuf 41:1559–1569

Clemons CM, Ibach RE (2004) Effects of processing method and moisture history on laboratory fungal resistance of wood-HDPE composites. Forest Prod J 54(4):50–57

Dikobe DG, Luyt AS (2017) Thermal and mechanical properties of PP/HDPE/wood powder and MAPP/HDPE/wood powder polymer blend composites. Thermochimica Acta. https://doi.org/10.1016/j.tca.2017.05.002

Dittenber DB, GangaRao HVS (2012) Critical review of recent publications on use of natural composites in infrastructure.Composites A 43:1419–1429

Du Y, Wu T, Yan N, Kortschot MT, Farnood R (2013) Pulp fiber-reinforced thermoset polymer composites: effects of the pulp fibers and polymer. Compos Part B. 48:10–17

Facca AG, Kortschot MT, Yan N (2006) Predicting the elastic modulus of natural fibrereinforced thermoplastics. Compos Part A. 37:1660–1671

Hung K-C, Yeh H, Yang T-C, Wu T-L, Xu J-W, Wu J-H (2017) Characterization of wood-plastic composites made with different lignocellulosic materials that vary in their morphology. Chem Compos Thermal Stabili Polym 9:726–736

Saiful IM, Hamdan S, Hasan M, Saleh AA, Rezaur RM (2012) Effect of coupling reactions on the mechanical and biological properties of tropicalwood polymer composites (WPC). Int Biodeterior Biodegradation 72:108–113

Jawaid M, Khalil HA (2011) Cellulosic/synthetic fibre reinforced polymer hybrid composites: a review. Carbohyd Polym 86(1):1–18

John M, Thomas S (2008) Biofibres and biocomposites. Carbohyd Polym 71:343–364

Kalaprasad G, Joseph K, Thomas S, Pavithran C (1997) Theoretical modeling of tensileproperties of short sisal fibre-reinforced low-density polyethylene composites. J Mater Sci 32:4261–4267

Kokta BV, Raj RG, Danesult C (1989) Use of wood flour as filler in polypropylene: studies on mechanical properties. Polym-Plast Technol Eng 28(3):247–259

Kokta BV, Chen R, Daneault C, Valade JL (1983) Use of wood fibers in thermoplastic composites. Polym Compos 4(4):229–232

Ku H, Wang H, Pattarachaiyakoop N, Trada M (2011) A review on the tensile properties ofnatural fiber reinforced polymer composites. Compos B. 42:856–873

Anaële L, Alain B, Christophe B (2015) Optimization of the mechanical performance of UD flax/epoxy composites by selection of fibres along the stem. Compos A 77:204–208

Yan L, Chouw N, Jayaraman K (2014) Flax fibre and its composites – a review. Compos B 56:296–317

Lu JZ, Wu Q-L, McNabb HS, Wu QL (2000) Chemical coupling in wood fiber and polymer composites: a review of coupling agents and treatments. Wood Fiber Sci 32(1):88–104

Lukkassen D, Meidell A (2007) Advanced Materials and Structures and Their Fabrication Processes. A book manuscript, Narvik University College

Machado JS, Knapic S (2017) Short term and long-term properties of natural fibre composites A book chapter in Advanced High Strength Natural Fibre Composites in Construction, Woodhead publishing, pp 447–458

Madsen, B, Kristofer, GE (2013) Wood versus plant fibers: similarities and differences in composite applications. Adv Mater Sci Eng Volume 2013, Article ID 564346

Marcovich NE, Ostrovsky AN, Aranguren MI, Reboredo MM (2009) Resin-sisal and wood flour composites made from unsaturated polyester thermosets. Compose Interfaces 16(7–9):639–657

Methacanon P, Weerawatsophon U, Sumransin N, Prahsarn C, Bergado DT (2010) Properties and potential application of the selected natural fibers as limited life geotextiles. Carbohydrate Poly 82:1090–1096

Michaeli W, Menges G (1989) Prediction of product properties in extrusion and injection molding. Adv Polym Technol 9(1):69–85

Migneault S, Koubaa A, Perre P, Riedl B (2015) Effects of wood fiber surface chemistry on strength of wood-plastic composites. Appl Surface Sci 343:11–18

Migneault S, Koubaa A, Erchiqui F, Chaala A, Englund K, Wolcott MP (2009) Effects of processing method and fiber size on the structure and properties of wood–plastic composites. Compos A. 40:80–85

Mohanty AK, Misra M, Drzal LT (2005) Thermoplastic wood fiber composites. A book chapter in Natural Fibers, Biopolymers, and Biocomposites. CRC Press

Moshibudi CK (2005) The processing properties of natural fiber higher α-olefin based thermoplastics. Master of Technology Thesis. Tshwane University of Technology, pp 1–85

Nourbakhsh A, Ashori A (2010) Wood plastic composites from agro-waste materials: analysis of mechanical properties. Bioresour Technol 101:2525–2528

Rong MZ, Zhang MQ, Liu Y, Yang GC, Zeng HM (2001) The effect of fiber treatment on the mechanical properties of unidirectional sisal-reinforced epoxy composites. Compos Sci Technol 61:1437–1447

Rowell RM (2005) Handbook of chemistry and wood composites. CRC Press, Boca Raton

Sain M, Pervaiz M (2008) Mechanical properties of wood-polymer composites. A book chapter in Wood–Polymer Composites. Wood head Publishing Series in Composites Science and Engineering, pp 101–117

Carlo S (2007) Impact properties of glass/plant fiber hybrid laminates. J Mater Sci 42:3699–3707

Stamboulis A, Baillie CA, Garkhail SK, Van Melick HGH, Peijs T (2000) Environmental durability of flax fibers and their composites based on polypropylene matrix. Appl Compos Mater 7:273–294

Stark NM, Matuana LM, Clemons CM (2004) Effect of processing method on surface and weathering characteristics of wood-flour/HDPE composites. J Appl Polym Sci 93:1021–1030

Thanate R, Kampanart T (2018) Alternative polymeric matrices for wood-plastic composites: effects on mechanical properties and resistance to natural weathering. Constr Build Mater 172:349–357

Paul W, Jan I, Ignaas V (2003) Natural fibers: can they replace glass in fiber reinforced plastics. Compos Sci Technol 63:1259–1264

Westman MP, Laddha SG, Fifield LS, Kafentzis TA, Simmons KL (2010) Natural fiber composites: a review, Report under National Technical Information Service, U.S. Department of Commerce, Springfield, VA 22161

Wong KJ, Yousif BF, Low KO (2010) Effects of alkali treatment on the interfacial adhesion of bamboo fibres. J Mater Des Appl 224:139–148

Yam KL, Gogoi BK, Lai CC, Selke SE (1990) Composites from compounding wood fibers with recycled high density polyethylene. J Poly Eng Sci 30(11):693–699

Yousif BF, El-Tayeb NSM (2009) Mechanical and wear properties of oil palm and glass fibres reinforced polyester composites. Int J Precis Technol 1:213–222

Yousif BF, Shalwan A, Chin CW, Ming KC (2012) Flexural properties of treated and untreated kenaf/epoxy composites. Mater Des. 40:378–385

Thermal Properties and Flammability of Wood Plastic Composites

Elammaran Jayamani and Vannethasrriy Balakrishnan

Abstract As the boundaries of engineering materials are pushed there is a wide increase in the demand for composite materials. In the past decade composite materials have been at the forefront of many engineering advancements. One of the category of composites which is wood plastic composites have emerged as a material with many uses especially in the construction industry. The combination of wood and thermoplastics form wood plastic composites which have unique mechanical properties which have been researched extensively. The usage of wood plastic composites as a building material has raised the concern on the thermal properties and flammability of wood plastic composite which this study is focusing on. The main reason behind this concern is due to the flammability of each separate material in the composite and if the combined composite is as flammable it could be catastrophic when used in the construction industry. The content of this study focuses on gathering previous research on the governing topics and forming better understanding on the thermal properties and flammability of wood plastic composites.

Keywords Wood plastic composites · Flammability · Thermal properties · Thermoplastics · Fibres

1 Composites

Composites have become a primary choice of material in various humankind actions and this is due to the improved nature of composites compared to its natural counterparts. Composites are a combination of components with different chemical and physical properties. Composites can be made by combining two or more

E. Jayamani (✉) · V. Balakrishnan
Faculty of Engineering, Computing and Science, Swinburne University of Technology, Sarawak Campus, Kuching, Sarawak, Malaysia
e-mail: ejayamani@swinburne.edu.my

V. Balakrishnan
e-mail: 100084512@students.swinburne.edu.my

S. Mavinkere Rangappa et al. (eds.), *Wood Polymer Composites*,
Composites Science and Technology, https://doi.org/10.1007/978-981-16-1606-8_8

artificial or natural elements which are stronger together compared to when they are individually separate (Anon 2020a, b, c, d). The individual materials combine and contribute their useful traits which produces an improved final product without the two materials completely blending and neither one of the components completely lose their individual characteristics. Traditional materials are replaced with composites as they have improved properties of their base materials and can easily be used in many situations.

There are many advantages to composites and the main one is that composites cost less compared to any other traditional materials with the exact same or similar properties. Furthermore, they are low weight and high strength, very durable, good electric insulators and can have high impact strength. Composites also provide design flexibility and have resistance to a wide range of chemical agents. Composites can have multiple properties that usually can't be found in a single material.

2 Wood Plastic Composites

A relatively recently developed composite is the wood plastic composite (WPCs). Wood plastic composite is a high-value product that comprises of thermoplastic polymers as matrices and wood flours as fillers and reinforcements (Kim et al. 2009). The driving force to produce this new composite was provided by the need to reduce the waste compiling in the landfill while exploring recycling technologies and the aim of exchanging petrochemical materials for increased bioderived sustainable materials, post-consumer plastics and wastepaper was seen as products that can be recycled and used for wood plastic composites (Oksman et al. 2008). Polymers such as polypropylene or polyethylene are blended with woody biomass from agriculture residue or fine powdered wood within an extruder which then is pelletised for further processing in the future or formed into sections such as planks profiles according to specific applications. The wood component of the composite typically makes up about 40 to 65 wt% (Kim et al. 2009).

Wood fibre, life fibre of greater aspect ratio and chopped bast are some of the wide range of variations and techniques used to enhance the properties of the composite by reinforced material rather than filled material (Oksman et al. 2008). Many researches were done to study the compatibility between wood fibre or flour and commonly used plastics to increase performance of the wood plastic composite (Anon 2020a, b, c, d). This composite is made to have better mechanical and thermal properties compared to the individual materials. wood plastic composites have advantages as listed below:

- Material is easy to maintain
- Has longer life span
- Water resistant
- Slip resistant
- Weather stability

3 Thermal Properties of Wood Plastic Composites

There are several factors affecting the service life of a multicomponent composite like the wood plastic composite. The thermal properties are one of the most important factors during the processing and usage of these composites. Many research have been done for the thermal properties of wood plastic composites and it primarily includes determination of crystallinity, heat deflection temperature, melting behaviour, thermal degradation, glass transition temperature, thermal expansion properties and flammability performance. There are several factors that affect the thermal properties of wood plastic composites such as the different types of wood species used, types of thermoplastics used and composition of the different materials in the composites.

3.1 Thermal Properties of Thermoplastics in Wood Plastic Composites

3.1.1 Polypropylene

Polypropylene is a thermoplastic which is the category of polymers that are mouldable at higher temperatures and solidify in cooler temperatures. Polypropylene is one of the most commonly used polymers in the world (Staff 2020). The structure of Polypropylene is a crystalline that is made up of propene monomers. One of the main reasons this polymer is widely used in various applications is due to the low manufacturing cost of this polymer. Besides the structure of the polymer allows for easy copolymerisation and this is beneficial in changing the mechanical properties of the polymer to be used in various engineering applications. When heat is applied to polypropylene, initially degradation is introduced to the polymers such as the formation of low molecular mass products and the break-up of macro-molecular chains (Fuad et al. 1995). Various inorganic compounds such as aluminosilicates, metallic oxides and zeolites are the compounds that activates the degradation of the polypropylene.

3.1.2 Polyethylene

Polyethylene are a variation of thermoplastic material which has gained traction as one of the most commonly used thermoplastics in multiple industries due to its unique crystalline structure and properties. The durability combined with its light weight properties give this material a unique property that is desired in various applications including as addition for wood plastic composites. Polyethylene can further be broken down into two main sub categories which are High-density

Polyethylene and Low-density polyethylene which both have similar mechanical properties such as high melting points and a great insulator which contributes to the high demand of this material.

3.2 Methods to Analyse Thermal Properties of Wood Plastic Composites

There have been many researches done wood plastic composites in the past few years and thermal properties of these wood plastic composites have been one of the factors that has been prioritised in the research. There are many types of thermal analysis methods that can be used to study and further understand the thermal properties of these composites, but the two main ones are the thermogravimetric analysis and the different scanning calorimetry analysis. These two methods have been widely used in many studies and researches that had been done in the past decade.

Thermogravimetric analysis measures the mass of a polymer as a function of time or temperature while the sample is subjected to a controlled temperature program in a controlled atmosphere (Earnest 1983). This analysis does not indicate the nature of the material lost but only provide a quantitative measurement of the change in mass of the sample. As the thermogravimetric analysis is mainly for polymer application, the temperature range for the analysis is typically ambient to more than 1000 °C. A purge gas flowing through the balance creates an atmosphere. The atmosphere can be oxidising, such as oxygen or air; inert, such as helium, nitrogen or argon; or reducing, such as foaming gas. In an oxidising atmosphere with slow heating rates, mas gain may be observed prior to degradation in polymers but generally they exhibit mass loss. Mass loss can be categorised as reaction products, volatile components or volatile degradation products. Information such as thermal stability and extent of cure can be yield from the characterisation of the mass loss processes by the thermogravimetric analysis (Prime et al. 2009).

The differential scanning calorimetry as stated previously is another method which is widely used in studies to investigate the thermal properties of a composite. It is a powerful method that is conventionally used to characterise the stability of a compound. The differential scanning calorimetry is a technique in which the heat flow that flows out or into a sample is measured as a function of time or temperature. This measurement is taken as the sample is exposed to a controlled temperature program. This thermal analysis method is used to evaluate many thermal properties of a composite such as, specific heat capacity, melting, transition temperature, crystallisation temperature and thermal stability. Data for a wide range of materials can be provided by the differential scanning calorimetry analysis which is perfect when using it for wood plastic composites as these composites are made of many different types of polymers and wood materials such as wood flour, sawdust and rice hulls with different composition.

3.3 Composition of Wood Plastic Composites

Wood and other natural fibres encounter three core obstacles which are hydrophilic characteristic, inherent incompatibility with non-polar thermoplastics and low thermal stability. Thermal stability of natural fibres and wood is a critical factor and low thermal stability can be a problematic issue. The thermoplastics such as high-density polyethylene (HDPE) and polypropylene (PP) in wood plastic composites have a much higher thermal stability compared to wood. For HDPE and PP, the mass loss is around 472 °C and 517 °C (Kim et al. 2004).

Previous studies done to investigate thermal stability of composites of thermoplastics and natural fibres have concluded that incorporating wood into thermoplastic composites results in reduces the thermal stability of the composites (Yang et al. 2005).

Alireza Kaboorani conducted a study on high density polyethylene composites with different formulation designs of wood, particle size and coupling agent contents to investigate the changes in thermal properties. Different scanning calorimetry (DSC) and thermogravimetric analysis (TGA) were used. Twenty-seven different formulations were used and it was concluded that the most influential factor for thermal stability of wood plastic composites is the wood content. Thermal stability of wood must be improved in order to have a better thermally stable composite. Particle size and coupling agent has a lesser impact on the composite's thermal stability when compared to wood. Coupling agent increased thermal stability by enhancement of interfacial adhesion between the polymer and wood. Smaller particles had higher chance to be embedded by matric polymer and this allowed the wood plastic composite to have a higher thermal stability regardless of the coupling agent content and wood content (Kaboorani 2010).

Another study was done to investigate the thermal performance of wood plastic composites with different compositions of fibre reinforcement which were wood fibre, floor sanding powder and rice husk, and for the matrix recycled high-density polyethylene. The results showed that the crystallinity and melting temperature of the wood plastic composite decreased as the wood fibre content increased. At a wood fibre content of 60%, it was noted that the dimensional stability of the wood plastic composites were constant and a wood fibre content which was any higher was not conducive for processing of the wood plastic composite (Guo et al. 2019).

3.4 Wood Species Variations in Wood Plastic Composites

The properties of wood plastic composite depend on various factors including the essential properties of the constituent materials. Wood plays a major role in wood plastic composite and there has been evidence that the species of wood can affect the thermal properties of wood plastic composites (Rogers and Simonsen 2005).

Jae-woo conducted a study on the effect of different wood species on the thermal and mechanical properties of wood plastic composite. In this study, various wood species such as pine, maple, red oak, Osage orange, walnut, eastern red cedar, hickory, yellow poplar, sweet gum and cherry were compounded with virgin isotactic polypropylene. The wood and thermoplastic had a 50 50 weight ratio. Wood plastic composite were made from all the different wood species and their thermal properties were evaluated. From the results of this study, the peak values from the dynamic mechanical analysis and the peak temperature indicated that the hickory and pine wood plastic composites had higher amorphous or void contents compared to the cherry and walnut wood plastic composites (Kim et al. 2009). There was a statistically significant difference in the mechanical properties compared to the thermal properties between the different species of woods.

In 2010, James S.Fabiyi conducted a research on thermal behaviours of high-density polyethylene based wood plastic composites with different wood species. The study was done to further understand the effect of hard and soft wood on the thermal and other properties of the wood plastic composite. There were five different wood species that were selected for this study which were ponderosa pine, Douglas-fir, hybrid poplar, black locust and white oak. Wood flour of 60% and high-density polyethylene of 40/5 was used to produce all five different types of wood plastic composites. Thermogravimetric analysis showed that the thermal stability of the wood plastic composite is species dependent. The final decomposition hemicelluloses in the softwoods occurred at a lower temperature compared to that of softwoods (Fabiyi and McDonald 2010).

Another study was done by Md Saiful on the thermal properties of wood plastic composites with various tropical woods. There were five different types of tropical light woods used in this study which are rubberwood, pulai, jelutong, batai and terbulan. The wood polymers were prepared by impregnating the different tropical woods with ethyl methacrylate that was combined with a cross-linker, hexamethylene diisocyanate. Differential scanning calorimetry and thermogravimetric analysis was used to evaluate the thermal properties of the wood plastic composites. The result showed that there were no significant difference in the thermal properties between the different wood plastic composite containing different tropical woods (Islam et al. 2012).

A.N Shebani conducted a study the effects of four different wood species which were eucalyptus, oak, acacia and pine on the thermal properties of wood plastic composites. The chemical composition and the morphological properties the distribution of the wood fibre length of each type of wood were carefully studied to further understand the effect of this investigation. Each wood species were prepared using polyvinyl alcohol co-ethylene as compatibilizer and made into wood plastic composites with liner low density polyethylene. The change in weight of each composite as a function of temperature was determined by conducting thermogravimetric analysis. The result disclosed that the wood with the superior thermal stability was the acacia. The higher lignin and cellulose contents including a favourable wood fibre length distribution allowed the wood plastic composite with acacia to have the best thermal stability (Shebani et al. 2009). The acacia composites showed a higher water absorption rate compared to the other three types of wood plastic composites.

3.5 *Improving Thermal Properties of Wood Plastic Composites*

3.5.1 Reinforcement of Polymers with Natural Fibres

Many studies have been done on polypropylene filled with different cellulosic fibres and the unanimous conclusion made by these studies is that natural fillers decline thermal properties of the wood plastic composite (Ismail et al. 2002). Due to large availability, low cost and biodegradability of fibres, using natural fibres as reinforcement in polymeric matrix has conquered commercial and academic interest (Xu et al. 2014). In the past decade, many studies have been done on using natural fibres to improve the thermal properties of wood plastic composites.

A research was done by Farhadinejad to study the thermal properties of wood plastic composites which were reinforced with cellulose micro fibril and nano inorganic fibre filler. Nano wollastonite was used by dispersing it into the composite in order to compensate for the composite's poor thermal characteristics. Wollastonite is a naturally occurring mineral that is hard and has a high impact ratio which also improves the performance and increase strength of the plastic. Oxidative induction time, differential scanning calorimetry, thermogravimetric analysis and limited oxygen index were the techniques used to study the thermal properties of the obtained wood plastic composite. The study found that due to the high specific surface area of the nano inorganic fibre filler, the thermal stability and crystallinity in the composite was increased as the nano wollastonite was introduced to the composite (Farhadinejad et al. 2012).

A study on the thermal properties of lignin/Polypropylene wood plastic composite was done to investigate the improvement of the change in thermal properties of this composite with the addition of flexible segment containing reactive compatibilizer. Lignin in powder form and polypropylene grains were used to prepare this wood plastic composite. After detailed investigation, the experimental results indicated that the addition of 2 wt% of flexible segment displayed the best performance with an increase of the thermal stability, degree of crystallinity and crystallisation rate of the wood plastic composite (Xu et al. 2014).

Introducing graphite as a filler in a wood plastic composite that consists of recycled polypropylene, kenaf core and maleic anhydride polypropylene resulted in a lower glass transition temperature of the composite compared to the wood plastic composite that did not incorporate graphite (Ramli et al. 2020). This shows that the addition of graphite to wood plastic composite improves the thermal property of the composite.

3.5.2 Compatibilizer

Natural fibres have a disadvantage of having low compatibility with the thermoplastic polymer in composites like the wood plastic composite. In order to overcome this disadvantage, many studies have suggested the used of compatibilizer. Researches that were done in this field concluded that incorporating compatibilizers into wood plastic composites can result in the improvement of both the mechanical and thermal characteristic of the achieved composite. Compatibilizer is a compound that consists of a nonpolar hydrocarbon head and a polar head, and are involved in the polymer matrix.

A study was done using oxidised polypropylene as a compatibilizer to investigate it's effects on the thermal properties of the wood plastic composites. This study also incorporated nano-clay into the composite to produce wood plastic composite with improved thermal properties. For this study, polypropylene polymer was oxidised to produce a compatibilizer and change its characterises such as surface tension, polarity and adhesion. Samples were produced using wood fibres, compatibilizer and polypropylene polymer. Thermal and differential analysis was conducted to evaluate the thermal properties of the composite. The result of the study showed that there was a significant improvement in the thermal properties of wood plastic composite due to the addition of oxidised poly propylene and Nano-clay particles (Nafchi et al. 2015). In addition to that, increasing the content of Nano-particles improved the thermal stability of the wood plastic composite.

Lu, John, Qinglin and Ioan conducted a study on two types of oxidised polyethylene as compatibilizer to investigate the improvement of the thermal resistance of wood plastic composites. This study used maleated polyethylene and oxidised polyethylene. The results showed that maleated polyethylene presented a better improvement on the thermal resistance of the wood plastic composite compared to oxidised polyethylene but it was concluded that the addition of a compatibilizer showed a general improvement in the thermal resistance of the wood plastic composite (Lu et al. 2005).

In 2015 a study was done on maleated polypropylene which was used as a compatibilizer to improve the thermal properties of a wood plastic composite that was made of wood wastes and recycled plastic. The study showed that there was a significant improvement in the thermal properties of the composite that included maleated polypropylene compared to the composite that only consisted of the wood waste and plastic (dos Santos et al. 2015). Another study used maleic anhydride as a compatibilizer of a polypropylene. Thermal gravimetric analysis was conducted to investigate the thermal properties of the wood plastic composite. The study resulted in an increase in the thermal resistance of the wood plastic composite with compatibilizer but there was a slight decrease in the composite's thermal stability (Mousavi et al. 2015).

Andre Luis did a study to further understand the influence of a compatibilizer in compatibility of wood plastic composite. In the study, maleated polyethylene was used as the compatibilizer. Andre Luis focused on improving the sustainability of the polymer industry by reusing lignocellulosic and plastic waste so the thermoplastic in the wood plastic composite used in this study was made with post-consumer HDPE. The resultant wood plastic composite with and without compatibilizer were compared by investigating the influence of the coupling agent, checking the thermal properties of the composites. Differential scanning calorimetry analysis was used to evaluate the temperatures, crystallization enthalpies and degree of crystallinity of the different composites. The results showed that coupling agent with the concentration of 3% in the polymer matrix showed the best results for the thermal properties of the wood plastic composite (Catto et al. 2014).

3.5.3 Coupling Agent

The use of natural oils as coupling agent in wood plastic composites were evaluated by Matheus Poletto to study the effects of these natural oils in the thermal and mechanical properties of the wood plastic composites. Four different natural oils that were used are hexanoic, decanoic, octanoic and dodecanoic acids. Polypropylene, wood flour, maleic anhydride-grafted polypropylene, maleic anhydride grafting and the different natural oils were used to produce four different wood plastic composites. Thermogravimetric analysis were used to evaluate the thermal properties of these different composites and it was clear that the natural oils significantly improved the thermal properties of the composites (Poletto 2019). Octanoic acid showed a better result compared to the wood plastic composites with the other three natural oils.

3.5.4 Reinforcing Fillers

Using reinforcing fillers is one of the many methods used to improve the overall properties if wood plastic composites. The incorporation of nanoparticles as reinforcing fillers has been used to improve the properties of lignocellulosic-thermoplastic composites such as montmorillonite and silica nanopowder (Zhao et al. 2008). Xiang Li conducted a study on the utilization of bamboo charcoal enhancement on the thermal properties of wood plastic composites. Bamboo charcoal countless small holes and this can be used to have a stronger interface between polymer matrix and filler. In this study, Bamboo powder was incorporated into low density polyethylene as the pores and gaps in the bamboo powder structure will lead to a better interfacial adhesion between the filler and the matrix. A series of bamboo charcoal wood plastic composites were prepared and it's thermal behaviours was examined using thermogravimetric. The results were compared between a wood plastic composite. The thermal properties of the bamboo charcoal wood plastic composite was higher than of the normal wood plastic composite (Li et al. 2014).

For many decades, chemical modification of wood has been heavily researched and studied. Among the various modification methods, impregnation under pressure or vacuum has been the most promising method. In 2011, Devi prepared wood plastic nanocomposite by impregnating wood with a polymer to study effects of in situ reinforcement in a wood plastic composite. The wood plastic composite was prepared by impregnating Simul wood with styrene-acrylonitrile copolymer, reactive polymerizable surfactant modified montmorillonite and glycidyl methacrylate. Thermogravimetric analyser was used in this study to measure the thermal properties of the composite. It showed that the thermal stability of the composite was enhanced on addition of glycidyl methacrylate and polymerizable surfactant modified montmorillonite (Devi and Maji 2011).

Another widely used reinforcement material for polymer composites is clay materials. Due to natural abundance and beneficial properties, montmorillonite is the most widely used type among all the other clay minerals. Ru Lui conducted a research on wood plastic composite that was made of polylactic acid and wood flour which was modified using sodium-montmorillonite and dodecyl dimethyl ammonium chloride. In this study, differential scanning calorimetry, stress relaxation, and thermogravimetric analysis was used to characterise the thermal stability of the composites with respect to their resistance against thermal decomposition and thermal deformation. The modification to the wood plastic composite improved the resistance against thermal deformation of the composite at suitable amount but when excess amount of the modified wood flour was added, the enhancements in the thermal stability of the composite diminishes (Liu et al. 2015). However, there was a negative effect on the resistance against thermal decomposition of the composites.

In 2016, Sumit Manohar conducted a study on the effects of modification of pristine nanoclay and it's application in wood plastic composites. A transition metal ion which was copper (II) Chloride was used to modify the pristine nanoclay to improve wood plastic composite's properties and to achieve better dispersion. The thermal properties of the wood plastic composite that had modified nanoclay as a reinforcing filler was studied. Wood flour, polypropylene and maleic anhydride grafted polypropylene coupling agent was used to make the pristine and modified nanoclay based wood plastic composite. Different concentration of the modified nanoclay and pristine was used as a reinforcing filler for the wood plastic composite. To study the thermal properties, thermogravimetric analysis was performed to determine the weight loss of the wood plastic composite as a function of temperature. The study concluded that the use of modified pristine nanoclay as a reinforcing filler in the wood plastic composite improved the composite's thermal properties (Yadav and Yusoh 2016).

Irene Bavasso did a study on incorporating Yerba mate residue into a polymer matrix to use it as a reinforcement material in a wood plastic composite. Yerba mate is processed from small stems and the leaves of Ilex paraguariensis and in primarily used as an infusion (Heck and de Mejia 2007). For this study, biocomposites was produced in polypropylene and high-density polyethylene at different yerba mate residue amounts. The thermal properties of this wood plastic composite was investigated by differential scanning calorimetry and thermogravimetric analysis. The conclusion made in this study was that the yerba mate can be a potential candidate as a reinforcing agent in wood plastic composites as in significantly improves the properties of the composite (Bavasso et al. 2019).

3.6 Thermal Conductivity of Wood Plastic Composites

As aforementioned with the vast uses of wood plastic composites it is a crucial element to understand the thermo-physical properties of the composites especially with regards to change in temperature. As wood plastic composites are more widely used in construction of buildings the effective thermal conductivity of various wood plastic composite becomes a important study as wood plastic composites could act as insulators in buildings. A thorough analysis on previous literature found a scarce amount of information published regarding the aforementioned topic with the exception of research done by Li et al. and Bourain et al. (Bourai). The research done by Bourai et al. produced insightful results on the effect of concentration and temperature on effective thermal conductivity of wood plastics.

The study took the approach of comparing numerical simulations based on theoretical models with experimental results found. The study concluded that the thermal conductivity of wood composites in solid state is almost independent of the temperature whereas as the polymer matrix approached its melting point there was a significant decrease in thermal conductivity. Furthermore, the study found a significant decrease in thermal conductivity as the wood content increase and this was significant in solid state.

Subsequently, wood plastic composites are being researched continuously as there are emerging studies supporting the idea of WPCs with higher thermal conductivity. As WPCs are widely used in homes and buildings an increased thermal conductivity could offer the benefit of heating through the composite as an option therefore this area is being researched and has a few published literature supporting the validity of this idea.

A study conducted by Xing Li Xhang et al. studied the idea of the possibility of adding minute amount of thermal fillers into the WPCs while having an objective of keeping this process economically competitive. The results of the study produced interesting findings and the main finding showed a longer thermal conductive pathway but with the drawback of decrease in interfacial adhesion for the composites (Xing Li Xhang). This results in pourer mechanical properties in the composite which could be decremental to the current usages of WPCs and therefore would need to be further researched and improved.

The thermal endurance of wood plastics is a area of research that has gained traction due to the widespread use of wood plastic composites. Sierra studied the thermal endurance and thermal degradation of wood plastic composites especially polypropylene composites. The research was done to compare the thermal degradation of pure Polypropylene powder and wood powder with Polypropylene wood composite. The findings showed a similarity between peak temperatures during decomposition of wood powder in comparison with the composite. However the

Polypropylene decomposition had a difference in temperature as the composite had higher decomposition temperature (Sierra et al. 2010).

4 Flammability of Wood Plastic Composites

In the new generation of engineering materials, wood plastic composites have emerged as a material in high demand for various exterior and interior applications which includes fencing, furniture, landscaping timbers, terraces and balconies. The high relative stiffness, durability to environment, high dimensional stability and low cost have caused wood plastic composite to be a prime construction material but the main drawback is the high flammability of wood plastic composites. Researchers have done various studies in the means of enhancing the flame retardancy of wood plastic composites and the main method currently used is the addition of flame retardants to the composite (Kalali et al. 2019).

As the growth of environmental awareness increases the demand for eco-friendly material has increased thus Ammonium Polyphosphate (APP) is the conventional flame retardant used in WPCs as its highly effective and is environmentally friendly. APP is a branched or linear polymeric compound with a variable degree of polymerization. At temperatures above 300 °C the long chain APP degrades which generates polyphosphoric acid and ammonia while at temperatures around 150 °C short chain APP decomposes. Thus, the temperature in which the material will be used is an important factor in choosing APP as a flame retardant. Besides APP substances such as magnesium hydroxide and aluminium hydroxide have been published to be used as flame retardants in WPCs (Kalali et al. 2019).

A study done by Irina Turku studied the flammability of wood plastic composites prepared from plastic waste compared to the commercial wood plastic composites which produced interesting results. The main motivation of this study was as the demand for wood plastic composite grows the use of recycled plastics would be able to aid in the reduction of global plastic wastage. The main concern was the presence of various particles in the recycled plastic may hinder the flame retardancy of the wood plastic composites although previous studies have shown composite manufactured from pristine Polypropylene to be less thermo-oxidatively stable then composite containing recycled Polypropylene. The study conducted by Irina Turku showed a higher thermostable composite in the recycled plastic variance compared to the pristine polymer composites. The main conclusion formed on the increase in thermal stability was the pre existing presence of various additives including fire retardants and pigment in the recycled plastic which influenced the increase in thermal stability.

As aforementioned the primary flame retardant is the APP however a study done by Ehsan studied the addition of phytic acid modified layered double hydroxide (Ph-LDH). In the field of flammability retardancy test on wood plastic composites the char formation is a critical study as the formation of char increases the fire resistance of a composite. In this study the residual char was studied and the observations were used in the results obtained. The results obtained from the study showed an earlier degradation of the knew composite further improved the char formation which in return increases the fire retardancy. Besides the addition of phytic acid modified layered double hydroxide showed a reduction in smoke formation while also causing no decrease in tensile properties of the wood plastic composite (Turku et al. 2017).

Subsequently, further studies done by Majid et al. tested the effect of adding nanomagnesium oxide in wood plastic composites with high-density polyethylene. The study tested the mechanical properties and flammability of the composite and the results showed a 3phc nanomagnesium addition caused the composite to have higher flexural strength. A positive effect was also observed for the flammability test as the char residue increased and the total smoke production decreased (Kiaei et al. 2017).

4.1 Plastic Effects on Flammability

As aforementioned the two main common plastics used in WPCs are Polyethylene and polypropylene which both are flammable components in WPCs. The most common ways wood plastic composites are characterised according to their flammability is through flame spread index (FSI) and for a common comparison average wood would have an FSI in the range between 100 to 200 whereas Inorganic cement board surface would be 0. There are a few classes in which materials are characterised based on FSI numbers and most wood plastic composites come under the Class C category which is FSI between 76 and 200.

The FSI numbers mentioned above have become a great concern in wood plastic composites used for the purpose of infrastructure buildings. One main research showed that as the number of home decks made of wood plastic composite increases the concern to make this material flame retardant has increased as the research shows brushfires often ignite a house via the decks. (reference). One of the easiest and cheapest ways manufacturers have found that could reduce the flammability of wood plastic components is by adding low-flammable plastics mainly Polyvinyl Chloride.

One of the main drawback to this is if the WPC catches on fire the PVC would release hydrogen chloride which is toxic and harmful. Besides with the addition of more plastics into the wood plastic composite the main concern is the changes in mechanical properties as the ratio of wood to plastic decreases greatly. These are the few main challenges in reducing the flammability of wood plastic composites that have to be continuously researched and improved.

4.2 Effect of Mineral Fillers on Flammability

Inert fillers such as talc, cellulose fibre, calcium carbonate, clay and glass fibre can reduce the flame spread. This is done my removing the "fuel" for the flame propagation. Flame retardants typically produce counter-ignition or counter-flame matters such as inflammable gases or water but these fillers don't thus they do not act as flame retardants. There fillers reduce heat generation by diluting the fuel in the solid phase.

4.3 Effect of Density on Flammability

A thorough study of the literature found no published data that studied the correlation of density of WPCs and their flammability. Nevertheless the theory in books suggest there will be a strong correlation between density of WPCs and the flammability as WPCs with lower density would be more porous and this would create space for oxygen to be trapped in the pores which in return would aid flame and increase the flame spread compared to higher density flame boards. A study done on various species of woods and its comparison in regards with ignition time is shown in the Table 1 below which supports the aforementioned hypothesis.

4.4 Composite Ignition and Mineral Fillers

Ignition occurs in the presence of oxygen and when the flammable material reaches the ignition temperature point. As most WPCs use polyethylene as the plastic in the composite this research would focus more on that specific plastic. Polyethylene as itself and not in a composite has a flammable temperature point of around 360–367 °C.

In order to offset the flammability risks caused by plastics certain inert fillers such as cellulose fibre, calcium carbonate, glass fibre and clay have been known to be used in the industry for its flame retardant qualities. Most of the times the common use of mineral fillers as flame retardants are for their charring favour

Table 1 Types, Density and Ignition time of wood

		Ignition time for heat flux	
Species	Density (c/cm^3	18 kW/m^2	55 kW/m^2
Red oak	0.66	930	13
Southern pine	0.51	740	5
Redwood	0.31	741	3
Basswood	0.31	183	5

which does not act as a counter flame measure but rather as a substance that stops the flame from its worst stages.

4.5 Flame Retardant Materials

As wood plastic composites are mainly composed of woods the general theory would assume similar flame retardants could be used as used in woods. Through research this hypothesis has been found to not be through as WPCs use very different flame retardants. Flame retardants are infused into wood plastic composites as solids during formulation and therefore the flame retardants would have to withstand high temperatures during the processing stage of the WPCs.

Flame retardants used in wood plastic composites can mainly be broken down into three categories which represent their physical actions and the divisions are water releasing and cooling, halogen forming and flame choking, and char forming of a protective layer. As aforementioned the inert fillers come under the category of char formation as they do not act as an active flame retardant and do not change the ignition point.

Active flame retardants work in such a way that they inhibit combustion through a few mechanisms and the first mechanism is through the release of water which prevents oxygen from aiding the flame compounds. Besides, the flame retardants form non-flammable gasses which act as a barrier that protects the flammable material and prevents oxygen from reaching it. Brominated flame retardants remove free radicals in the gas phase and as a result prevent or slow down the burning process by reducing heat generation and by producing flammable gases. Furthermore, the third category is formation of char in which when heated the materials form a solid surface which cotes the flammable material which prevents the flame from feeding on the material.

5 Conclusion

The extensive study on previously published literature gave insightful understandings on the thermal properties and flammability of wood plastic composites. The main takeaway from an extensive literature review shows that the field of material engineering is ever evolving and in the case of wood plastic composites there is constant changes on the addition of materials to the composites to alter the thermal properties and with advancement in technologies there is a huge area of interest to research and improve the thermal properties of wood plastic composites. As mentioned multiple times with wood plastic composites becoming a staple material in construction the continuous research on improving flame resistance in wood plastic composites has great benefits for the safety and improvement of buildings.

6 Future Perspective

Upon completion of analysing previous literature and understanding the properties of wood plastic composites it has shown that there are areas where wood plastic composites could be improved. As aforementioned in the conclusion section in order to increase the usage of wood plastic composites in the industry the first step has to be by conducting more research to understand and improve every aspect of the composite. Furthermore, looking at the current data it shows that wood plastic composites are mainly only used to replace wood in the construction industry and this would change in the future as all industries shift to using more composite materials. With the future increase in demand and the ongoing research on wood plastic composite the wood plastic composite industry could see multiple new iterations of the composite consisting of various new combinations of wood and polymers which improve the thermal and flammability properties.

References

Anon (2020a) Differential Scanning Calorimetry (DSC) Analysis, viewed 30 August, 2020. https://www.intertek.com/analysis/dsc/

Anon (2020b) What Are Composites? - Composites 101, viewed 30 August, 2020. https://compositeslab.com/composites-101/what-are-composites/

Anon (2020c) What is a Composite Material? (A Definitive Guide), viewed 30 August, 2020. https://www.twi-global.com/technical-knowledge/faqs/what-is-a-composite-material#WhataretheDifferentTypes

Anon (2020d) Wood-Plastic Composite - an overview | ScienceDirect Topics, viewed 30 August, 2020. https://www.sciencedirect.com/topics/materials-science/wood-plastic-composite

Bavasso I, Bracciale M, Sbardella F, Tirillò J, Sarasini F, Di Palma L (2019) Effect of yerba mate (Ilex paraguariensis) residue and coupling agent on the mechanical and thermal properties of polyolefin-based composites. Polym Compos 41(1):161–173

Catto A, Stefani B, Ribeiro V, Santana R (2014) Influence of coupling agent in compatibility of post-consumer HDPE in thermoplastic composites reinforced with eucalyptus fiber. Mater Res 17(suppl 1):203–209

Devi R, Maji T (2011) Preparation and characterization of wood/styrene-acrylonitrile copolymer/MMT nanocomposite. J Appl Polym Sci 122(3):2099–2109

dos Santos L, Trombetta E, Flores-Sahagun T, Satyanarayana K (2015) Effect of domestic compatabilizer on the performance of polypropylene-sawdust composites. J Compos Mater 50 (10):1353–1365

Earnest C (1983) Thermal analysis of hectorite. Part I. Thermogravimetry. Thermochim Acta 63 (3):277–289

Fabiyi J, McDonald A (2010) Effect of wood species on property and weathering performance of wood plastic composites. Compos Appl Sci Manuf 41(10):1434–1440

Farhadinejad Z, Ehsani M, Khosravian B, Ebrahimi G (2012) Study of thermal properties of wood plastic composite reinforced with cellulose micro fibril and nano inorganic fiber filler. Eur J Wood Wood Prod 70(6):823–828

Fuad M, Mustafah J, Mansor M, Ishak Z, Omar A (1995) Thermal properties of polypropylene/rice husk ash composites. Polym Int 38(1):33–43

Guo Y, Zhu S, Chen Y, Li D (2019) Thermal properties of wood-plastic composites with different compositions. Materials 12(6):881

Heck C, de Mejia E (2007) Yerba mate tea (ilex paraguariensis): a comprehensive review on chemistry, health implications, and technological considerations. J Food Sci 72(9):R138–R151

Islam M, Hamdan S, Sobuz H, Rahman R, Ahmed A (2012) Thermal and decay-resistance properties of tropical wood–plastic composites. J Compos Mater 47(12):1493–1500

Ismail H, Edyham MR, Wirjosentono B (2002) Bamboo fibre filled natural rubber composites: the effects of filler loading and bonding agent. Polym Test 21:139–144

Kaboorani A (2010) Effects of formulation design on thermal properties of wood/thermoplastic composites. J Compos Mater 44(18):2205–2215

Kalali E, Zhang L, Shabestari M, Croyal J, Wang D (2019) Flame-retardant wood polymer composites (WPCs) as potential fire safe bio-based materials for building products: Preparation, flammability and mechanical properties. Fire Saf J 107:210–216

Kiaei M, Moghdam Y, Kord B, Samariha A (2017) The effect of Nano-MgO on the mechanical and flammability properties of hybrid nano composites from wood flour-polyethylene. Maderas. Ciencia y tecnología, no. ahead

Kim H-S, Yang H-S, Kim HJ, Park H-J (2004) Thermogravimetric analysis of husk flour filled thermoplastic polymer composites. J Therm Anal Calorim 76:395–404

Kim J, Harper D, Taylor A (2009) Effect of wood species on the mechanical and thermal properties of wood-plastic composites. J Appl Polym Sci 112(3):1378–1385

Li X, Lei B, Lin Z, Huang L, Tan S, Cai X (2014) The utilization of bamboo charcoal enhances wood plastic composites with excellent mechanical and thermal properties. Mater Des 53:419–424

Liu R, Peng Y, Cao J (2015) Thermal stability of organo-montmorillonite-modified wood flour/poly(lactic acid) composites. Polym Compos 37(7):1971–1977

Lu J, Wu Q, Negulescu I (2005) Wood-fiber/high-density-polyethylene composites: coupling agent performance. J Appl Polym Sci 96(1):93–102

Mousavi S, Arjmand O, Talaghat M, Azizi M, Shooli H (2015) Modifying the properties of polypropylene-wood composite by natural polymers and eggshell nano-particles. Polym Renewable Resour 6(4):157–173

Nafchi H, Abdouss M, Najafi S, Gargari R, Mazhar M (2015) Effects of nano-clay particles and oxidized polypropylene polymers on improvement of the thermal properties of wood plastic composite. Maderas. Ciencia y tecnología, no. ahead

Oksman Niska K, Sain M (2008) Wood-polymer composites. CRC Press, Boca Raton

Poletto M (2019) Natural oils as coupling agents in recycled polypropylene wood flour composites: Mechanical, thermal and morphological properties. Polym Polym Compos 28 (7):443–450

Prime B, Bair H, Vyazovkin S, Gallagher P, Riga A (2009) Thermal analysis of polymers: fundamentals and applications, pp. 241–317

Ramli R, Zulkifli M, Rabat N (2020) Effect of Graphite on mechanical thermal and morphological properties of kenaf recycle polypropylene wood plastic composites. Mater Sci Forum 981:144–149

Rogers J, Simonsen J (2005) Interfacial shear strength of wood–plastic composites: a new pullout method using wooden dowels. J Adhes Sci Technol 19(11):975–985

Shebani A, Van Reenen A, Meincken M (2009) The effect of wood species on the mechanical and thermal properties of wood—LLDPE composites. J Compos Mater 43(11):1305–1318

Sierra J, Cardona E, Ospina S, Maldonado A (2010) Study of the thermal endurance and the thermal degradation kinetics of a Polypropylene wood composite. In: 68th annual technical conference of the society of plastics engineers 2010, vol 2, pp 1029–1033

Staff C (2020) Everything You Need To Know About Polypropylene (PP) Plastic. https://www.creativemechanisms.com/blog/all-about-polypropylene-pp-plastic. Accessed 30 Aug 2020

Turku I, Kärki T, Puurtinen A (2017) Flammability of wood plastic composites prepared from plastic waste. Fire Mater 42(2):198–201

Xu X, He Z, Lu S, Guo D, Yu J (2014) Enhanced thermal and mechanical properties of lignin/polypropylene wood-plastic composite by using flexible segment-containing reactive compatibilizer. Macromol Res 22(10):1084–1089

Yadav S, Yusoh K (2016) Modification of pristine nanoclay and its application in wood-plastic composite. e-Polymers 16(6): 447–461

Yang H-S, Wolcott MP, Kim H-S, Kim H-J (2005) Thermal properties of lignocellulosic fillers-thermoplastic polymer bio-composites. J Therm Anal Calorim 82:157–160

Zhao R, Wang X, Yuan J, Lin J (2008) Investigation of feasibility of bamboo charcoal as solid-phase extraction adsorbent for the enrichment and determination of four phthalate esters in environmental water samples. J Chromatogr a 1183(1–2):15–20

Tribology of Wood Polymer Composites

Karthikeyan Subramanian, Senthilkumar Krishnasamy, Chandrasekar Muthukumar, Suchart Siengchin, Kalusuraman Gnaniar, and Arunprasath Kanagaraj

Abstract Wood is a solid material, and it is derived from shrubs and trees. The wood is composed of cellulose (40% to 50%), hemicellulose (15% to 25%), and lignin (15% to 30%). It can be classified as (i) softwood or (ii) hardwood, and some of the advantages include biodegradable nature, non-corrosive, high load-bearing capacity, etc. Though the wood has many benefits, it has a significant disadvantage of porous nature, poor resistance against abrasion, and delamination. Furthermore, the wood is susceptible to (i) light, (ii) heat, and (iii) ultraviolet (UV) radiations.

K. Subramanian
Department of Automobile Engineering, Kalasalingam Academy of Research and Education, Anand Nagar, Krishnankoil 626 126, Tamilnadu, India
e-mail: skarthikeyan@klu.ac.in

S. Krishnasamy (✉)
Center of Innovation in Design and Engineering for Manufacturing (CoI-DEM), King Mongkut's University of Technology North Bangkok, 1518, Wongsawang Road, Bangsue 10800, Bangkok, Thailand
e-mail: kmsenthilkumar@op.kmutnb.ac.th

C. Muthukumar
School of Aeronautical Sciences, Hindustan Institute of Technology and Science, Padur, Kelambakkam, Chennai 603103, Tamil Nadu, India
e-mail: chandrasm@hindustanuniv.ac.in

S. Siengchin
Department of Materials and Production Engineering, The Sirindhorn, International Thai-German Graduate School of Engineering (TGGS), King Mongkut's University of Technology North Bangkok, 1518 Wongsawang Road, Bangsue 10800, Bangkok, Thailand
e-mail: suchart.s.pe@tggs-bangkok.org

K. Gnaniar
Department of Agricultural Engineering, Kalasalingam Academy of Research and Education, Anand Nagar, Krishnankoil 626 126, Tamilnadu, India
e-mail: kalusunrk@klu.ac.in

A. Kanagaraj
Department of Mechanical Engineering, Kalasalingam Academy of Research and Education, Anand Nagar, Krishnankoil 626 126, Tamilnadu, India
e-mail: arunprasath@klu.ac.in

S. Mavinkere Rangappa et al. (eds.), *Wood Polymer Composites*,
Composites Science and Technology, https://doi.org/10.1007/978-981-16-1606-8_9

Thus, this chapter examines that the issues related to the tribological behavior of wood plastic-based composites. The term 'tribology' is dealt with the subject of (i) wear, (ii) friction, and (iii) lubrication characteristics. Furthermore, any tribo-materials are developed by analyzing their friction and wear characteristics by varying (i) applying a load, (ii) sliding velocity, (iii) sliding distance, and (iv) working temperature.

Keywords Wood polymer composites · Tribology · Thermoset · Thermoplastics · Filler

1 Introduction

Wood-polymer composites (WPC) are widely used due to its easy availability, high strength, low density, and low cost. Thus, the wood polymer-based composites are now extended to many applications: marine, automotive, aircraft, electronic packing, pharmaceutical industries, to mention but a few (Dale Ellis 2000). Furthermore, combining polymer with wood materials possessing many advantages includes mechanical properties. Wood as an individual material having disadvantages of moisture absorbing property which makes not suitable for most of the heavy structural material (Tuntsev et al. 2017). Based on the applications, the woods are being combinedly used with additives and binder materials. The usage of stabilizers, antioxidants, coupling agents makes the wood as a suitable material.

The wood fiber properties majorly depend on the wood type, matrix, filler, fiber treatment, additives, and coupling agent used (Kumar et al. 2019). All the additives help to optimize the material's performance and make it suitable for particular applications. Besides, the usage of polymers and the woods make them a more robust material. The interphase adhesion and bonding between the material and the polymer matrix is an essential factor to be considered (Jeong and Park 2019). Most wood polymer composite materials like polypropylene (PP), polyester based materials, exhibit good bonding properties. The silane treated fiber materials increase the material's bonding ability; some of the surface modification techniques were also used to enhance the bonding characteristics (Koohestani et al. 2017).

The emergence of polymer matrix composite in wear application was increasing during the last few decades. The tribological applications such as gears, cams, clutches, rollers, and bearings using natural fiber reinforced composite showed improved results than conventional materials (Jena 2019). Natural materials as filler components for various composite materials using wood dust, wood charcoal, apple shell, pineapple wood, coconut shell, etc., increases due to its unique wear resistance properties (Friedrich 2018). This chapter explains the tribological behavior of wood fiber-based composites using thermoset and thermoplastic-based resins. Furthermore, the morphological characterization and wood filler filled composites are discussed.

2 Tribological Behavior of Thermoset/Wood Fiber-Reinforced Composites

Wood fiber reinforced composite exhibits comparable results when compared to conventional based materials. Thus the usages of wood fiber-based composites are being increased in various fields: (i) automobile, (ii) mining, (iii) biomedical, (iv) marine, and (v) aerospace industries (Kumar and Singh 2020). Generally, the polymers are classified into two major types: (i) thermoplastic polymer and (ii) thermosetting polymer. The most commonly used thermosetting polymers are polyester, epoxy, vinyl ester, phenolic, and polyimide (Singh et al. 2018; Samyn and Zsidai 2017). Figure 1 shows the usages of polymers in different fields, whereby the thermoset based polymer exhibits a significant contribution than the rest of the polymers (Renard 2005).

Wood fibers are obtained from the recovered sawdust, and these are one of the industrial waste materials abundantly available from civil and mechanical industries. The ability of wood powder can be easily molded into any desired shape by mixing and extruding processes. Due to the difference in the shape of cells' growth in wood, heterogeneous tissue morphology can occur, which affects the physical, mechanical, and machining properties. Researchers examined the tribological behavior of wood flour/epoxy matrix composites. They reported that the wood fiber loading increases the load-carrying capacity of the tribomaterials. However, the wear resistance of the composites was found to be decreased; it was ascribed to the tribo layer formation on the contacting surfaces. But in the case of abrasive wear, the wear resistance was increased due to increasing the fiber loading. Wood flour/ epoxy matrix composites showed that the specific wear rate (Ws) in the order

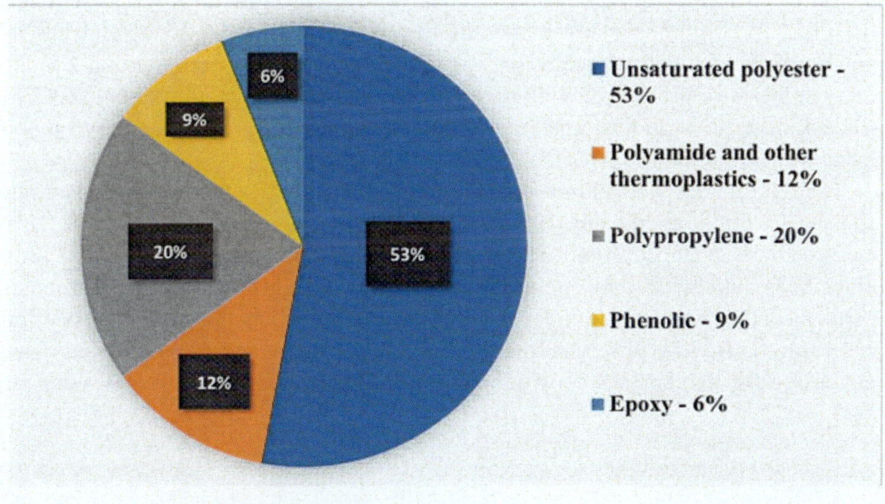

Fig. 1 Thermosetting resin used in composite markets

of $\sim 10^{-14}\ \frac{m^3}{Nm}$ in sliding condition and $10^{-10}\ \frac{m^3}{Nm}$ in abrasive wear condition (Dwivedi and Chand 2008).

Researchers investigated the tribological properties of wood flour/carbon nanotube/polyester matrix composites (Nabinejad et al. 2019). The addition of a smaller quantity of 0.2 wt.% of the wood flour to a multi-wall carbon nanotube enhanced the tribological properties. Lalit Ranakoti et al. (2019) analyzed the tribological behavior of wood/glass fiber/epoxy matrix composites. They reported that the wear rate of the hybrid composites was increased; it was ascribed to the addition of wood fillers in the epoxy matrix composites. The authors also conducted the design of experiments for ranking the order of composite preferences. Karthik Babu et al. (2000) studied the wear properties of epoxy composite with different fiber dosages of pine wood dust powder (0, 5, and 10%). The addition of pine wood dust powder enhanced the sliding wear resistance of the epoxy matrix composites. Shakuntala et al. (2014) conducted the tribological behavior of wood apple shell/epoxy matrix composites. It was observed that the incorporation of the wood apple shell powder enhances the mechanical and erosion wear property of composites.

3 Tribological Behavior of Thermoplastics/Wood Fiber-Reinforced Composites

Wood polymer composites are one of the renewable resources of materials which are obtained from waste wood. The mixture of wood fiber and plastic with binding agents is combined to produce the composites stronger and denser. The wood-based polymeric matrix composites are commonly fabricated by injection molding technique using thermoplastic resins such as high-density polyethylene (HDPE), polyvinyl chloride (PVC), polypropylene (PP), and low-density polyethylene (LDPE). The significant advantages of the wood polymer-based composites are non-corrosive, biodegradable, high strength, high stiffness, low thermal conductivity and ease of production with recyclability, etc. The application of the WPC's is (i) deck floors, (ii) rail fencings, (iii) doors, (iv) windows, (v) frames and (vi) indoor furniture (Borah and Kim 2016).

The addition of wood flour with thermoplastic polymer increases the hardness and the frictional coefficient. However, the addition of wood flour content (i) increases the water absorption properties and (ii) deteriorates the mechanical properties (Mysiukiewicz and Sterzyński 2017). In another work, the researchers performed the wear study under dry sliding conditions with varying wood flour percentages. It was observed that the wear performance was improved when increasing the percentages of wood flour in the thermoplastic matrix. Besides, the wear rate was lowered with 40wt.% of wood flour at 100N loading conditions (Beer et al. 2007). Ibrahim et al. studied the tribological performance of wood flour/PP matrix composites by varying the wood flour content from 5 wt.% to 55 wt.% (order of 10 wt.%). They observed that the adhesion of wood flour was improved in maleated polypropylene (MAPP); it also helped to enhance the interfacial interaction and wettability. Furthermore, the wear resistance and the mechanical behavior of these composites were improved by increasing the wood flour content (Ibrahim et al. 2019).

Xiang et al. fabricated the Polyoxymethylene (POM)/low-density polyethylene (LDPE)/wood fibers using injection molding technique, and the composites were subjected to friction and wear studies using reciprocating sliding tribotester. They reported that 5 wt.% of LDPE acted effectively since the internal lubricant for the POM. The wear performance of the composites was found to be improved by the addition of wood fiber; however, the coefficient of friction was experienced lesser effect due to the presence of LDPE. Furthermore, the researchers suggested that the combination of POM/LDPE/wood fiber composites could be acted environmentally friendly and cost-effective for tribo-based applications (Xiang et al. 2012). In another work, Waßmann and Ahmed reported that the lowest friction and low wear observed for the Beeswax-impregnated and compressed wood samples. Furthermore, the friction coefficient was comparable to ultra-high-molecular-weight-polyethylene (UHMWPE) and Polytetrafluoroethylene (PTFE). The wear was observed to be lower than 290 and 3 times compared to the UHMWPE and PTFE (Waßmann and Ahmed 2020).

Olga and Tomasz fabricated a wood flour/PP and wood flour/PLA composites, and all these polymeric matrix composites were subjected to wear study. The wear performance was analyzed by varying the wood flour content from 0 wt.% to 30 wt.% (order of 10 wt.%). Based on the favorable coefficient of friction results, these polymeric composites have been suggested as bearing material (Mysiukiewicz and Sterzyński 2017). Kumar et al. (2019) examined the wear performance of PP-based wood-plastic composites by varying the wood dust proportion (i.e., 10 wt.% to 20 wt.%) of Mango, Sheesham, Mahogany, and Babool. They reported that the Mango and Mahogany based PP matrix composites exhibited the better wear performance than the rest of the wood dust composites. Furthermore, they reported that the wear resistance of wood-plastic composites was decreased with increasing the content of wood dust particles. Jeamtrakull et al. (2012) explored the wear behavior of *Xylia kerrii* Craib & Hutch/E-chopped strand glass fiber/PVC, *Hevea brasiliensis* Linn/ E-chopped strand glass fiber/PVC, and *Mangifera indica* Linn/ E-chopped strand glass fiber/PVC composites. The wear behavior of all the wood fiber-based composites was examined by varying (i) the wood fiber

Table 1 Tribo properties of wood flour based thermoplastic composite

Wood polymer (WP)/Thermo Plastic combinations	Tribo-properties																				
	Hardness			Co-efficient of friction			Dry sliding wear			Wet sliding wear			Abrasive wear			Wear resistance			Frictional wear		
	G	M	E	G	M	E	G	M	E	G	M	E	G	M	E	G	M	E	G	M	E
WP/PTFE	*					*	*				*		*			*					*
WP/PVDF		*			*				*			*		*			*		*		
WP/PTFE			*			*			*			*			*			*			*
WP/PVC	*			*			*			*			*				*			*	
WP/HDPE		*			*		*				*		*					*	*		
WP/LDPE		*				*			*				*			*					*
WP/PP	*			*			*				*			*			*		*		
WP/PLA		*		*					*		*			*		*			*		

*Good-G, Moderate-M, Excellent-E

content and (ii) the sliding distance. The result analysis reported that the addition of glass fiber content enhanced the wear properties. Furthermore, the greater the specific wear rate was observed for all types of wood fiber-based composites when increasing the sliding distances. The tribo properties such as hardness, coefficient of friction, dry sliding wear, abrasive wear, etc. are tabulated in Table 1 for different types of wood fiber based thermoplastic matrix composites.

PTFE = Polytetrafluoroethylene; PVDF = Polyvinylidene fluoride; PVC = Polyvinyl chloride; HDPE = High density poly ethylene; LDPE = Low density poly ethylene; PP = Polypropylene; PLA = Polylactic acid.

4 Morphology Characterization of the Worn Surface of the Wood Polymer Composites

Yin et al. (2016) classified the wood surface as wenge with intact surface and tamarisk with natural voids on the surface as shown in Fig. 2(a–b). As the sliding ball contacts and moves over the surface, the material removal is different based on the wenge or tamarisk morphology and the applied load. In the case of the wood surface with tamarisk (Fig. 2(f)), the contact area decreases, and the groove could store the removed debris from the test, hence resulting in less friction co-efficient whereas, in the case of the wood surface like wenge (Fig. 2(e)), the removed debris remains on the wood surface and moves along with the sliding ball to create more abrasive leading to high friction coefficient. Similarly, the wear deformation due to the applied load is dissimilar for the wood surface with wenge and tamarisk, as shown in Fig. 2(c–d).

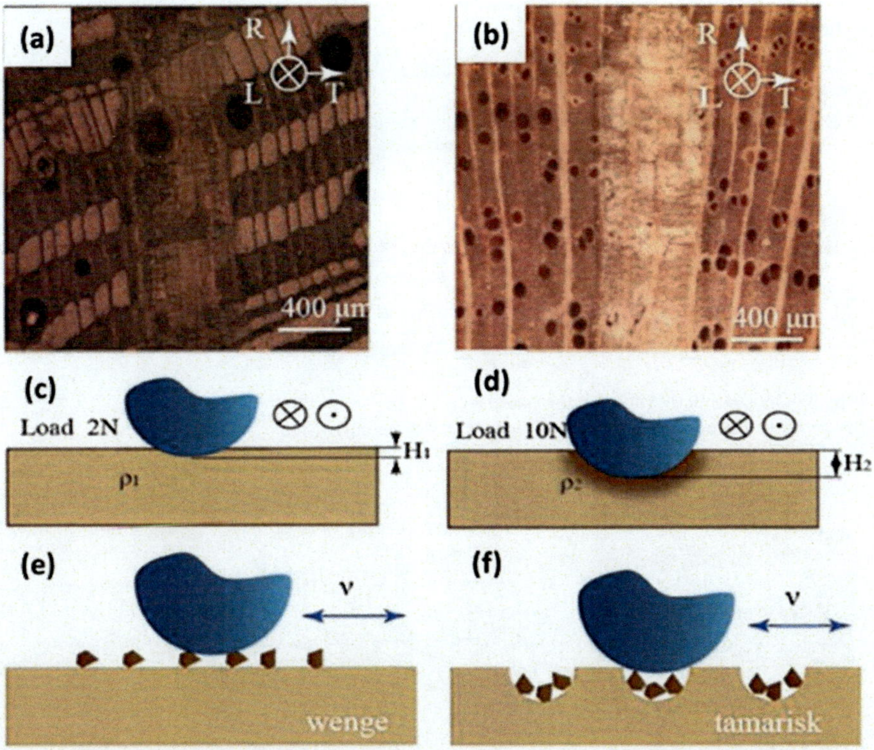

Fig. 2 Morphology of the wood and wood deformation due to the sliding wear (a) microstructure of the wenge, (b) microstructure of the tamarisk, (c) Wear deformation due to the applied load of 2 N, (d) Wear deformation due to the applied load of 2 N, (e) worn surface and grinding trace in the wenge and (f) worn surface and grinding trace in the tamarisk (Yin et al. 2016) [Re-used with permission from Elsevier, License Number- 4890830538277]

Wood polymer composite consists of the plywood bonded with adhesive or wood flour blended with a polymer. The wear deformation mechanism of the WPC is different from that of the wood surface. For the WPC, polymers are exposed to the sliding contact initially. With the increase in applied load, sliding distance, and sliding velocity, the wear tracks were formed, and polymers chip away from the wear area, exposing the wood fiber. The polymer removal to the sliding interface is irregular, as shown in Fig. 3(a–b). No other visible changes were observed in the wear area with respect to the fiber loading for wood flour/PP composites. The wear mechanism is explained as three-body abrasive wear in which the debris removed from the surface acts as a third body along with the sliding interface. This debris, combined with the applied load, increases the material removal rate at higher sliding velocities (Ibrahim et al. 2019).

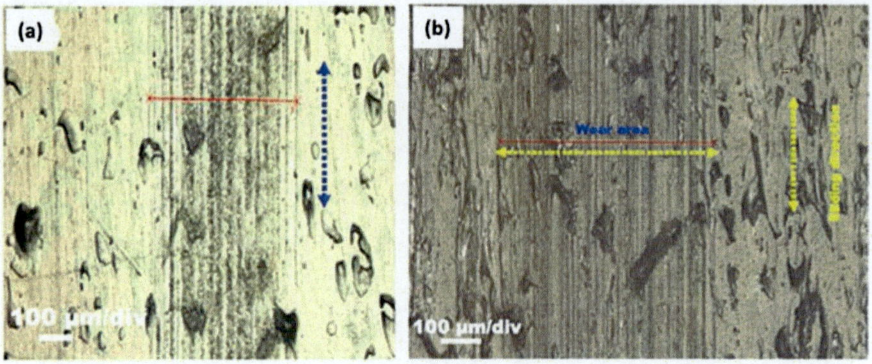

Fig. 3 Optical micrograph of the worn surface of wood flour/PP composite (a) 5 wt.% and (b) 15 wt.% (Ibrahim et al. 2019) (Open Access Content)

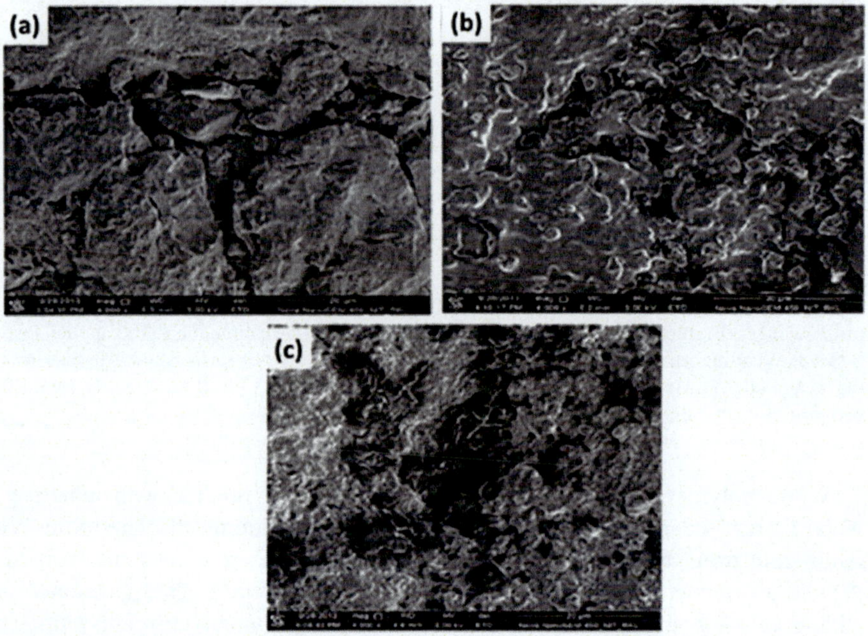

Fig. 4 Microstructure of the worn surface of the wood-apple shell/epoxy composite subjected to erosion wear at an impact velocity of 48 m/s, and impact angle of 45° (a) 5 wt.% filler, (b) 10 wt.% filler, and (c) 15 wt.% filler (Shakuntala et al. 2014) (Open Access Content)

The influence of fiber loading on the erosion wear was investigated in a study. At lower wood fiber weight, cracks were formed on the polymer surface in addition to the material removal (Fig. 4(a)) while such cracks were absent at higher fiber loading (Fig. 4(b–c)). The addition of fiber weight improves the wear resistance of the composites; however, the material removal was prominent at the higher fiber

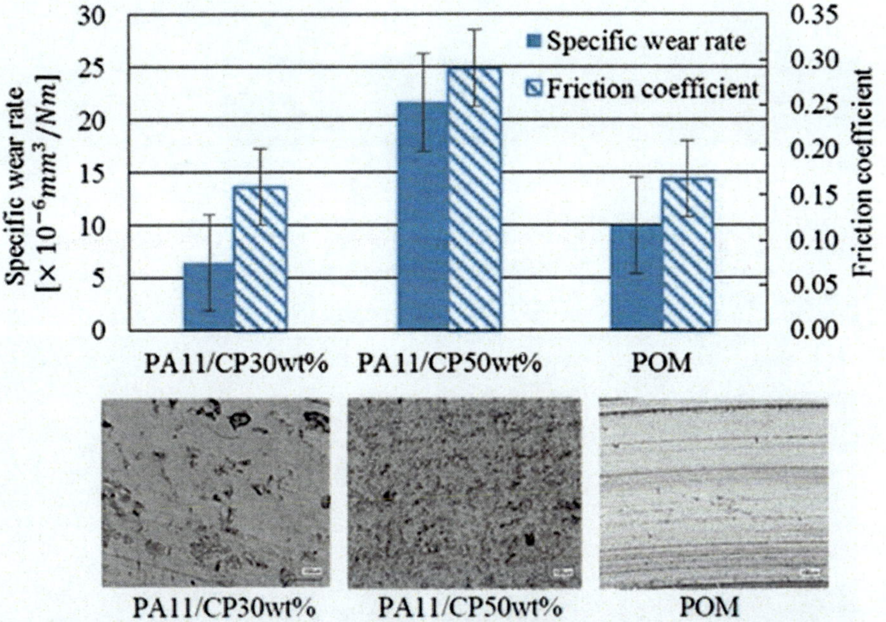

Fig. 5 Correlation of material removal from the morphology and specific wear rate-friction coefficient for the polyacetal resin and the cellulose powder (CP) reinforced with polyamide resin (PA11) (Araki et al. 2016) (Open Access Content)

weight, leading to a higher specific wear rate. The weak interfacial adhesion between the wood flour and polymer was responsible for more material removal at the higher fiber loading (Shakuntala et al. 2014).

Araki et al. (2016) investigated the tribological performance of polyacetal resin and the cellulose powder reinforced with the polyamide resin composite (CP/PA11) at 30 wt.% and 50 wt.% of CP. Morphological changes in the surface after the dry sliding test (speed of 50 $\frac{mm}{s}$ at room temperature for 3 h) are presented along with the specific wear rate and friction coefficient in Fig. 5. Material removal was more prevalent for the PA11/CP50 wt.% as confirmed by the rough surface is following the higher friction coefficient and larger specific wear rate. The softening of the composite surface during the sliding test and heat generation due to the friction was responsible for material removal from the surface. A higher rate of material removal for composite with higher fiber weight % implies the poor interfacial bonding between the CP and PA11, which led to more material removal.

According to Brostow et al. (2016) the film transfer phenomenon over the ball surface sliding on the test specimen during the sliding test increases the material removal leading to a higher specific wear rate in the wood/HDPE composite. It can be noticed in Fig. 6(a–b) that the film surface area formed on the ball surface is

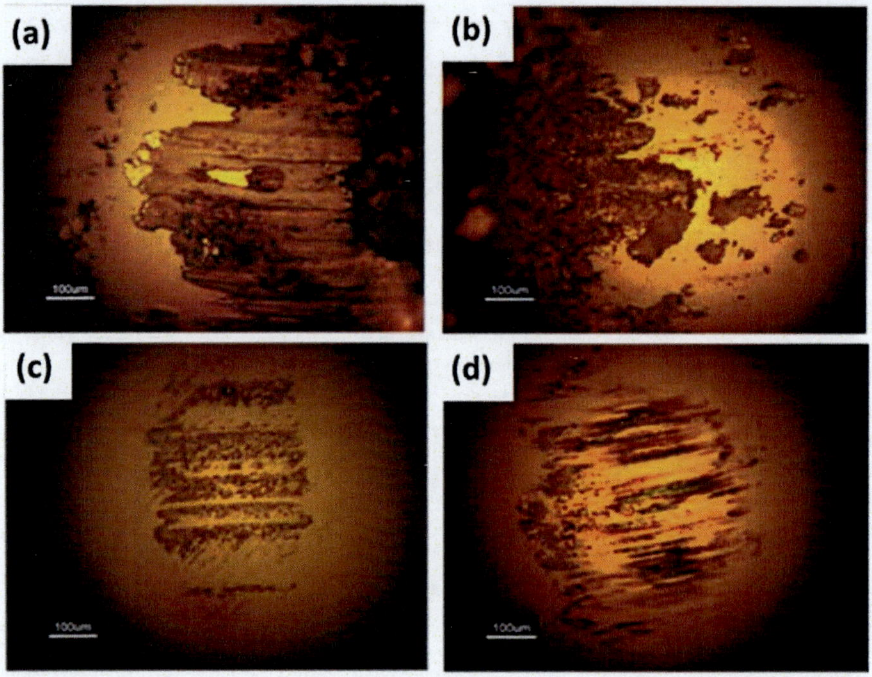

Fig. 6 Film transfer mechanism (a) untreated wood/HDPE, (b) silane treated wood/HDPE, (c) HDPE and (d) wood/HDPE/ sol–gel silica (Brostow et al. 2016) [Re-used with permission from Elsevier, License Number-4892340836953]

higher for the composite with untreated wood and silane treated wood. The HDPE and wood/HDPE blended with the sol–gel silica (Fig. 6(c-d) showed lower film formation, which corresponds directly to the lower material removal and lower friction during the sliding test.

5 Tribological Behavior of Wood Filler Reinforced Polymer Matrix Composites

In general, the polymers are mixed up with fillers (except adhesives) to obtain a favorable property. The filler filled composites exhibited better performance than the unfilled polymeric matrix. On the other side, the composites reinforced with fillers are expected to exhibit superior performance than the unfilled polymers, for the former cheapening agents are chosen, and for the latter, reinforced fillers are preferred (Kokta et al. 1989; Bledzki et al. 1998; Kim and Pal 2010). In recent days, fillers' usage is increased due to the increasing price of oil and monomer prices (Thiagamani et al. 2019; Kumar et al. 2020).

The significant role of the tribologist needs to select the right combination of (i) polymers, (ii) fillers, and (iii) additives for obtaining favorable properties based on the intended application. In some cases, the wear characteristics (i.e., wear-resistance and friction co-efficient) may not play a vital role; however, the fabricated materials need to function appropriately depends upon the system requirement. For instance, brake pads and clutches need low wear with a higher coefficient of friction. In some cases, it is necessary to develop a material that possesses low wear and low friction (e.g., gears and bearings applications) characteristics (Friedrich et al. 2005; Parikh and Gohil 2015; Omrani et al. 2016; Joseph et al. 2020; Milosevic et al. 2020).

Researchers reported that for enhancing wear and frictional behavior of polymeric matrix composites, the adhesion between the material and its counterpart needs to be reduced; also, the respective materials' (i) hardness, (ii) compressive strength, and (iii) stiffness are required to improve. All the above-said statements can be possible by employing special fillers. However, all kinds of fillers may not be benefitted from the wear performance of the polymeric matrix composites. The fillers should (i) decompose and (ii) produce reaction products, which would help to improve the bonding between the counterface and the transfer film; consequently, the wear resistance gets improved. On the other side, the fillers produce more discontinuities within the material, thus gives a negative impact, i.e., a decreasing trend in wear resistance. Therefore, it is necessary to understand the wear mechanisms by knowing exactly (i) growth, (ii) bonding, and (iii) the loss of transfer films. However, the mechanical and chemical interactions of transfer films are challenging; therefore, they are suggested to explore current and future research (Bahadur 2000; Jintang 2000; Friedrich et al. 2005; Karthikeyan et al. 2017). Table 2 summarizes the significant findings from the various forms of filler/ polymeric matrix composites subjected to tribological studies.

Kranthi and Satapathy analyzed (2010) the wear performance of pine wood dust /epoxy matrix composites using a fascinating computational tool called 'artificial neural network'. They varied the filler contents in the ranges of 0–15 wt.%. The experimental study of dry sliding wear test was conducted based on the design of experiments; they also reported some influencing factors that determine the wear rates of wood dust matrix composites. The result analysis established that the wear resistance of the virgin epoxy matrix was enhanced by incorporating the pine wood dust particles. Furthermore, the specific wear was influenced by varying the primary factors: (i) normal load, (ii) pinewood dust content and, (iii) sliding velocity. In another study (Mysiukiewicz and Sterzyński 2017), the coefficient of friction of wood flour filler was analyzed after mixed with two different types of resins: Polypropylene (PP) and poly (lactic acid) (PLA). The increasing addition of fillers from 0 to 30% within the PP and PLA, showed a decreasing trend in the coefficient of friction; it was ascribed to (i) hardening of polymer composites, (ii) variations in surface geometry or heat dissipations during the wear test.

Table 2 Examples of tribological studies of various filler reinforced composites with their findings

Type of composite	Type of wear study	Significant findings	References
• Mahogany/PP, Babool/PP, Mango/PP, Sheesham/PP • Fillers were varied in the order of 10, 15, and 20%, respectively	• Pin-on-disc	• Wear (g) was found to be decreased by increasing the filler content • The highest wear (10.5 g) was observed at mango/PP composites • The lowest wear (5.5 g) was observed at mango/PP and Sheesham/PP composites	(Kumar et al. 2019)
• Wood flour/polypropylene (PP) • Variation of wood flour content: 0–55%	• Dry sliding • ball-on-disc sliding tester	• 35% of wood flour loaded composites exhibited lesser in friction coefficient • 15% of wood flour loaded composites showed lesser in wear area (mm) • Roughness: 340.7 nm	(Ibrahim et al. 2019)
• Wood apple shell fillers • Variation of fillers from 5–20% (order of 5%)	Erosion test	• The impingement angle was varied from 30 °C to 90 °C • Irrespective of filler loading, the erosion wear (g/g) was found to be increased at 45 °C • Between 60–90 °C, there was no increase in erosion wear	(Shakuntala et al. 2014b)
• Phoenix dactylifera tree/glue epoxy rapid matrix; • Date stones fillers (DS), date flesh (DF)	• Two-body abrasion study Rotating cylindrical drum tester	• When increasing the content of DS, DF, the abrasive wear resistance was decreased for the former, and for the latter, it was increased	(Ruggiero et al. 2016)

Mogaji et al. (2017) incorporated wood sawdust with the polystyrene, whereas the filler content was varied from 5 to 30% (order of 5%). It was observed that the wear loss (%) was increased with increasing the wood dust content. Among the composites, 5% of sawdust loaded polystyrene matrix composites showed the best wear performance that the rest of the samples. Achebe et al. (2019) successfully developed a brake pad for automobile vehicles using cane wood/palm fruit/epoxy matrix composites, whereas 6.52%:3.48%:30% were selected among the various combinations. Further, the wear rate of these composites was found to be $4.13\frac{mg}{m}$.

Tasdemir (2017) examined the static and dynamic coefficient of friction of Olive pit powder/PP and Almond shell powder/PP composites by varying the load from 2 to 7 N ranges. From both types of composites, the filler content was varied from 10 wt.% to 40 wt.% (order of 10). The wear performance showed that the friction coefficient was increased with increasing the load and the filler content. It was ascribed to (i) the more fiber pull-out and (ii) poor adhesion between the filler and the matrix. Nabinejad et al. (2019) studied the wear performance of wood flour/ multi-walled carbon nanotubes/polyester matrix composites by varying the filler loadings ranging from 0 wt.% to 40 wt.%. The improved tribological behavior was noticed even at 10wt.% addition of wood flour content, which was attributed to the formation of a soft layer; also, it acted as self-lubricating material.

6 Conclusion

The wear behavior of wood fiber-based composites has been reviewed based on the thermoset matrix and thermoplastic matrix. Based on the reviewed works, the thermoset based polymers are being used for many applications compared to the thermoplastic-based polymers.

The wood polymer-based composites possessed many advantages: non-corrosive, biodegradable, high strength, high stiffness, low thermal conductivity and ease of production with recyclability, etc. The application of the wood polymer composites is (i) deck floors, (ii) rail fencings, (iii) doors, (iv) windows, (v) frames, and (vi) indoor furniture.

The wood fiber properties depend on many factors, such as wood, matrix, filler, fiber treatment, additives, and coupling agent. It is known that the treated fiber-reinforced composites could demonstrate enhanced wear characteristics than the untreated fiber composites. Though, there was a lack of wear-based studies reported on the fiber treatments.

The wear performance of polymeric matrix composites could have improved by incorporating wood fillers. Thus, fillers' usage has been increased in recent days due to the increasing price of oil and monomer prices.

Acknowledgements We hereby acknowledge and sincerely appreciate unalloyed supports from the management of the Kalasalingam Academy of Research and Education, Krishnankoil 626126, Tamilnadu, India. This research was also, supported by King Mongkut's University of Technology North Bangkok (KMUTNB), Thailand through Grant No. KMUTNB-64-KNOW-007.

Conflict of Interest None declared.

References

Achebe CH, Obika EN, Chukwuneke JL, Ani OE (2019) Optimisation of hybridised cane wood–palm fruit fibre frictional material. Proc Inst Mech Eng Part L J Mater Des Appl 233:2490–2497. https://doi.org/10.1177/1464420719863445

Araki K, Hamabe G, Sano Y et al (2016) Study of friction and abrasion properties of wood plastic composites. High Perform Optim Des Struct Mater II 166:403

Bahadur S (2000) The development of transfer layers and their role in polymer tribology. Wear 245:92–99. https://doi.org/10.1016/S0043-1648(00)00469-5

Beer P, Gogolewski P, Klimke J, Krell A (2007) Tribological Behaviour of Sub-micron Cutting-ceramics in Contact with Wood-based Materials. Tribol Lett 27:155–158. https://doi.org/10.1007/s11249-007-9212-2

Bledzki AK, Reihmane S, Gassan J (1998) Thermoplastics reinforced with wood fillers: a literature review. Polym-Plast Technol Eng 37:451–468

Borah JS, Kim DS (2016) Recent development in thermoplastic/wood composites and nanocomposites: a review. Korean J Chem Eng 33:3035–3049

Brostow W, Datashvili T, Jiang P, Miller H (2016) Recycled HDPE reinforced with sol–gel silica modified wood sawdust. Eur Polym J 76:28–39

Bunsell AR, Renard (2005) Fundamentals of Fibre Reinforced Composite Materials. CRC Press

Dale Ellis W (2000) Wood-polymer composites: review of processes and properties. Mol Cryst Liq Cryst Sci Technol Sect A Mol Cryst Liq Cryst 353:75–84. https://doi.org/10.1080/10587250008025649

Dwivedi UK, Chand N (2008) Influence of wood flour loading on tribological behavior of epoxy composites. Polym Compos 29:1189–1192. https://doi.org/10.1002/pc.20548

Friedrich K (2018) Polymer composites for tribological applications. Adv Ind Eng Polym Res 1:3–39. https://doi.org/10.1016/j.aiepr.2018.05.001

Friedrich K, Zhang Z, Schlarb AK (2005) Effects of various fillers on the sliding wear of polymer composites. Compos Sci Technol 65:2329–2343. https://doi.org/10.1016/j.compscitech.2005.05.028

Ibrahim MA, Hirayama T, Khalaf D (2019) An investigation into the tribological properties of wood flour reinforced polypropylene composites. Mater Res Express 7. https://doi.org/10.1088/2053-1591/ab600c

Jeamtrakull S, Kositchaiyong A, Markpin T et al (2012) Effects of wood constituents and content, and glass fiber reinforcement on wear behavior of wood/PVC composites. Compos Part B Eng 43:2721–2729

Jena H (2019) Study of Tribo-Performance and Application of Polymer Composite BT - Automotive Tribology. In: Bhattacharya S, Patel VK, Kumar V (eds) Katiyar JK. Springer Singapore, Singapore, pp 65–99

Jeong B, Park B-D (2019) Effect of molecular weight of urea–formaldehyde resins on their cure kinetics, interphase, penetration into wood, and adhesion in bonding wood. Wood Sci Technol 53:665–685. https://doi.org/10.1007/s00226-019-01092-1

Jintang G (2000) Tribochemical effects in formation of polymer transfer film. Wear 245:100–106. https://doi.org/10.1016/S0043-1648(00)00470-1

Joseph J, Munda PR, Kumar M et al (2020) Sustainable conducting polymer composites: study of mechanical and tribological properties of natural fiber reinforced PVA composites with carbon nanofillers. Polym Technol Mater 59:1088–1099

Karthik Babu NB, Ramesh T, Muthukumaran S (2020) Physical, tribological and viscoelastic behavior of machining wear debris powder reinforced epoxy composites. J Clean Prod 272:122786. https://doi.org/10.1016/j.jclepro.2020.122786

Karthikeyan S, Rajini N, Jawaid M et al (2017) A review on tribological properties of natural fiber based sustainable hybrid composite. Proc Inst Mech Eng Part J J Eng Tribol 231:1616–1634

Kim JK, Pal K (2010) Overview of wood-plastic composites and uses. In: recent advances in the processing of wood-plastic composites. Springer, pp 1–22

Kokta BV, Raj RG, Daneault C (1989) Use of wood flour as filler in polypropylene: studies on mechanical properties. Polym Plast Technol Eng 28:247–259

Koohestani B, Ganetri I, Yilmaz E (2017) Effects of silane modified minerals on mechanical, microstructural, thermal, and rheological properties of wood plastic composites. Compos Part B Eng 111:103–111. https://doi.org/10.1016/j.compositesb.2016.12.021

Kranthi G, Satapathy A (2010) Evaluation and prediction of wear response of pine wood dust filled epoxy composites using neural computation. Comput Mater Sci 49:609–614. https://doi.org/10.1016/j.commatsci.2010.06.001

Kumar S, Singh KK (2020) Tribological behaviour of fibre-reinforced thermoset polymer composites: A review. Proc Inst Mech Eng Part L J Mater Des Appl 1464420720941554. https://doi.org/10.1177/1464420720941554

Kumar S, Vedrtnam A, Pawar SJ (2019) Effect of wood dust type on mechanical properties, wear behavior, biodegradability, and resistance to natural weathering of wood-plastic composites. Front Struct Civ Eng 13:1446–1462. https://doi.org/10.1007/s11709-019-0568-9

Kumar TSM, Senthilkumar K, Chandrasekar M, et al (2020) Influence of fillers on the thermal and mechanical properties of biocomposites: an overview. In: Biofibers and biopolymers for biocomposites. Springer, pp 111–133

Milosevic M, Valášek P, Ruggiero A (2020) Tribology of natural fibers composite materials: an overview. Lubricants 8:42

Mogaji PB, Ayodeji SP, Olatise AD, Oladele IO (2017) Investigation of the properties and production of sawdust ceiling tile using polystyrene as a binder. African J Sci Technol Innov Dev 9:655–659. https://doi.org/10.1080/20421338.2017.1352158

Mysiukiewicz O, Sterzyński T (2017) Influence of water on tribological properties of wood-polymer composites. Arch Mech Technol Mater 37:79–84. https://doi.org/10.1515/amtm-2017-0013

Nabinejad O, Liew WYH, Debnath S et al (2019) Tribological behavior of unsaturated polyester hybrid composites containing wood flour and carbon nanotubes. SN Appl Sci 1:1–9. https://doi.org/10.1007/s42452-019-0792-x

Omrani E, Menezes PL, Rohatgi PK (2016) State of the art on tribological behavior of polymer matrix composites reinforced with natural fibers in the green materials world. Eng Sci Technol an Int J 19:717–736

Parikh HH, Gohil PP (2015) Tribology of fiber reinforced polymer matrix composites - a review. J Reinf Plast Compos 34:1340–1346. https://doi.org/10.1177/0731684415591199

Ranakoti L, Gupta MK, Rakesh PK (2019) Analysis of mechanical and tribological behavior of wood flour filled glass fiber reinforced epoxy composite. Mater Res Express 6: https://doi.org/10.1088/2053-1591/ab2375

Samyn P, Zsidai L (2017) Temperature effects on friction and wear of thermoset polyester fabric composites. Polym Plast Technol Eng 56:1003–1016. https://doi.org/10.1080/03602559.2016.1247281

Shakuntala O, Raghavendra G, Samir Kumar A (2014) Effect of filler loading on mechanical and tribological properties of wood apple shell reinforced epoxy composite. Adv Mater Sci Eng 2014. https://doi.org/10.1155/2014/538651

Singh AK, Siddhartha GP, Singh PK (2018) Evaluation of mechanical and erosive wear characteristics of TiO2 and ZnO filled bi-directional e-glass fiber based vinyl ester composites. SILICON 10:309–327. https://doi.org/10.1007/s12633-016-9447-3

Tasdemir M (2017) Polypropylene/olive pit & almond shell polymer composites: Wear and friction. IOP Conf Ser Mater Sci Eng 204. https://doi.org/10.1088/1757-899X/204/1/012015

Thiagamani SMK, Krishnasamy S, Siengchin S (2019) Challenges of biodegradable polymers: an environmental perspective. Appl Sci Eng Prog x:1. https://doi.org/10.14416/j.asep.2019.03.002

Tuntsev D V, Sattarova ZG, Galiev IM (2017) Multi-layer wood-polymer composite. In: Materials engineering and technologies for production and processing III. Trans Tech Publications Ltd, pp 47–52

Waßmann O, Ahmed SIU (2020) Slippery wood: low friction and low wear of modified beech wood. Tribol Lett 68

Xiang XH, Xiang DH, Fang W, Ma JL (2012) Friction and wear behavior of POM composites filled with LDPE and wood fibers. In: Advanced materials research. Trans Tech Publ, pp 293–296

Yin W, Liu Z, Tian P et al (2016) Tribological properties of wood as a cellular fiber-reinforced composite. Biotribology 5:67–73

Water Absorption and Swelling Behaviour of Wood Plastic Composites

Sabarish Radoor, Jasila Karayil, Jyothi Mannekote Shivanna, and Suchart Siengchin

Abstract Wood polymer composite (WPC) is an environmentally friendly and sustainable material which has been exploited in the filed such as building and construction marine, packaging, house wares, aerospace and automotive industry. In recent years, there is a rapid growth in the usage of WPC as it possesses low maintenance cost. The properties of WPC depend on the interaction between wood and polymer. WPC has characteristic water uptake and swelling properties. A knowledge of water absorption and swelling behaviour property of WPC is essential to tune the WPC for different application. Hence, in this chapter we give a brief outlook about the water absorption and swelling behaviour of wood plastic composites. The effect of processing method, fiber content, orientation, matrix type, wood content, coupling agent, crosslinking, immersion time and temperature on the water absorption and swelling of WPC is briefly described. This chapter also discuss different swelling test of wood polymer composite.

Keywords Wood plastic composites · Water absorption · Swelling · Coupling agent · Wood fiber

S. Radoor (✉) · S. Siengchin
Department of Mechanical and Process Engineering, The Sirindhorn International Thai-German Graduate School of Engineering (TGGS), King Mongkut's University of Technology North Bangkok, 1518 Wongsawang Road, Bangsue, Bangkok 10800, Thailand

J. Karayil
Department of Chemistry, Government Women's Polytechnic College, Calicut, Kerala, India

J. M. Shivanna
Center for Nano and Material Sciences, Jain University, Jain Global Campus, Bangalore 562112, Karnataka, India

S. Mavinkere Rangappa et al. (eds.), *Wood Polymer Composites*, Composites Science and Technology, https://doi.org/10.1007/978-981-16-1606-8_10

1 Introduction

The demand for cheap and environmental-friendly material is increasing day by day. So, it is not surprising that in the recent years there is significant interest in the development of novel composite material using renewable resources (Bismarck et al. 2006). Wood is a cheap and abundantly available material which is widely used in the construction of several household and industrial materials. Its salient features include low density, specific mechanical properties, easy processability and is ability to act as good reinforcement for polymer composite (Harper and Wolcott 2004; Li and Wolcott 2004; Singh and Mohanty 2007). Therefore, in the last few years, wood has been used to reinforce polymer matrix and the composite so generated is known by the name wood plastic composite (WPC). The wood used to develop WPC will be either in the form of short fibers, flour or particulates. The thermoplastics selected to manufacture WPC should possess soften temperature less than the thermal degradation of wood (Sliwa 2012; Sudár 2014). Examples for such polymers are PS (poly styrene), LDPE (low density polyethylene), HDPE (high density polyethylene), PP (polypropylene) etc. The development of WPC can be considered as a cleaner technology as it employed naturally available raw material, wood. Moreover, report shows that recycled polymer could be used for generating WPC (Gramlich et al. 2006; Vardai et al. 2019; Zhang et al. 2019). WPC has widespread application and have been used in construction of furniture, automotive parts, transportation materials and so on (Renner et al. 2010). However, it suffers from several drawbacks which limit its application. One of the major drawbacks of WPC is its high moisture absorption. The components in wood such as cellulose, hemicellulose and lignin are responsible for enhancing the water uptake capacity of the WPC (Hosoya et al. 2007). The non-uniform dispersion of wood fillers in the polymer matrix is another problem which is encountered while developing WPC (Medupin 2013). The poor dispersion of wood in the matrix will negatively affect the overall performance of the composite. WPC also suffers from compatibility issues. As wood is polar in nature, it has a poor compatibility with non-polar polymer. The weak incompatibly between polymer and wood adversely effect the properties of WPC especially its mechanical properties (Ashori et al. 2014). The interfacial interaction between polymer and wood are generally improved by chemical treatment (alkali treatment, silane, mercerization, acetylation etc.), plasma treatment or by introducing compatibilizers or coupling agents (maleic anhydride, acrylic anhydride etc.) (Borysiak 2013; Donath 2006; Roussel et al. 2001; Rowell 2006; Salehian 2013).Wood is easily degraded when exposed to environmental conditions such as radiation, moisture etc. Hence, wood polymer composite is susceptible to weathering and necessary treatment method should be implemented to enhance the life of WPC (Catto 2016).

2 Water Absorption of Wood Plastic Composite

Water absorption is one of the characteristic features of WPC; a proper knowledge of water absorption is essential to design WPC for different applications. Previous studies show that water uptake and swelling property of WPC affects the overall performance of the composite (Mishra and Naik 1998; Panthapulakkal and Sain 2016; Yang et al. 2006). Water absorption adversely affect the interfacial adhesion between wood and the polymer (Dányádi et al. 2010). It also creates favourable environment for the growth of microorganism thus making WPC susceptible to fungal attack (Chen et al. 2019). Hence insecticides and fungicides are often incorporated in WPC to protect the composite form microorganism. Diffusion is considered to be the main mechanism of water absorption in WPC. The water molecules exist mainly in cell wall, interfacial gaps and lumen of cell (Li et al. 2012).

Several researchers have evaluated the water absorption behaviour of WPC (Mrad 2018; Temiz et al. 2006). Butylina et al. (2010) compared the water absorption studies of wood-PLA (polylactic acid) and wood-PP (polypropylene) composite. The result shows that wood-PLA composite has high water absorption than wood-PP composites and is attributed to the low density and high-water affinity of PLA. Composite with low density have high water absorption and thickness swelling. Owing to its high-water absorption, wood-PLA composite exhibits low mechanical properties. Hemicellulose is an amorphous polymer and is one of the constituents of cellulosic fiber. It is hydrophilic in nature and is mainly responsible for the high-water absorption of wood fiber. Several extraction techniques have been suggested to remove hemicelluloses from wood fiber. Gwon et al. (2010) employed sodium hydroxide treatment to remove hemicellulose from wood fiber. The sodium hydroxide treated composite displayed low water absorption which could be due to the removal of hemicellulose from the surface of wood. Above observation was confirmed by SEM analysis (Fig. 1). The smooth surface of sodium hydroxide treated sample indicates the removal of hemicellulose and lignin from the wood fiber. The authors also monitored the effect of talc treatment on the water absorption behaviour of the composite. Their studies show that the talc treatment enhances the water absorption and is ascribed to the hydrophilic nature of the talc. Meanwhile, Hosseinaei et al. (2012) employed hot water extraction method to remove hemicellulose from the surface of wood. Removal of hemicelluloses from wood enhances its hydrophobic character which in turn improves its adhesion with the polymer matrix. Thus, the treated composite restricts the penetration of water molecule and therefore has low water absorption value. In their work, authors also report the effect of treatment temperature on the water absorption behaviour of the composite. The experimental result suggests that treatment temperature highly influences the water absorption value of the composite. The composite treated at high temperature (170 °C) displayed low water absorption (Fig. 2).

Islam et al. (2013) employed acrylonitrile (AN) treatment to improve the physical and mechanical properties of WPC. The water absorption test revealed that

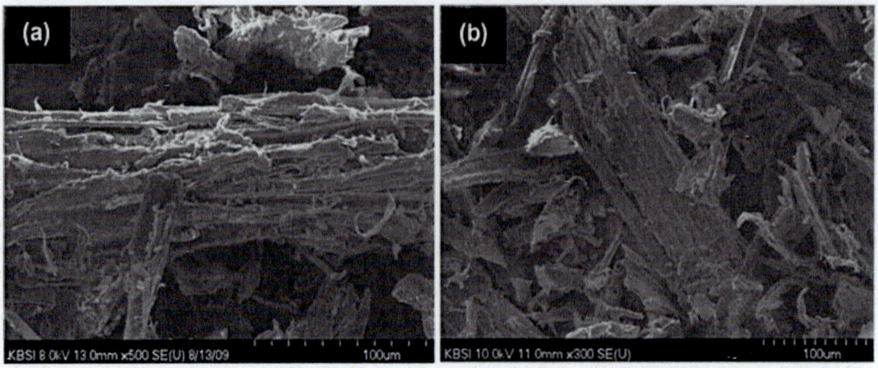

Fig. 1 SEM photographs of fibers: (a) without NaOH treatment and (b) with NaOH treatment [Reproduced with permission from Elsevier, License Number: 4854271145944 (Gwon et al. 2010)]

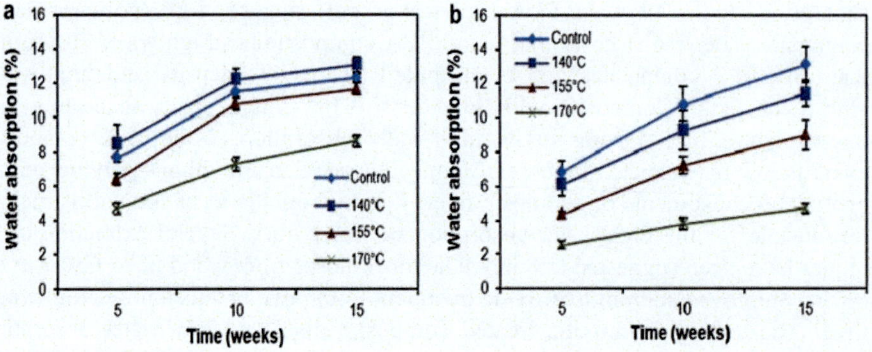

Fig. 2 Water absorption property of WPCs: (a) uncoupled samples, and (b) coupled samples [Reproduced with permission from Elsevier, License Number: 4854271433340 (Hosseinaei et al. 2012)]

AN treated composite has low water absorption than untreated one. This is due to chemical interaction between AN and the hydroxyl group of wood. Wei et al. (2013) investigated the effect of esterification on the properties (weathering, water absorption, thermal properties, tensile property etc.) of WPC. During esterification reaction, the –OH group in wood fiber is replaced by an alkyl group. This reduced the hydrophilic character of wood and consequently treated WPC absorb less water than untreated sample. The increasing order of water absorption is as follows: untreated > acetylated > propionated > benzoated. This is ascribed to the high hydrophobic character of benzoated sample. The enhancement in hydrophobicity improves the compatibility of wood fiber with polymer matrix, which is reflected in the mechanical parameter (tensile and flexural strength).

Coupling agent is usually added to control the water absorption behaviour of WPC. Maleic anhydride is one of the frequently used coupling agents. Patil et al. (2000) developed WPC using different agro waste such as wheat straw, cane bagasse and teak wood sawdust. In their work, they utilized maleic anhydride (MA) treatment to modify wood fiber. The water absorption of MA treated and untreated composite was compared. The wheat straw containing WPC has highest amount of water absorption, while cane bagasse/teak wood sawdust displayed lowest. This may be due to difference in the arrangement of chemical structure. Due to the loose structure, wheat straw is prone to absorb water. However, the compact structure of cane bagasse and teak wood sawdust restrict the entry of water into the WPC. Meanwhile, Nourbakhsh et al. (2009) used PE-g-MA (polyethylene grafted maleic anhydride) to improve properties of WPC. They observed that 2% and 4% PE-g-MA treated poplar fiber has low water absorption than untreated WPC. Mishra and co-worker (2004) reported the water absorption of MA treated novolac/ natural fiber (banana, hemp, and sisal fibers). Their studies show that MA treated composite has low affinity for water. Among the different fiber they studied, the hemp reinforced composite displayed high-water absorption both in treated and untreated form. Low water absorption was noted for sisal fiber. The authors concluded that MA treated composite has low water absorption, irrespective of the fiber they chose. Espert et al. (2004) suggested EVA (Ethylene–vinyl acetate) as one of the good coupling agent to solve the compatibility issue in WPC. EVA is the polymer of acetate with one part bearing a hydrophilic acetate group and the other part a hydrophobic polymer chain. This enables it to connect with both natural fiber as well as with the polymer. Thus, EVA treated composite displayed strong interfacial compatibility. They also observed that upon increasing the fiber content, the capacity to absorb water also increases. However, the matrix type has no differentiating effect on the water absorption process. The effect of absorption temperature was also evaluated and the analysis shows that high temperature disfavour water absorption. Thus, the authors conclude that water absorption of WPC is governed by fiber content, matrix type and temperature. Migneault et al. (2009) investigated the effect of processing method and fiber diameter on the performance of WPC. Their studies show that IM (injection moulding) composite have low water absorption than extruded composite. This could be due to the difference in the surface quality of the composite with processing method. The exposed fiber of extruded composite promotes the diffusion of water whereas the composite processed by IM treatment has hydrophobic surface thereby restricting the entry of water molecule into the composite. The fiber orientation and its length/diameter ratio are also found to play a significant role in controlling the water absorption process. Long fiber tends to increase the voids within the composite and thus enhance the high-water absorption (Fig. 3). Turku et al. (2017) utilized recycled plastic waste and spruce wood flour particle to develop WPC. Highest water absorption was observed for composite made using municipal mixed plastic waste, probably due to the weak fiber/matrix interaction. This statement was confirmed by SEM analysis (Figs. 4 and 5). Composite with poor interfacial bonding displayed several voids and interfacial gaps, which is favourable for water absorption.

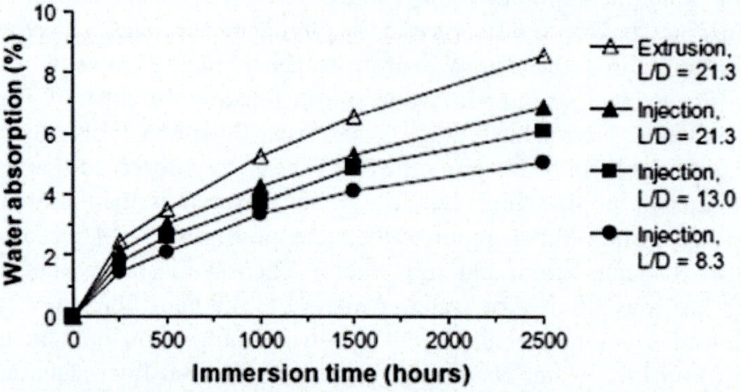

Fig. 3 Water absorption with different time for extruded and injection molded WPC [Reproduced with permission from Elsevier, License Number: 4854240695207 (Migneault et al. 2009)]

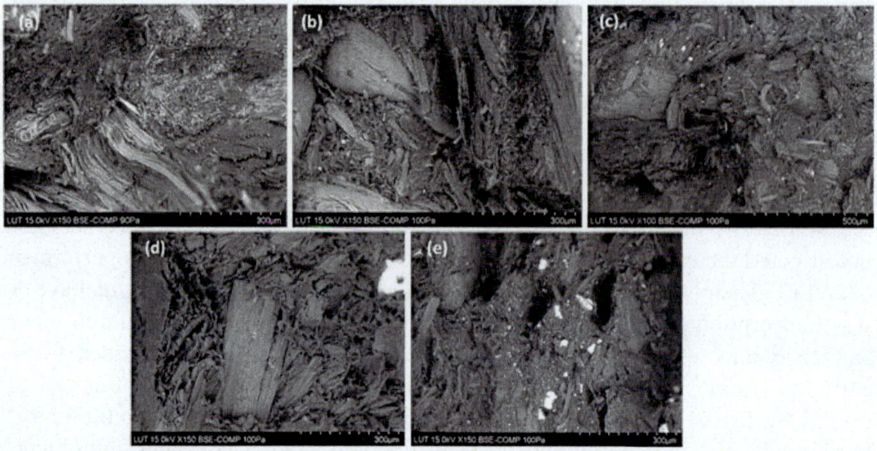

Fig. 4 SEM micrographs of the composite samples: (a) reference; (b) PB_C2; (c) PB_C5; (d) PB_C9; (e) PB_C10. (PB_C2 and PB_C5-construction plastic waste, PB_C9-flexible packaging material, PB_C10 - municipal mixed plastic waste) [Reproduced with permission from Elsevier, License Number: 4854241158976 (Turku et al. 2017)]

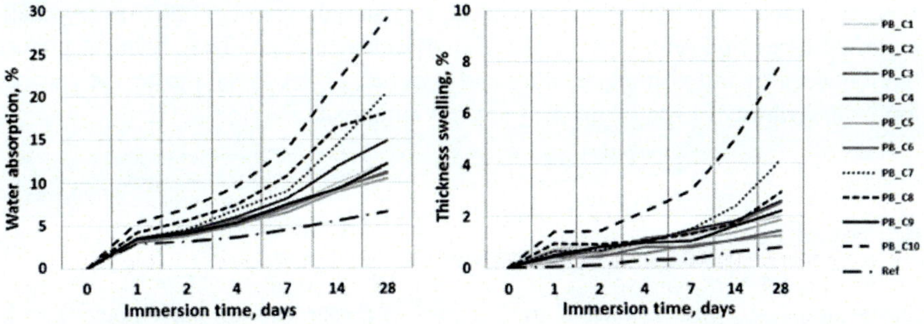

Fig. 5 Water absorption of composite samples: (PB_C1 -PB_C9-construction plastic waste, PB_C7 consisted of hard non-packaging material PB_C9-flexible packaging material, PB_C10 - municipal mixed plastic waste) [Reproduced with permission from Elsevier, License Number: 4854241158976 (Turku et al. 2017)]

Najafi et al. (2016) studies show water absorption property of WPC obtained from sawdust and recycled plastics (HDPE and PP) could be enhanced by increasing the proportion of plastics. The authors also compared the water adsorption of composite obtained from recycled plastics and virgin plastics. It was observed that composite obtained from recycled plastics has high water absorption than those obtained from virgin plastics. This is attributed to presence of voids and gaps in the interphase of recycled plastic and fiber. The authors thus suggested that the poor adhesion between recycled plastics and wood fiber could be improved by incorporating a compatibilizer. Biochor (BC) is a hydrophobic material which is obtained as by-product during the synthesis of gas or bio-oil. Devallance et al. (2016) for the first time introduce biochar into PP/wood fiber composite and evaluated its mechanical, water adsorption and thermal properties. Based on the experimental studies, they report that composite with high amount biochor resist the water absorption and thus displayed lowest water absorption and thickness swelling percentage. Their observation is strengthened by SEM observation. The poor interfacial bonding in composite without BC is evident from the SEM analysis. Incorporation of BC into the composite reduces the gap between BC particle and PP and thus decreases its affinity for water. The authors thus suggest BC as an alternative for wood in composite material. Zhang et al. (2011) studied the effect of different compatibilizer on the properties of PP/wood flour composite. The composite treated with compatibilizer possess superior moisture resistance than untreated sample. This could be due to the reduction in the number of exposed hydroxyl group in treated composite. SEM analysis strongly supports the water absorption behaviour. Aref et al. (2013) pre-treated wood fiber using hot or cold water and employed it to fabricate a WPC. The water uptake of the composite made using pre-treated wood is much lower than that of untreated one. The pre-treatment improves the compatibility between the fiber and the matrix and thereby reduce the hygroscopicity of the composite. Devi et al. (Council of Scientific & Industrial Research (India). Publications & Information Directorate. & Indian National

Science Academy., 1994; "Indian journal of materials science," 2013) chemically modified pinewood with styrene and glycidyl methacrylate. The chemical interaction between hydroxyl group of fiber and styrene - GMA, is said to be the cause of low water affinity of the chemically treated composite.

3 Swelling Test

Dimensional stability, is one of the significant constraints for any material to be used for its definite applications. In case of wood research and wood aimed applications, dimensional stability is attributed to its shape, size and swelling behaviour adopted by wood at specified environment. However, the hygroscopic nature of wood upshot the dimensional instability due to its swelling and shrinking nature (Rowell 2006; Williams 2005). Wood mainly constitutes of lignin, cellulose and hemicellulos having enormous amount of free hydroxyl functionalities. These functionalities are known to absorb the water bestowing to climatic conditions inducing stress in surface and internal physique. The induced stress causes cracks and swelling of wood (Evans et al. 2008; Evans 2007; Feist 1983; Jebrane et al. 2009). Measuring alteration in dimensional stability is crucial and knowledge on the assessments for the same is equally important to rectify the issues associated with composite. Despite of availability of numerous tests, they suffer a consistency, yielding compromised results depending on availability of groundwork facilities, type of woods, time and environmental conditions. Mathematically, the moisture content (MC) in a wood either in free vapour form or in bound water form, it could be calculated by given Eq. (1).

$$MC = (Mw/Md - 1) \tag{1}$$

Where Md and Mw are masses of wood in dry and wet conditions respectively. Present section of the chapter deals with the availability of various investigations to measure the dimensional stability corresponding to swelling nature of wood with an aid of two very basic principles, dealing with swelling due to water and due to water vapour.

3.1 *Tests for Wood Swelling Caused by Direct Contact of Water*

As the title indicates, the wood was allowed for swelling with direct water contact. Wherein, it involves two major categories. 1) Allowing wood and water in contact for a longer period i.e, more than 24 h (Hill 2007, 2011); 2) Exposing wood to water for shorter periods (Sargent 2019; Virta et al. 2006). The first category is basically for the applications involving maximum contact of water, like outdoor applications. Widely used method for the first category is water saturation test and is tranquil, do not demand special laboratory equipment (Skaar 2012). It include repetitive water soaking and calculation of anti-shrinking effect (ASE) (Deka and Saikia 2000; Elvy et al. 1995). The test will be repeated with standard references to confirm the swelling behaviour. However, the results are not comparable with other materials since the set of standards are not available yet. Also, when used for outdoor application where water is in direct contact, nor the fully oven dried wood or completely saturated wood serves the authenticate values due to non-linearity in the swelling behaviour but emphasis the predicted values (Ma et al. 2005; Noack et al. 1973). As of now, there are no standard test protocol for the water soak test and the variations are many. One can vary the size from 18 × 40 × 15 mm (Bak and Németh 2012)to as small as 20 × 20 × 5 mm (Hill et al. 2004), time lengths for shorter and longer durations, variations in soaking cycles (Rowell and Ellis 2007) or soaking them to reach the constant dimensions (Šefc et al. 2009), while presenting the values as the ratio of changes in tangential and radial dimensions (Ohmae et al. 2002). Along with these parameters, one can note the changes in an anisotropy affecting the dimensional stability of wood (Rafsanjani et al. 2014; Rowell and Dickerson 2014).

In the past few decades' ago, the rewetting or re-saturation of wood was not considered much and one used to simply re-wet the wood until it reaches beyond fibre saturation point. This method was widely used and considered to be greener approach, imposing no effect on the stability studies (Naderi and Hernandez 2007). However, the variable behaviours of woods dried at different conditions were noted (Stevens 1963). In 1965, Covington found the reduced mechanical strength of the re-wetted wood samples from never dried ones (Naderi and Hernandez 2007). With enormous observations and investigations another study reported a creation of permanent cracks and creeps on re-wetting wood sample in 14% of initial moisture content (Hernández and Bizoň 2007).Moreover, two stage drying process were all encountered by some of the studies (Hill 2007; Liu and Simpson 1999), the transport diffusion co-efficient were accurate for red oak tree for multiple tests (Liu and Simpson 1999). Whatever the soaking method and number of cycles, many findings suggested that, very first cycle is not representative as it always results is higher ASE values than the consequent sub-cycles (Čermák et al. 2015; Rowell 2007).

A method of sample preparation as well as techniques of measurements play a vital role. Grain direction variation for different grains results in uncommon ASE values. It is important to choose the samples with no defects and having growth rings perfectly parallel to at least one edge of the target sample (Lanvermann et al. 2014; Rafsanjani et al. 2012). Direction of specimen is normally measured in calipers. After water adsorption, the volume of the specimen would be calculated by displacement of water (Jankowska et al. 2017) and could be compared with the ones taken individually using calipers with reduced errors (Sargent 2019).Specimens having no dimensional accuracy are recommended to dip in paraffin first, and then measuring the water displaced subsequently, however, this step can be applied only at the end after all measurements (Jankowska et al. 2017).

Moving on to swelling test involving short term soaking, Swellometer tests are commonly adopted (Feist 1979; Lukowsky et al. 1997; Williams 1999). In this step, generally the thin wood pieces will be equilibrated at relative humidity of 65%. Then the pieces will be soaked in water for shorter time (about 30 min). Changes in tangential direction are usually noted down (Murínová et al. 2014). Many times, a flatbed micrometer will be used to measure the swelling (Rowell et al. 2007; Youngquist et al. 2007). Window and door manufactures from United States had found another method which is similar to above method but simpler one (Sargent 2019). Where wood pieces were slightly lesser in thickness and specimens were equilibrated at higher temperatures and humidity. Another group used Wilhelmy plate method to study the swelling behaviour of veeners (Son and Gardner 2007). The study uses non swelling liquid probe octane for the measurement. The study concluded the higher efficiency of Wilhelmy plate method for non-rectangular specimens over calliper methods due to the measured changes in micrometer level. Computational study involving comparison of standard swellometers was performed by Oh et al. (Oh 2000). Any of the adopted measures wood depict the swelling behaviour when water and wood are in contact. However, wood swelling due to humid air is not negligible since, the wood tends to decay due to humid air.

3.2 *Tests for Wood Swelling due to Humidity*

Humidity is another factor, gains significant due to unavoidable climatic changes and is applicable for all kinds of wood (indoor and outdoor). Manly, there are two types of testing for this category. Equilibrium test involving the longer period, where samples are allowed to reach the equilibrium and come to specified weight at various humidity conditions (Gaff and Gašparík 2013; Sargent 2019). The other one is the shorter one where samples are kept under humidity for a set of periods and the changes on swelling are measured before and after exposure. In the farmer method, wood samples will be kept in humidity environment to obtained the equilibrium in mass and then the dimensions are measured. Again, the equilibrium is attained at second stage humidity exposure and dimensions are noted. The cycle would be repeated for many times with higher and lower levels of humidity. The step also involves an oven dry cycle either in the beginning or the end of the overall test (Bekhta and Niemz 2003; Lande et al. 2004). Watching the weights in every cycle is most important in all these tests. These are time dependent and for some samples it goes up to few months to attain the equilibrium (Barkas 1949; Ding et al. 2011). Conditions for the humidity test also vary as in water swelling test. The temperature conditions start from 20 to 30 °C (Choong and Achmadi 2007; Ding et al. 2011). Sometimes a single study involves numerous humidity conditions for the relative comparison (31, 44).

Unlike, water swelling, humidity swelling possesses a major drawback. As humidity related changes occurs eventually under humid environmental conditions, the process is very slow and hence understanding the behavioural changes of wood is very important. These tests suffer a lack of standard test procedure and prolonged time. Short term humidity exposure suffers some of the contradictions. One of the work reports the time taken by thermally treated wood was longer than the unmodified one (Pfriem 2011). Whereas, the other work reports equal time taken for both woods (Bak and Németh 2012). It is also noted that, exposure to humidity even for a shorter time induce loads of changes and hence study on short-term exposure response is stressed out. But unfortunately, vey less number of tests are available for the same. On closing note, it is understood that, all together four types of tests are available. Each test possesses few drawbacks. Performances are usually compared between modified and unmodified ones; fast and repetitive checks are ideal. The standardization of test methods is still lacking and future could focus on long term tests having both humidity and water contacts.

4 Swelling Behaviour of Wood Polymer Composites

The increase in demand and supply of economic chain is challenging the traditional building methods and materials. Various places, wherein a wood alone or a polymer alone was used are getting replaced by fibre reinforced polymers. Wood/polymer composites offer a high strength, low weight and great economic value. Some of the specified advantages depending on polymer and wood, such as anticorrosion, water-proof, fire retarding and many more are the interesting factors of wood/ polymer composites (Yousif and El-Tayeb 2007). Low density, hazardless or minimum hazardous, biodegradability, relatively high tensile strengths and flexural modulus and abundance are some more attractive and catchy advantages of wood polymer composites (Wambua et al. 2003). However, there is raise in the concern about hydrophilic nature of natural fibres or woods, which come out with a problem of dimensional stability. The moisture accumulation lead to degradation of wood also even corrosion where wood is contact with metals (Schmidt 2006; Zelinka and Stone 2011). These complications could be avoided by keeping the moisture content of wood below 15% (Jakes et al. 2013). Various modifications have been performed with reference to wood and polymer to increase the stability with respect to swelling behaviours. People have used coupling agent like maleic anhydride modified polypropylene (MAPP) to improve the stability of wood polymer composites, however, there is a limit on the performance due to water uptake of wood and its swelling (Kazemi Najafi et al. 2008).

The best way to improve the compatibility between polymer and wood is the use of block copolymer compatibilizers. The principle has come purely from two copolymers line inducing the miscibility between two polymers (Sjoerdsma et al. 1980). However, this case could not achieve the complete miscibility but ensure the reduced interfacial energy and the better compatibility. Some of the most commonly used copolymers in wood polymer composites are styrene ethylene/butylene styrene (SEBS), MAPP, ionomer polymer. Low-density polyethylene – wood flour (LDPE–WF) composite system has utilized (maleic-anhydride) MA grafted SEBS copolymer. The location of copolymer was analysed and found to present at the interface between PE and wood matrix (Oksman et al. 1998). Y. Xu studied natural fiber reinforced PVC composites with same copolymer (Xu et al. 2008). The nonlinearity in the swelling behaviour of different test specimens was observed with varying weight percentages. Kamal B et al., proposed a composite material of Pinus radiata WF, recycled and virgin polyolefin with HDPE and PP (Adhikary et al. 2008). Water adsorption for composites increased with wood content and immersion time irrespective of type and form of the polymers. Incorporation of coupling agents like MAPP reduced the swelling behaviour and showed linear relationship. Composites of bamboo and natural rubber gave a reduced swelling ability compared to bamboo alone (Thygesen et al. 2010). Phenol formaldehyde and hexamethylene tetramide were used as bonding agents and the increased concentration of them has reduced the swelling behaviour of the composites. Oil palm trunks with phenol formaldehyde and urea formaldehyde were made composites with resin (North Carolina State University. Department of Wood and Paper Science.). The dimensional stability of the composite with urea formaldehyde was very less comparatively with the adsorption of more amount of water. Moisture sorption ability for maple, red oak and southern pine were tested by saturating with acrylic and methacrylic monomers (Mitchell et al. 2010). Few monomers penetrated deeply and gave swelling range of 2 to 21% at 105 °C emphasizing the importance of penetration of additive into wood cells walls. Polymers suffer hydrostatic swelling pressure during swelling of them due to mechanical constraints (Barkas 1942). With increasing in swelling, the force to push against the mechanical constraints will also increase and there exists a pressure. Chemical or physical cross linking as mentioned above or molecular entanglements would serve as inherent for such mechanical constraints at molecular scale. Overall, the swelling in wood polymer composites is decided not just by the addition of monomers, coupling agents, resins or polymers but by the penetration of those into wood cells walls. With these, the potential of mechanical constraints is also vital. However, the results vary from different wood compositions and composite materials with their concentration as well as time of contact.

5 Conclusion and Future Perspective

Wood is a promising green material to reinforce several polymer composites. Wood polymer composite is considered as an alternative for conventional polymer composite, which often contains toxic chemicals. Water absorption and swelling are one of the characteristic features of wood polymer composite and a knowledge of this parameter will help in the design of composites for various applications. This chapter deals with the water absorption and swelling behaviour of wood polymer composite. A brief literature review of related topic is discussed in the current chapter. Based on the light of previous reports, we can conclude that the wood absorption and swelling affects the overall performance of wood fiber composite.

These properties could be modified by several chemical and physical treatments, introduction of coupling agents, compatibilizer etc. The length, diameter and orientation of wood fiber are also found to play a major role in deciding the water absorption and swelling behaviour of wood polymer composite. In addition to this, immersion time and processing methods also affects the water absorption and swelling behaviour of wood fiber composites.

Acknowledgements The authors are thankful for the financial support by the King Mongkut's University of Technology North Bangkok (KMUTNB), Thailand and grant funded the Post-Doctoral scholarship Thailand (Grant No. KMUTNB-63-Post-03 to SR) and (Grant No. KMUTNB-BasicR-64-16).

References

Adhikary KB, Pang S, Staiger MP (2008) Long-term moisture absorption and thickness swelling behaviour of recycled thermoplastics reinforced with Pinus radiata sawdust. Chem Eng J 142 (2):190–198. https://doi.org/10.1016/j.cej.2007.11.024

Aref I, Nasser R, Ali I, Al-Mefarrej H, Al-Zahrani S (2013) Effects of aqueous extraction on the performance and properties of polypropylene/wood composites from Phoenix dactylifera and Acacia tortilis wood. J Reinf Plast Compos 32(7):476–489. https://doi.org/10.1177/0731684412454462

Ashori A, Babaee M, Jonoobi M, Hamzeh Y (2014) Solvent-free acetylation of cellulose nanofibers for improving compatibility and dispersion. Carbohyd Polym 102:369–375. https://doi.org/10.1016/j.carbpol.2013.11.067

Bak M, Németh R (2012) Changes in swelling properties and moisture uptake rate of oil-heat-treated poplar (Populus × euramericana cv. Pannonia) wood. BioResources 7(7):5128–5137

Barkas WW (1942) Wood water relationships—VII. Swelling pressure and sorption hysteresis in gels. Trans. Faraday Soc. 38(0):194–209. https://doi.org/10.1039/tf9423800194

Barkas WW (1949) The swelling of wood under stress

Bekhta P, Niemz P (2003) Effect of high temperature on the change in color, dimensional stability and mechanical properties of spruce wood. Holzforschung 57(5):539–546

Bismarck A, Baltazar-Y-Jimenez A, Sarikakis K (2006) Green composites as panacea? socio-economic aspects of green materials. Environ Dev Sustain 8(3):445–463. https://doi.org/10.1007/s10668-005-8506-5

Borysiak S (2013) Fundamental studies on lignocellulose/polypropylene composites: effects of wood treatment on the transcrystalline morphology and mechanical properties. J Appl Polym Sci 127(2):1309–1322. https://doi.org/10.1002/app.37651

Butylina S, Martikka O, Kärki T (2010) Comparison of water absorption and mechanical properties of wood–plastic composites made from polypropylene and polylactic acid. Wood Mat Sci Eng 5(3–4):220–228. https://doi.org/10.1080/17480272.2010.532233

Catto AL, Montagna LS, Almeida SH, Silveira RMB, Santana RMC (2016) Wood plastic composites weathering: Effects of compatibilization on biodegradation in soil and fungal decay. Int Biodeterior Biodegradation 109:11–22. https://doi.org/10.1016/j.ibiod.2015.12.026

Čermák P, Rautkari L, Horáček P, Saake B, Rademacher P, Sablík P (2015) Analysis of dimensional stability of thermally modified wood affected by re-wetting cycles. BioResources 10(2):3242–3253

Chen Y, Guo X, Peng Y, Cao J (2019) Water absorption and mold susceptibility of wood flour/polypropylene composites modified with silane-wax emulsions. Polym Compos 40(1):141–148. https://doi.org/10.1002/pc.24616

Choong ET, Achmadi SS (2007) Effect of extractives on moisture sorption and shrinkage in tropical woods. Wood Fiber Sci 23(2):185–196

Council of Scientific & Industrial Research (India). Publications & Information Directorate., & Indian National Science Academy (1994). Indian journal of engineering and materials sciences. In New Delhi: Publications & Information Directorate, CSIR in association with Indian National Science Academy

Dányádi L, Móczó J, Pukánszky B (2010) Effect of various surface modifications of wood flour on the properties of PP/wood composites. Compos A Appl Sci Manuf 41(2):199–206. https://doi.org/10.1016/j.compositesa.2009.10.008

Deka M, Saikia C (2000) Chemical modification of wood with thermosetting resin: effect on dimensional stability and strength property. Biores Technol 73(2):179–181

DeVallance DB, Oporto GS, Quigley P (2016) Investigation of hardwood biochar as a replacement for wood flour in wood–polypropylene composites. J Elastomers Plast 48(6):510–522. https://doi.org/10.1177/0095244315589655

Ding T, Gu L, Li T (2011) Influence of steam pressure on physical and mechanical properties of heat-treated Mongolian pine lumber. Eur J Wood Wood Prod 69(1):121–126

Donath S, Militz H, Mai C (2006) Creating water-repellent effects on wood by treatment with silanes. Holzforschung 60(1):40–46. https://doi.org/10.1515/hf.2006.008

Elvy SB, Dennis GR, Ng L-T (1995) Effects of coupling agent on the physical properties of wood-polymer composites. J Mater Process Technol 48(1–4):365–371

Espert A, Vilaplana F, Karlsson S (2004) Comparison of water absorption in natural cellulosic fibres from wood and one-year crops in polypropylene composites and its influence on their mechanical properties. Compos A Appl Sci Manuf 35(11):1267–1276. https://doi.org/10.1016/j.compositesa.2004.04.004

Evans P, Urban K, Chowdhury M (2008) Surface checking of wood is increased by photodegradation caused by ultraviolet and visible light. Wood Sci Technol 42(3):251–265

Evans PD (2007) A note on assessing the deterioration of thin wood veneers during weathering. Wood Fiber Sci 20(4):487–492

Feist WC (1979) Protection of wood surfaces with chromium trioxide, vol 339. Forest Products Laboratory, Forest Service, US Department of Agriculture

Feist WC (1983) Weathering and protection of wood

Gaff M, Gašparík M (2013) Shrinkage and stability of thermo-mechanically modified aspen wood. BioResources 8(1):1136–1146

Gramlich WM, Gardner DJ, Neivandt DJ (2006) Surface treatments of wood–plastic composites (WPCs) to improve adhesion. J Adhes Sci Technol 20(16):1873–1887. https://doi.org/10.1163/156856106779116623

Gwon JG, Lee SY, Chun SJ, Doh GH, Kim JH (2010) Effects of chemical treatments of hybrid fillers on the physical and thermal properties of wood plastic composites. Compos A Appl Sci Manuf 41(10):1491–1497. https://doi.org/10.1016/j.compositesa.2010.06.011

Harper D, Wolcott M (2004) Interaction between coupling agent and lubricants in wood–polypropylene composites. Compos A Appl Sci Manuf 35(3):385–394. https://doi.org/10.1016/j.compositesa.2003.09.018

Hernández RE, Bizoň M (2007) Changes in shrinkage and tangential compression strength of sugar maple below and above the fiber saturation point. Wood Fiber Sci 26(3):360–369

Hill CA (2007) Wood modification: chemical, thermal and other processes, vol 5. Wiley, Hoboken

Hill CA (2011) Wood modification: an update. BioResources 6(2):918–919

Hill CA, Farahani MM, Hale MD (2004) The use of organo alkoxysilane coupling agents for wood preservation. Holzforschung 58(3):316–325

Hosoya T, Kawamoto H, Saka S (2007) Cellulose–hemicellulose and cellulose–lignin interactions in wood pyrolysis at gasification temperature. J Anal Appl Pyrol 80(1):118–125. https://doi.org/10.1016/j.jaap.2007.01.006

Hosseinaei O, Wang S, Taylor AM, Kim J-W (2012) Effect of hemicellulose extraction on water absorption and mold susceptibility of wood–plastic composites. Int Biodeterior Biodegradation 71:29–35. https://doi.org/10.1016/j.ibiod.2011.12.015

Indian journal of materials science (2013). In (pp. 1 online resource). New York, NY: Hindawi Pub. Corp

Islam MS, Hamdan S, Hassan A, Talib ZA, Sobuz HR (2013) The chemical modification of tropical wood polymer composites. J Compos Mater 48(7):783–789. https://doi.org/10.1177/0021998313477894

Jakes JE, Plaza N, Stone DS, Hunt CG, Glass SV, Zelinka SL (2013). Mechanism of transport through wood cell wall polymers. J Forest Prod Ind 2(6):10–13

Jankowska A, Drożdżek M, Sarnowski P, Horodeński J (2017) Effect of extractives on the equilibrium moisture content and shrinkage of selected tropical wood species. BioResources 12 (1):597–607

Jebrane M, Sèbe G, Cullis I, Evans PD (2009) Photostabilisation of wood using aromatic vinyl esters. Polym Degrad Stab 94(2):151–157

Kazemi Najafi S, Kiaefar A, Tajvidi M (2008) Effect of bark flour content on the hygroscopic characteristics of wood–polypropylene composites. J Appl Polym Sci 110(5):3116–3120

Lande S, Westin M, Schneider M (2004) Properties of furfurylated wood. Scand J For Res 19 (sup5):22–30

Lanvermann C, Wittel FK, Niemz P (2014) Full-field moisture induced deformation in Norway spruce: intra-ring variation of transverse swelling. Eur J Wood Wood Prod 72(1):43–52

Li TQ, Wolcott MP (2004) Rheology of HDPE–wood composites. I. Steady state shear and extensional flow. Compos Part A Appl Sci Manuf 35(3):303–311 https://doi.org/10.1016/j.compositesa.2003.09.009

Li Y, Wu Q, Li J, Liu Y, Wang X-M, Liu Z (2012). Improvement of dimensional stability of wood via combination treatment: swelling with maleic anhydride and grafting with glycidyl methacrylate and methyl methacrylate. Holzforschung 66(1) https://doi.org/10.1515/hf.2011.123

Liu JY, Simpson WT (1999) Two-stage moisture diffusion in wood with constant transport coefficients. Drying Technol 17(1–2):258–267. https://doi.org/10.1080/07373939908917528

Lukowsky D, Peek R, Rapp A (1997) Water-based silicones on wood. International Research Group on Wood Protection, IRG/Wp, 97-30144

Ma E-N, Zhao G-J, Cao J-Z (2005) Hygroexpansion of wood during moisture adsorption and desorption processes. Forestr Stud China 7(2):43–46

Medupin R (2013) Mechanical properties of wood waste reinforced polymer matrix composites. Am Chem Sci J 3(4):507–513. https://doi.org/10.9734/acsj/2013/5637

Migneault S, Koubaa A, Erchiqui F, Chaala A, Englund K, Wolcott MP (2009) Effects of processing method and fiber size on the structure and properties of wood–plastic composites. Compos A Appl Sci Manuf 40(1):80–85. https://doi.org/10.1016/j.compositesa.2008.10.004

Mishra S, Naik JB (1998) Absorption of water at ambient temperature and steam in wood-polymer composites prepared from agrowaste and polystyrene. J Appl Polym Sci 68(4):681–686. https://doi.org/10.1002/(sici)1097-4628(19980425)68:4<681::Aid-app20>3.0.Co;2-t

Mishra S, Naik JB, Patil YP (2004) Studies on swelling properties of wood/polymer composites based on agro-waste and novolac. Adv Polym Technol 23(1):46–50. https://doi.org/10.1002/adv.10073

Mitchell MR, Link RE, Zelinka SL, Glass SV (2010) Water vapor sorption isotherms for southern pine treated with several waterborne preservatives. J Test Eval 38(4). https://doi.org/10.1520/jte102696

Mrad H, Alix S, Migneault S, Koubaa A, Perré P (2018) Numerical and experimental assessment of water absorption of wood-polymer composites. Measurement 115:197–203. https://doi.org/10.1016/j.measurement.2017.10.011

Murínová T, Daňková J, Mec P (2014). Utilization of wood modification for the purpose of moisture volume changes reduction. Paper presented at the Advanced Materials Research

Naderi N, Hernandez RE (2007) Effect of re-wetting treatment on the dimensional changes of sugar maple wood. Wood Fiber Sci 29(4):340–344

Najafi SK, Kiaefar A, Hamidina E, Tajvidi M (2016) Water absorption behavior of composites from sawdust and recycled plastics. J Reinf Plast Compos 26(3):341–348. https://doi.org/10.1177/0731684407072519

Noack D, Schwab E, Bartz A (1973) Characteristics for a judgment of the sorption and swelling behavior of wood. Wood Sci Technol 7(3):218–236

North Carolina State University. Department of Wood and Paper Science. Bioresources. In. Raleigh, N.C.: Dept. of Wood and Paper Science, College of Natural Resources

Nourbakhsh A, Ashori A (2009) Preparation and properties of wood plastic composites made of recycled high-density polyethylene. J Compos Mater 43(8):877–883. https://doi.org/10.1177/0021998309103089

Oh Y-S, Sites LS, Sellers J, Nicholas DD (2000) Computerized dynamic swellometer evaluation of oriented strand products. Forest Prod J 50(3):35

Ohmae K, Minato K, Norimoto M (2002) The analysis of dimensional changes due to chemical treatments and water soaking for hinoki (Chamaecyparis obtusa) wood. Holzforschung 56 (1):98–102

Oksman K, Lindberg H, Holmgren A (1998) The nature and location of SEBS-MA compatibilizer in polyethylene-wood flour composites. J Appl Polym Sci 69(1):201–209. https://doi.org/10.1002/(sici)1097-4628(19980705)69:1<201::Aid-app23>3.0.Co;2-0

Panthapulakkal S, Sain M (2016) Studies on the water absorption properties of short hemp—glass fiber hybrid polypropylene composites. J Compos Mater 41(15):1871–1883. https://doi.org/10.1177/0021998307069900

Patil YP, Gajre B, Dusane D, Chavan S, Mishra S (2000) Effect of maleic anhydride treatment on steam and water absorption of wood polymer composites prepared from wheat straw, cane bagasse, and teak wood sawdust using Novolac as matrix. J Appl Polym Sci 77(13):2963–2967. https://doi.org/10.1002/1097-4628(20000923)77:13<2963::Aid-app20>3.0.Co;2-0

Pfriem A (2011). Alteration of water absorption coefficient of spruce (Picea abies (L.) Karst.) due to thermal modification. Drvna industrija: Znanstveni časopis za pitanja drvne tehnologije 62 (4): 311–313

Rafsanjani A, Derome D, Wittel FK, Carmeliet J (2012) Computational up-scaling of anisotropic swelling and mechanical behavior of hierarchical cellular materials. Compos Sci Technol 72 (6):744–751

Rafsanjani A, Stiefel M, Jefimovs K, Mokso R, Derome D, Carmeliet J (2014) Hygroscopic swelling and shrinkage of latewood cell wall micropillars reveal ultrastructural anisotropy. J R Soc Interface 11(95):20140126

Renner K, Kenyó C, Móczó J, Pukánszky B (2010) Micromechanical deformation processes in PP/wood composites: particle characteristics, adhesion, mechanisms. Compos A Appl Sci Manuf 41(11):1653–1661. https://doi.org/10.1016/j.compositesa.2010.08.001

Roussel C, Marchetti V, Lemor A, Wozniak E, Loubinoux B, Gérardin P (2001) Chemical modification of wood by polyglycerol/maleic anhydride treatment. Holzforschung 55(1):57–62. https://doi.org/10.1515/hf.2001.009

Rowell RM (2006) Chemical modification of wood: a short review. Wood Mat Sci Eng 1(1):29–33. https://doi.org/10.1080/17480270600670923

Rowell RM (2007) Chemical modification of wood. Handbook of engineering biopolymers, homopolymers, blends, and composites. Hanser Gardner Publications, Inc., Cincinnati, pp 673–691

Rowell RM, Dickerson JP (2014) Acetylation of wood. In: Deterioration and protection of sustainable biomaterials pp 301–327. ACS Publications

Rowell RM, Ellis WD (2007) Determination of dimensional stabilization of wood using the water-soak method. Wood Fiber Sci 10(2):104–111

Rowell RM, Imamura Y, Kawai S, Norimoto M (2007) Dimensional stability, decay resistance, and mechanical properties of veneer-faced low-density particleboards made from acetylated wood. Wood Fiber Sci 21(1):67–79

Salehian P, Karimi K, Zilouei H, Jeihanipour A (2013) Improvement of biogas production from pine wood by alkali pretreatment. Fuel 106:484–489. https://doi.org/10.1016/j.fuel.2012.12.092

Sargent R (2019) Evaluating dimensional stability in solid wood: a review of current practice. J Wood Sci 65(1):1–11

Schmidt O (2006) Wood and tree fungi. Springer

Šefc B, Trajković J, Hasan M, Katović D, Bischof Vukušić S, Frančić M (2009) Dimensional stability of wood modified by citric acid using different catalysts. Drvna Industrija: Znanstveni Časopis Za Pitanja Drvne Tehnologije 60(1):23–26

Singh S, Mohanty A (2007) Wood fiber reinforced bacterial bioplastic composites: Fabrication and performance evaluation. Compos Sci Technol 67(9):1753–1763. https://doi.org/10.1016/j.compscitech.2006.11.009

Sjoerdsma SD, Dalmolen J, Bleijenberg ACAM, Heikens D (1980) Dynamic mechanical properties of polystyrene/low density polyethylene blends. Polymer 21(12):1469–1474. https://doi.org/10.1016/0032-3861(80)90149-4

Skaar C (2012) Wood-water relations. Springer Science & Business Media

Sliwa F, El Bounia N-E, Charrier F, Marin G, Malet F (2012) Mechanical and interfacial properties of wood and bio-based thermoplastic composite. Compos Sci Technol 72(14):1733–1740. https://doi.org/10.1016/j.compscitech.2012.07.002

Son J, Gardner DJ (2007) Dimensional stability measurements of thin wood veneers using the Wilhelmy plate technique. Wood Fiber Sci 36(1):98–106

Stevens W (1963) The transverse shrinkage of wood. For Prod J 13:386–389

Sudár A, Burgstaller C, Renner K, Móczó J, Pukánszky B (2014) Wood fiber reinforced multicomponent, multiphase PP composites: structure, properties, failure mechanism. Compos Sci Technol 103:106–112. https://doi.org/10.1016/j.compscitech.2014.08.018

Temiz A, Terziev N, Jacobsen B, Eikenes M (2006) Weathering, water absorption, and durability of silicon, acetylated, and heat-treated wood. J Appl Polym Sci 102(5):4506–4513. https://doi.org/10.1002/app.24878

Thygesen LG, Tang Engelund E, Hoffmeyer P (2010) Water sorption in wood and modified wood at high values of relative humidity. Part I: Results for untreated, acetylated, and furfurylated Norway spruce. Holzforschung 64(3). https://doi.org/10.1515/hf.2010.044

Turku I, Keskisaari A, Kärki T, Puurtinen A, Marttila P (2017) Characterization of wood plastic composites manufactured from recycled plastic blends. Compos Struct 161:469–476. https://doi.org/10.1016/j.compstruct.2016.11.073

Vardai R, Lummerstorfer T, Pretschuh C, Jerabek M, Gahleitner M, Pukanszky B, Renner K (2019) Impact modification of PP/wood composites: a new approach using hybrid fibers. Express Polym Lett 13(3):223–234. https://doi.org/10.3144/expresspolymlett.2019.19

Virta J, Koponen S, Absetz I (2006) Modelling moisture distribution in wooden cladding board as a result of short-term single-sided water soaking. Build Environ 41(11):1593–1599

Wambua P, Ivens J, Verpoest I (2003) Natural fibres: can they replace glass in fibre reinforced plastics? Compos Sci Technol 63(9):1259–1264

Wei L, McDonald AG, Freitag C, Morrell JJ (2013) Effects of wood fiber esterification on properties, weatherability and biodurability of wood plastic composites. Polym Degrad Stab 98 (7):1348–1361. https://doi.org/10.1016/j.polymdegradstab.2013.03.027

Williams RS (1999) Effect of water repellents on long-term durability of millwork treated with water-repellent preservatives. For Prod J 49(2):53

Williams RS (2005) Weathering of wood. In Handbook of wood chemistry and wood composites, 7, 139–185

Xu Y, Wu Q, Lei Y, Yao F, Zhang Q (2008) Natural fiber reinforced poly(vinyl chloride) composites: effect of fiber type and impact modifier. J Polym Environ 16(4):250–257. https://doi.org/10.1007/s10924-008-0113-8

Yang H-S, Kim H-J, Park H-J, Lee B-J, Hwang T-S (2006) Water absorption behavior and mechanical properties of lignocellulosic filler–polyolefin bio-composites. Compos Struct 72 (4):429–437. https://doi.org/10.1016/j.compstruct.2005.01.013

Youngquist JA, Krzysik A, Rowell RM (2007) Dimensional stability of acetylated aspen flakeboard. Wood Fiber Sci 18(1):90–98

Yousif B, El-Tayeb N (2007) Tribological evaluations of polyester composites considering three orientations of CSM glass fibres using BOR machine. Appl Compos Mater 14(2):105–116

Zelinka SL, Stone DS (2011) Corrosion of metals in wood: Comparing the results of a rapid test method with long-term exposure tests across six wood treatments. Corros Sci 53(5):1708–1714

Zhang J, Li Y, Xing D, Wang Q, Wang H, Koubaa A (2019) Reinforcement of continuous fibers for extruded wood-flour/HDPE composites: effects of fiber type and amount. Constr Build Mater 228:116718. https://doi.org/10.1016/j.conbuildmat.2019.116718

Zhang X, Yang H, Lin Z, Tan S (2011) Polypropylene hybrid composites filled by wood flour and short glass fiber: effect of compatibilizer on structure and properties. J Thermoplast Compos Mater 26(1):16–29. https://doi.org/10.1177/0892705711417030

Long Term Natural Weathering of PP Based WPCs: The Effect of TiO_2 on Selected Color, Physical, Mechanical, Morphological and Chemical Properties

Fatih Mengeloğlu and Vedat Çavuş

Abstract Long-term weathering performance of polypropylene based wood plastic composites manufactured using a combination of extrusion and injection molding processes were evaluated. The effect of micronized titanium dioxide (TiO_2) content (3-6-9 wt%) on the color changes (Spectrophotometer), selected mechanical properties, surface morphology (SEM), and surface chemical changes (FTIR) of the neat (PP) and composites samples (PP/WF) were investigated after 6, 12, 18 and 24 months of weathering in the City of Izmir, Turkey. Addition of TiO_2 has reduced ΔE^* and ΔL^* values from 35 to near 10 indicating higher resistance to weathering. Morphological study showed that distinguishable micro-cracks were present on all 12 months weathered samples (without TiO_2). Presence of TiO_2 has increased the number of cracks but they were less severe due to the reduced length and depth. After 24 months of weathering, large cracks and delamination of PP layer were present on the surfaces. However, this phenomenon was also severe and less pronounced in groups having TiO_2 compared to others. FTIR study showed that TiO_2 slightly reduced the intensity of the all bands associated with weathering process. Study showed that TiO_2 can be utilized to some extent in the formulations of PP and PP/WF against detrimental weathering effects.

Keywords Fourier transform infrared (FTIR) spectroscopy · Wood plastic composites · Titanium dioxide · Mechanical properties · Lightness and color properties

F. Mengeloğlu
Faculty of Forestry, Department of Forest Industry Engineering, Kahramanmaras Sutcu Imam University, Kahramanmaras, Turkey
e-mail: fmengelo@ksu.edu.tr

F. Mengeloğlu
Institute of Natural and Applied Sciences, Department of Materials Science and Engineering, Kahramanmaras Sutcu Imam University, Kahramanmaras, Turkey

V. Çavuş (✉)
Forest Faculty, Forest Industry Engineering, Izmir Katip Celebi University, Izmir, Turkey
e-mail: vedat.cavus@ikcu.edu.tr

S. Mavinkere Rangappa et al. (eds.), *Wood Polymer Composites*, Composites Science and Technology, https://doi.org/10.1007/978-981-16-1606-8_11

1 Introduction

Wood Plastic Composite (WPCs) is the general name of the composites having thermoplastic matrix and lignocellulosic-based reinforcer/filler. WPCs can be manufactured using equipment similar to ones utilized in thermoplastic polymer processing (Matuana and Heiden 2004; Hietala 2013; Çavuş and Mengeloğlu 2020). A wide range of the polymeric matrix such as polyethylene, polypropylene, and polyvinyl chloride (PVC) and filler like sawdust and wood fiber are utilized in various WPC products (Smith and Wolcott 2006; Behravesh et al. 2010). During production of WPCs, various additives are utilized in order to ease the processing conditions, to give an aesthetic appearance to composites, and to improve their resistance against physical, chemical and weathering effects (Mengeloğlu and Çavuş 2019). WPCs have been used for a wide range of application areas due to their desirable physical and mechanical properties. They are especially preferred for outside applications such as cladding, decking, fencing, paneling, and furniture (Çavuş 2019).

The main advantages of WPCs over wood material can be listed as good dimensional stability, durability against fungi and insects, and their low maintenance requirements during their service life (Peng et al. 2014). However, especially in outdoor applications, weathering is still a major disadvantage of these materials (Stark et al. 2004; Bouza et al. 2011; Butylina et al. 2012; Teaca 2013; Peng et al. 2014). In outdoor applications, the performance of WPCs is generally lower than other polymer based composites. Weather conditions such as changes in temperature (high and low), sunlight, moisture, oxygen, and other atmospheric contaminants play an imported role on the durability of WPCs (Stark and Rowlands 2003; Matuana et al. 2011; Peng et al. 2014; Krehula et al. 2014; Chen et al. 2016). The UV exposure of WPCs plays a key role on the changes of durability properties of WPCs (Stark et al. 2004). Polymeric materials can experience environmental degradation by physical, chemical, and biological processes or a combination of them. Several factors such as temperature (thermal degradation), air (oxidative degradation), moisture (hydrolytic degradation), microorganisms (biodegradation), light (photo degradation), high-energy radiation (UV, irradiation), chemical agents (corrosion), and mechanical stress might affect this process leading irreversible changes in the material (Shah et al. 2008; Tokiwa et al. 2009; Chen et al. 2016; Brebu 2020). These changes usually occur in two steps; disintegration and mineralization. Previous one covers the deterioration of appearance (e.g., color change), physical, and morphological properties (e.g., mass change, crystallinity) or mechanical properties of material (e.g., embrittlement, fragmentation), while later one includes the complete breakdown of material into water, carbon dioxide, and other simple inorganic compounds (Brebu 2020).

It is known that the chemical composition of wood material consist of cellulose, hemicellulose, lignin and extractives. They all have significant effect on physical and mechanical properties of wood. Wood polymer composites (WPC) is a general term that used for lignocellulosic material filled polymer composites. The hygroscopic nature of wood particles (swelling and absorbing water) plays a key role on micro crack generation at the surface of composites. This might contribute to the loss of modulus and strength properties in composites. Moisture also degrades the wood-plastic interface and decreases stress transfer efficiency from matrix to fiber (Chen et al. 2016). Lignin, which is, absorbs from 80 to 95% of UV light; found to be more sensitive to degradation than other wood components (Kallakas et al. 2015). UV radiation degrades the WPCs surface, further causing it to become cracked and flaky. Synergism between UV radiation and water also contribute to mechanical property losses of WPCs. It is reported that the surface of the weathered WPC is eroded through water spray, making new surfaces to degrade (Chen et al. 2016).

There are different options to quantify the effect of UV light on plastic and composite materials; environmental chamber, weatherometer, among others. Of course, they do not precisely matches the spectrum of natural UV light, but they provide all four major components of oxidative degradation of materials at conditions of natural exposure: UV light, heat, water, and oxygen. It should be taken under consideration that equivalency of artificial weathering and natural weathering period will be different for any geographical location and for any color of the material studied (Klyosov 2007).

There are several studies conducted to evaluate the effect of weathering on selected properties of WPCs. Most of these studies focussed on the changes of color, mechanical and chemical properties of weathered WPCs (Falk 2000; Stark 2007; Stark and Matuana 2007; Stark and Müeller 2008; Beg and Pickering 2008; Matuana et al. 2011; Krehula et al. 2014; Kallakas et al. 2015; Eshraghi et al. 2016; Catto et al. 2016; Cesarino et al. 2019; Kumar et al. 2019; Chao et al. 2020). Many of these studies assessed the changes on the properties of weathered WPCs after either short-term natural weathering or UV or Arc-Xenon accelerated aging. There is limited information on mechanical, chemical and color properties of long-term natural weathered WPCs. The aim of this study was to determine the selected color, physical, mechanical, morphological, and chemical properties of 6, 12, 18 and 24 months weathered WPCs having 3, 6 and 9% micronized titanium dioxide (TiO_2).

2 Materials and Methods

2.1 Materials

Turkish red pine (*Pinus brutia* Ten.) wood particles (density 0.543 g/cm^3) which were supplied by workshop in Kahramanmaraş city (Turkey) were turned into flour form with Wiley mill. The flours were classified and particles staying between 40 mesh-size and 60 mesh-size wires were used in this study. Before manufacturing, moisture content of the selected wood flours (WF) were reduced to below 1% by dessicating in oven for 24 h at 103 °C (±2).

Table 1 Experimental design of the study

ID	Polymer (%)	WF (%)	TiO_2 (%)	MAPP (%)	Wax (%)
PP	94	0	0	3	3
PP-3	91	0	3	3	3
PP-6	88	0	6	3	3
PP-9	85	0	9	3	3
PP/WF	54	40	0	3	3
PP/WF-3	54	40	3	3	3
PP/WF-6	48	40	6	3	3
PP/WF-9	45	40	9	3	3

Polypropleyne was obtained from Petkim Petrochemical Co. (melting point: 163 °C, melt flow index (MFI): 4.5 g/10 min, and density: 0.905 g/cm^3). Micron size titaniumdioxide (Purity: 92.5%, Particle Size: 26.6 µm, Surface area: 50 ± 15 m^2/g; Mean density: 0.23 g/cm^3, Color: White and pH: 7) as a colorant. Paraffin wax (K.130.1000) was used as a lubricant (density: 0.93 g/cm^3, softening point: 56 to 58 °C). The MAPP obtained from Clariant was used as a coupling agent (density: 0.91 g/cm^3, softening point: 156 °C). Experimental design of the study is given in Table 1.

2.2 Methods

Depending on the manufacturing recipe (Table 1), dried wood flours (WF), polypropylene, MAPP, lubricant and titanium dioxide were mixed in a mixer at 900–1000 rpm speed for 2 min for homogeneous blend. The mixed materials were compounded in a single-screw extruder at 40 rpm screw speed and temperatures of 170–200 °C. The extruded strands were dipped into a water pool (23 °C ± 2). Strands were cooled, pelletized and dried to under 1% moisture content in oven at 103 °C (±2) for 24 h.

The injection molding machine with 5–6 MPa injection pressure, 80 mm/s injection speed and 30sn cooling time were used to manufacture test samples. Manufacturing parameters of single-screw extruder and injection molding machine was presented in Fig. 1.

Tensile, and impact properties were determined based on ASTM D 638 and ASTM D 256, respectively. Tensile testing was performed on Zwick 10KN while a HIT5.5P by Zwick™ was used for impact property testing for notched samples. The notches were added using a Polytest notching cutter by RayRan™. For both physical and mechanical properties, the test samples were conditioned at %65 ± 5 relative humidity and 23 ± 2 °C for at least 15 days prior to testing. Five test samples for each group were tested.

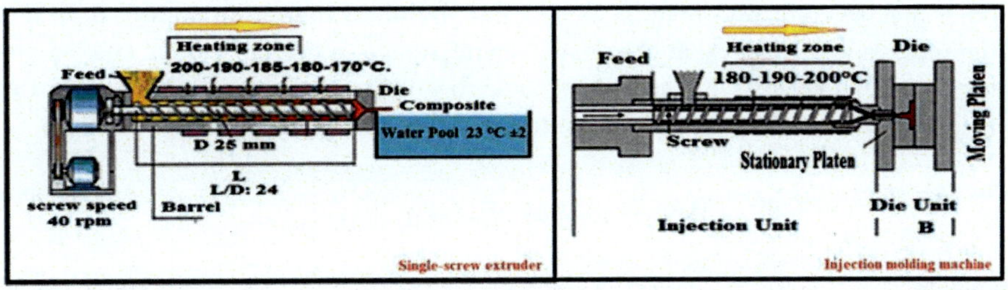

Fig. 1 Single-screw extruder and injection molding machine manufacturing parameters for the study

Table 2 Average climate data of Izmir province

Date range	Temperature (°C)	Humidity (%)	Precipitation (kg/m^2)	Wind speed (m/sn)
6 months	14.15	67.09	76.72	4.097
12 months	23.31	51.38	31.78	2.957
18 months	12.30	63.75	2.772	4.166
24 months	24.97	47.91	30.15	3.157
Avarege	18.68	57.53	35.34	3.594

According to ASTM D 1641 standards, the samples were exposed to open air conditions as 0–6–12–18– and 24 months in the location 38°26′25.3″ north and 27° 10′29.9″ east of İzmir city. The climatic data of İzmir was obtained from İzmir Meteorology Regional Directorate from August 24th of 2015 to August 24th of 2017 (24 months). Average values were presented in Table 2.

The test samples were placed on special stands at an angle of 45° to the ground and samples were always facing to south. The test samples were adjusted so that the height of the stand was 50 cm above the ground. Special attention was paid to keep the water away from stands where the test samples were placed. For this purpose, near-by organic materials such as grass were removed. The samples were visually inspected regularly every month. After each weathering period, the test samples were conditioned in the air conditioning cabinet until they reached equilibrium moisture at 23 ± 2 °C and 50 ± 5% relative humidity. Tensile, impact properties and color measurements were performed to determine the effects of natural weathering on the properties of wood-plastic composites. To determine the amount of color change, the test samples were numbered and first measurements were made. Surfaces of the test samples were cleaned during the measurements of the aged test samples. To determine the effects of aging on the color of test samples were analyzed according to CIE L, a, b Color space system with a Konica Minolta CM–3600 spectrophotometer according to the guidelines set forth in ASTM-D 2244. This method is a comparison of three coordinates that are L, a, and b; and other precise color information such as hue, lightness, and saturation. L represents lightness; a and b represent chromaticity directions, +a goes to the red and −a goes to the green, +b goes to the yellow and −b goes to the blue direction. The total color variation ($\Delta E^*_{a^*b^*}$) can be estimated using the Eq. 1 (Ecco et al. 2017).

$$\Delta E^* = \sqrt{(\Delta L)^2 + (\Delta a^*)^2 + (\Delta b^*)^2}$$

where:

ΔL *; The values Δa * and Δb * are the changes that occur between the initally (i) and the final (f) colors.
ΔE *; shows the total color changes in the L, a and b directions of the colors. The low value of ΔE * indicates that the color does not change or changes slightly. The highest value indicates the highest color change.

The morphology of test samples were investigated using scanning electron microscopy (SEM). For this purpose, ZEIS machine model EVO LS10 was used. Before starting the experiment, samples were dippped into liquid nitrogen, and then were broken to half with a hammer. The fructured surface of the test samples were coated with a gold layer (in 10 mA and in 120 s) to provide conductive surfaces thus avoiding electrostatic charging during observation.

Surface chemical changes of the unweathered and 12 month weathered PP, PP-3TiO_2, PP/WF and PP/WF-3TiO_2 samples were studied through Attenuated total reflectance Fourier-transform infrared (ATR-FTIR) spectroscopy. Agilent Cary 630 model ATR-FTIR was used and the spectra of the samples were recorded between 4000 cm^{-1} and 1000 cm^{-1} wave frequencies. Baseline and ATR corrections for the penetration depth and frequency variations were applied using the microlab software supplied with the equipment.

3 Result and Discussion

The effect of TiO_2 rates on density and selected mechanical properties of neat polypropylene (PP) and their wood flour filled composites (PP/WF) were investigated in 6 months apart during 24 months of yearlong natural weathering period. After each weathering period, surface color changes of all manufactured samples was measured using Konica Minolta CM–3600 spectrophotometer. After color measurements, density, tensile strength (TS), tensile modulus (TM), elongation at break (EatB), impact strength (IS) and hardness (H) values were determined. Surface morphology of unweathered and 12- and 24 month-weathered PP, PP-3TiO_2, PP/WF, PP/WF-3TiO_2 samples were studied using scanning electron microscopy (SEM). Surface chemistry of unweathered and 12-month weathered PP, PP-3TiO_2, PP/WF; PP/WF-3TiO_2 samples were also examined using FTIR-ATR Spectroscopy.

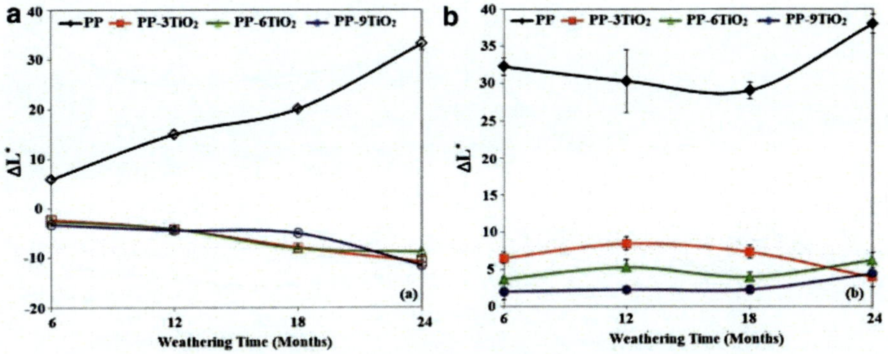

Fig. 2 Lightness (ΔL) of PP: (a) and PP/WF composites: (b)

3.1 *Color Changes of the Samples During Weathering*

During 24 months of weathering, test specimens were exposed to an average of 18.7 °C temperature, 57.5% humidity, 35.3 kg/m^2 precipitation and 3.59 m/s wind speed (Table 2). Considering the effects of these conditions on specimens, it is inevitable to observe color changes and degradation at the sample surface. Additionally, rain washing of the sample-surface might remove some water-soluble extractives that impart color (Stark and Rowlands 2003; Matuana 2005).

Generally, UV light causes color changes, or lightness (fading) of polymer and polymer based composite materials. This indicates an oxidative degradation of the material at the surface and may lead to a damage of the bulk of the composite material. Studies and observation show that fading and surface cracking phenomena always go together. Early and more rapid crack growth was seen in materials fading quickly compared to ones fading slowly. The surface crack growth eventually leads to a measurable reduction in strength properties (Klyosov 2007).

Lightness (ΔL^*) and color changes (ΔE^*) of all samples were presented in Fig. 2 and Fig. 3, respectively. In those figures, (a) denotes PP groups and (b) shows PP/WF groups. The unweathered reference composite samples (LPP/WF) had initial lightness values of 33.2 ± 0.99. The addition of micronize TiO_2 increased the lightness of the composites to 70.7 ± 1.36, 77.7 ± 4.0 and 80.0 ± 0.98 for LPP/WF-3TiO_2, LPP/WF-6TiO_2 and LPP/WF-9TiO_2, respectively. It is commented that if the lightness (ΔL^*) has the most influence on ΔE^*, the trends of ΔL should follow that of ΔE (24). Figure 2 and Fig. 3 show almost identical trends of ΔE^* and ΔL^*.

PP and PP/WF specimens had the highest ΔE^* and ΔL^* values, indicating poor resistance against degredation. It should be noted that while ΔE^* and ΔL^* values of PP gradually increased with weathering period, those values of PP/WF were almost constant. This is an indication that most of the fading was occurred in the first 6 months period (LPP/WF increased from 33.2 ± 0.99 to 66.3 ± 2.16). Photochemical degradation by UV light occurs primarily in the lignin component, which is also largely responsible for fading of wood and WPC materials (Stark et al.

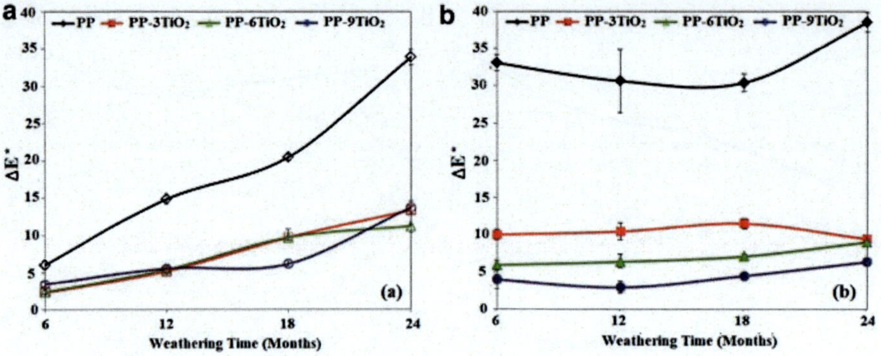

Fig. 3 Color changes (ΔE) of PP: (a) and PP/WF composites: (b)

2004; Klyosov 2007). It has been reported in previous studies that there is a strong correlation between the color changes and the lignin decay in wood (Klyosov 2007; Stark and Matuana 2007; Fabiyi et al. 2011). Removal of degraded lignin components from the sample surface leads to the exposure of the wood filler at the composite surface (Chang et al. 1982; Adhikary 2008). It has been reported in previous studies that color changes followed surface cracking. When composite materials expose to UV light, polymer chains and other organic compounds were broken due to the high-energy UV light. During this process, free radicals are formed and propagated until the free radicals are terminated by themselves (Klyosov 2007).

Test specimens having 3, 6 an 9% TiO_2 had ΔE^* values below 10 for both PP and PP/WF samples. Samples having TiO_2 in the formulations had the lowest ΔE^* and ΔL^* values in PP and PP/WF composites, showing higher resistance to degredation. Butylina et al. (2016) studied various pigments including TiO_2 for PP based WPCs. They reported that among all composites, the ones having titanium pigment showed the best results (the lowest lightness and colour change). It is reported that inorganic pigments often protect the WPC surface from fading, but accelerate the WPC oxidative degradation in the bulk (Klyosov 2007). In this study, it is believed that titanium dioxide pigment masking caused a sharp reduction in light transmission resulting in a delayed degradation. Similar result of sharp reduction in light transmission was also reported by Butylina et al. 2016.

3.2 *Density and Selected Mechanical Properties of the Samples*

The density graphs of the PP and PP/WF composites were presented in Fig. 4a and 4b, respectively. Addition of both WF and various concentration of TiO_2 increased the density of test specimens. This is generally explained by the rule of mixture;

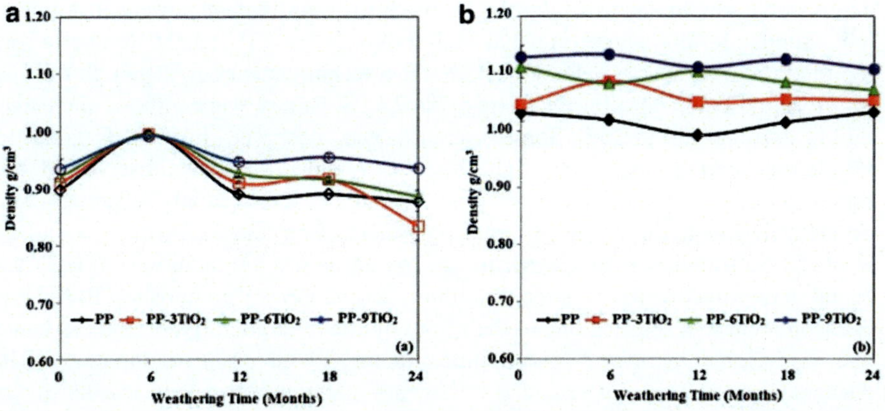

Fig. 4 Density values of PP: (a) and PP/WF: (b) specimens

incorporation of high-density filler into a low-density polymer results in a composite with higher density compared to polymer itself (Çavuş and Mengeloğlu 2020). The increased density of the composites with the rising percentage of fillers and additives are reported in the literature (Clemons 2002; Klyosov 2007; Mengeloğlu et al. 2015; Çavuş and Mengeloğlu 2016). Weathering time also affected the density of the specimens. In the first six months, especially for the groups without WF, there was some increase in density values of the test specimens (Fig. 4a). For those samples, in the following 12, 18 and 24 months of weathering period, density was gradually decreased. The changes on density during weathering periods can be due to the gaps, breaks and degradations occurring on polymer chains. Reduced density values were mostly caused by visually apparent deteriorated surfaces (Stark and Matuana 2007; Yang et al. 2015).

The tensile properties of the manufactured samples were studied and their interaction graphs of tensile strength (TS), tensile modulus (TM) and elongation at break (EatB) values were presented in Fig. 5. TS values of PP and PP/WF composites were shown in Fig. 5a and 5b, respectively. In unweathered samples, TS values were reduced noticeably with the addition of WF and various concentration of TiO_2. Similar reduction with WF presence was reported by others (Çavdar et al. 2016; Çavuş and Mengeloğlu 2020). Lowered TS values were also reported for composites having TiO_2 (Butylina et al. 2016). Increased TS values of PP with nano-size TiO2 were also reported by Esthappan et al. (2012). In the case of weathered samples, large amount of reduction on TS values of PP were observed. TS values were redued from 29.3 ± 0.2 MPa to 12.2 ± 0.8, 14.4 ± 0.3 MPa, 11.0 ± 0.3 and 5.1 ± 0.7 MPa at the end of the 6-, 12-, 18- and 24-month weathering periods, respectively (Fig. 5a). This reduction was due to the chain scission mechanism of polymer caused by photodegration (Soccalingame et al. 2015, 2016). Compared to neat PP sample, PP/WF ones had more resistance to weathering effects and retain its strength (Fig. 5b). It is believed that mechanical degradation was hindered by the high amount of antioxidant present in lignin of WF. Similar results suggesting that WF stabilizes the PP against weathering are reported (Peng et al. 2014; Peng et al. 2015; Soccalingame et al. 2016). In the case of PP and PP/WF specimens having TiO2, TS values were almost unchanged during weathering period. Some increase was observed in first 6 month of weathering period and then slight and slow reduction was observed during remainder of the weathering periods (Fig. 5a and 5b). Reduced tensile strength after accelerated weathering were reported by others (Beg and Pickering 2008; Devi et al. 2013; Butylina et al. 2016). In the present study, the addition of TiO_2 was found to improve the tensile strength of both PP and PP/WF composites. Rutile type TiO_2 pigment was selected due to their high tinting strength, high brightness values and surface treatment with aluminum, silicon and zirconium compounds. This particular product commercialized as having maximum weather resistance and suitable for outdoor applications. Butylina et al. (2016) reported that TiO_2 pigment can mask discolouration to some degree by causing a sharp reduction in light transmission. It is possible that TiO_2 with high tinting and high brightness values might have reduced the light transmission hindering the occurring deep and propagating cracks at the surface. It should be noted that TS values of samples with TiO_2 were increased with weathering. Additional study is needed to understand mechanism of how pigment works and interacts.

TM values of PP and PP/WF composites were shown in Fig. 5c and 5d, respectively. Similar to TS results, TM values of PP were reduced with addition of TiO_2. In previous studies, TiO2 presence in PP increased the TM values of the unweathered samples (Deka and Maji 2011; Esthappan et al. 2012; Butylina et al. 2016). Addition of WF, on the other hand, increased TM values. Lignocellulosic filler usually improves TM of the thermoplastic composites and usually explaines by the rule of mixtures (Matuana et al. 1998).

In weathered samples, TM values of PP were gradually reduced from 470 $\pm$ 27 MPa to 405 $\pm$ 6 MPa and 211 $\pm$ 30 MPa at the end of the 12- and 24-month weathering period, respectively (Fig. 5c).

For TiO_2 added PP samples, TM values, on the other hand, were increased up to 18 months weathering time and then started decreasing. Best result was achieved with samples having 9% TiO_2. TM values of PP/WF were slightly reduced first 6 month period, increased in the second 6 month period, and started reducing after 3rd 6 month period. Addition of TiO_2 in PP/WF composites improved TM values considerably for samples having 6% and 9% TiO_2.

Elongation at break (EatB) values of the PP and PP/WF composites were shown in Fig. 5e and 5f, respectively. Addition of both TiO_2 and WF reduced the EatB values. WF makes the polymer matrix more brittle causing lowered EatB values (Yeh et al. 2013). At the end of the 24 months of weathering period, All PP test specimens had significantly decreased EatB values compared to unweathered ones.

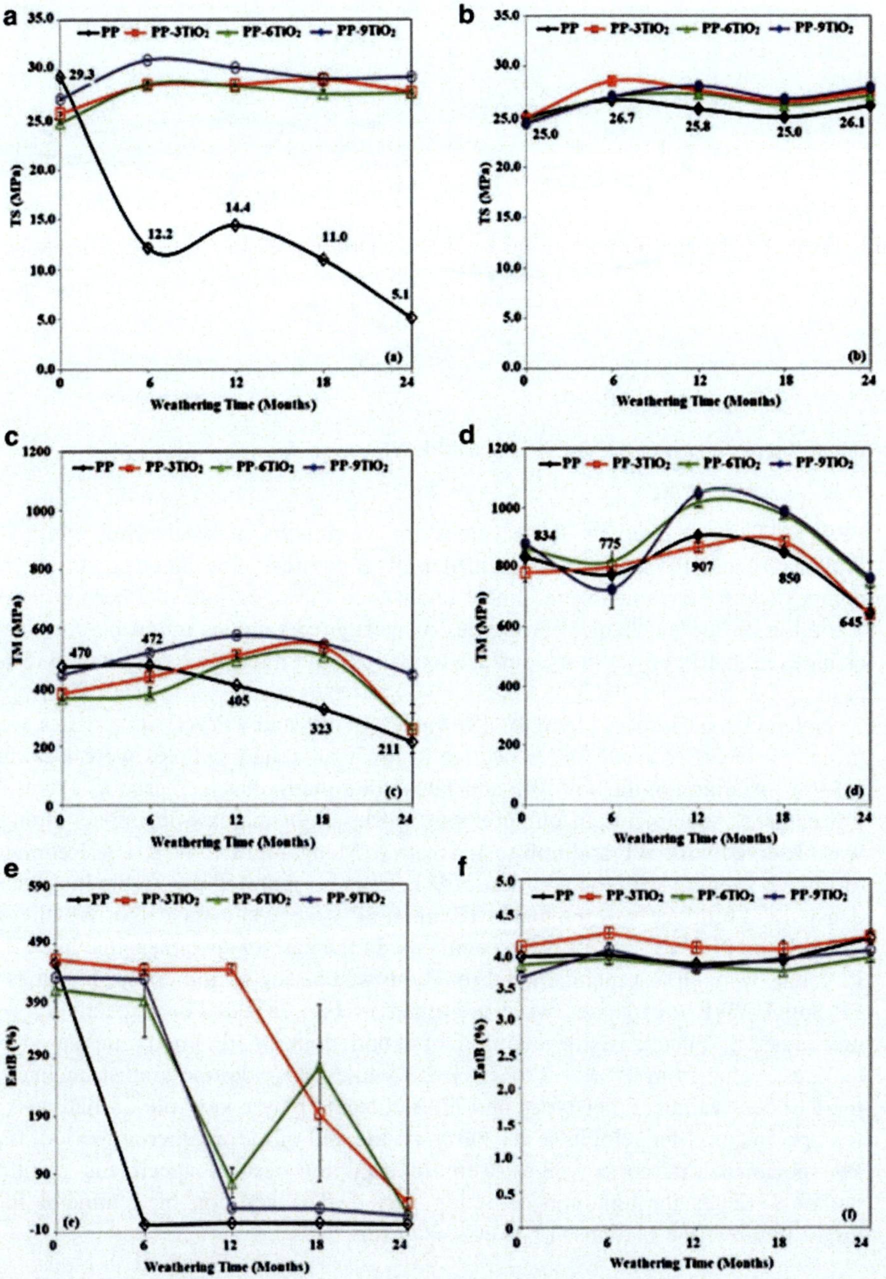

Fig. 5 Tensile strength, tensile modulus and elongation at break values of PP: a, c, e, and PP/WF composites: b, d, f

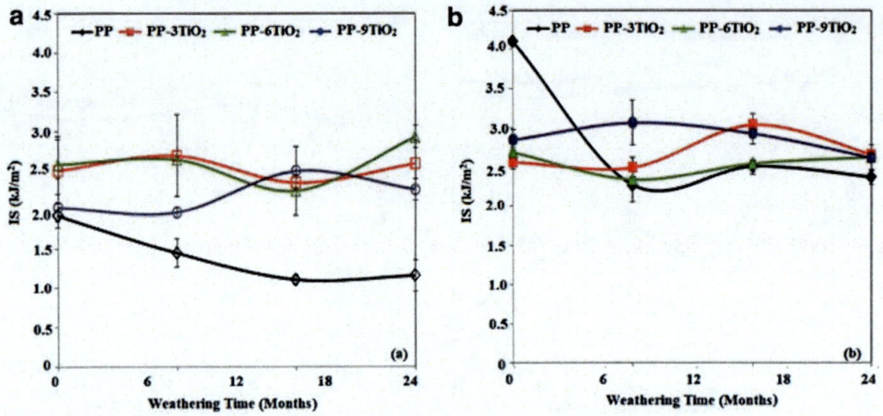

Fig. 6 Impact strength values of PP: (a) and PP/WF composites: (b)

PP/WF-3TiO_2 retained its EatB values for 12 months of weathering while PP/WF-6TiO_2 and PP/WF-6TiO_2 retained only 6 months of weathering. In PP/WF composites, EatB values were almost unchanged througout the weathering period. Variation on the mechanical properties of weathered samples might be due to the changes in matrix crystallinity, surface oxidation, and interfacial degradation (Stark et al. 2004; Matuana 2015).

Notched izod impact strength (IS) values of PP and PP/WF composites were presented in Fig. 6a and 6b, respectively. IS of all tested samples were less than 5 kJ/m^2. Addition of the both WF and TiO_2 in PP matrix has increased IS values for unweathered specimens. In our previous work, slight increase in impact strength was observed with WF addition to PP matrix (Mengeloğlu et al. 2015). Reduction of impact strength with the addition of WF was also reported due to the increase in density and brittleness of the matrix (Zaini et al. 1996; Matuana 2015). Weathering reduced the IS of PP and PP/WF specimens. In the case of specimens having TiO_2, IS values were almost maintained throughout weathering period. Shore D values of PP and PP/WF composites were presented in Fig. 7a and 7b, respectively. For unweathered specimens, the addition of wood flour in PP matrix increased the hardness value from 64 ± 1.7 to 75 ± 2.3 while TiO_2 addition almost unchanged the hardness values. Filler types and filler contents play a key role on hardness of test specimens (Mengeloğlu et al. 2007). At the end of the weathering period, only PP specimens' hardness values were reduced but rest of specimens maintain hardness values throughout weathering period. The addition of a mineral filler might improve the hardness of WPCs (Klyosov 2007).

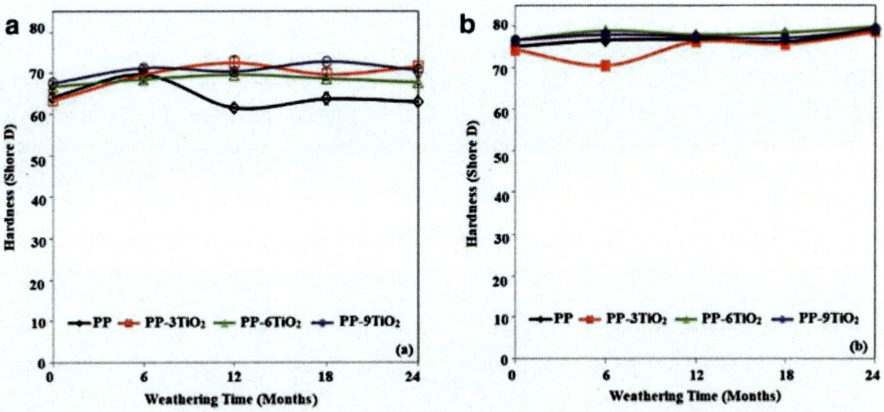

Fig. 7 Shore D values of PP: (a) and PP/WF composites: (b)

3.3 Surface Morphology of the Samples

Surface morphology of 12 and 24 months of naturally-weathered samples were studied. Micrographs of unweathered and naturally weathered PP, PP-3TiO_2, PP/WF and PP/WF-3TiO_2 samples are shown in Fig. 8. SEM images of all unweathered samples showed regular and smooth surfaces with no significant defects. For composites, this is an indication of a sufficient encapsulation of wood flour by a layer of PP matrix. After 12 months of weathering, clearly distinguishable microcracks were present on all weathered samples. It is believed that these cracks were related to polymer chain scission as a result of seasonal changes in moisture and temperature during weathering period.

Similar results were also reported for artifically weathered polymer composites (Stark and Matuana 2007; Fabiyi et al. 2008; Peng et al. 2014). Compared to PP surfaces, more cracks were observed on PP/WF surfaces, indicating that the presence of wood fiber (WF) accelerated the photodegradation of PP (Darie et al. 2013; Peng et al. 2014).

In addition, protruded wood particles on the composite surface were also present. Butylina et al. (2016) also observed that craks in composites started around lignocellulosic or mineral fillers near the surface. It should also be noted that presence of TiO_2 in samples has changed the shape of the surface cracks. Number of cracks were increased but their length and depth were less severe. Butylina et al. (2016) reported similar findings in their study where they used zinc oxide, iron oxide and titanium dioxide as pigments for PP based composites. They suggested that this might be due to the UV shielding ability of the studied pigments. After

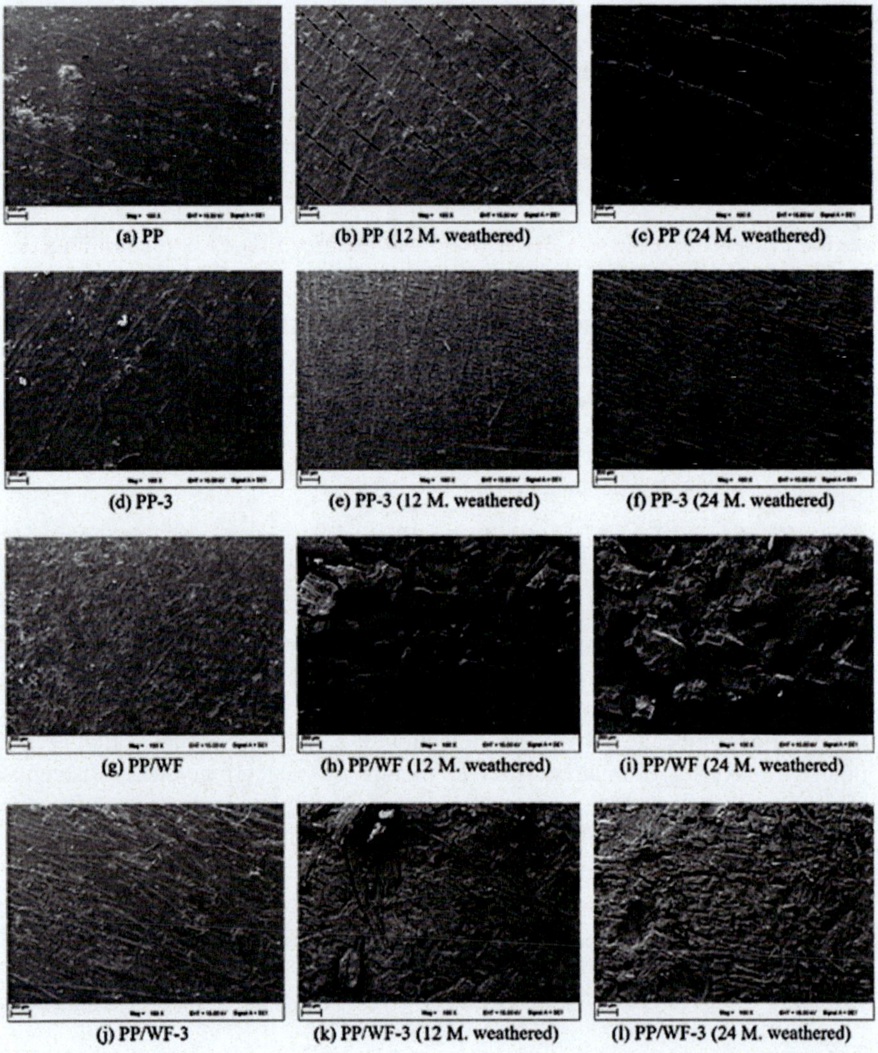

Fig. 8 SEM micrographs of unweathered (a, d, g, j), 12 months weathered (b, e, h, k) and 24 months weathered (c, f, i, l) samples (×100 magnification)

24 months of weathereing, large cracks and delamination of PP layer were present on the surfaces of all samples. However, this phenomenon was less severe and less pronounced in groups having TiO_2 compared to other groups (Peng et al. 2014).

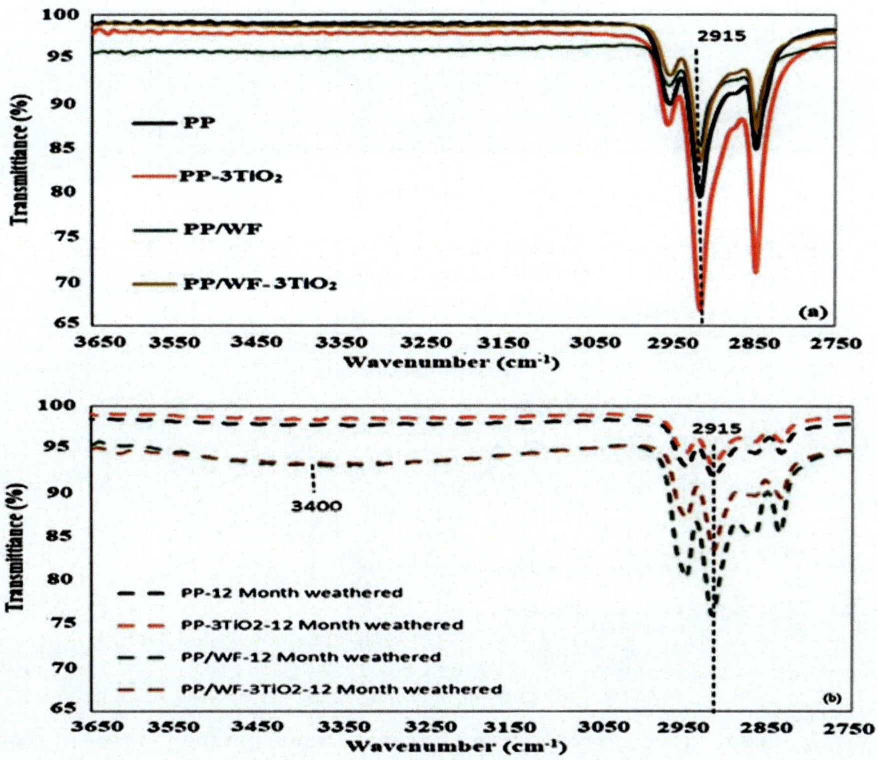

Fig. 9 FTIR spectra (3650–2750) of manufactured composites before (a) and after 12 months (b) of weathering

3.4 Surface Chemical Changes Due to Weathering

Surface chemical changes of the unweathered and 12 month weathered PP, PP-3TiO_2, PP/WF and PP/WF-3TiO_2 samples were studied through FTIR-ATR spectroscopy analysis. Figure 9a and 9b present the FTIR spectra in the region from 3750 to 2750 cm^{-1}. A strong band can be observed at 2915 cm^{-1} in all unweathered samples (Fig. 9a), which is assigned to alkane CH vibrations of methylene groups of PP (Stark and Matuana 2007; Peng et al. 2014; Chen et al. 2016). After 12 months of weathering, intensity of this peak was decreased for both PP and PP/WF samples. Peng et al. (2014) reported that this peak was almost disappeared for the cellulose filled and combination of cellulose and lignin filled PP based composites. After weathering, the peaks assigned to free hydroxyl (–OH) groups were seen at $\sim$3400 cm^{-1} for PP/WF and PP/WF-3TiO_2 indicating that more WF was exposed on the weathered surfaces.

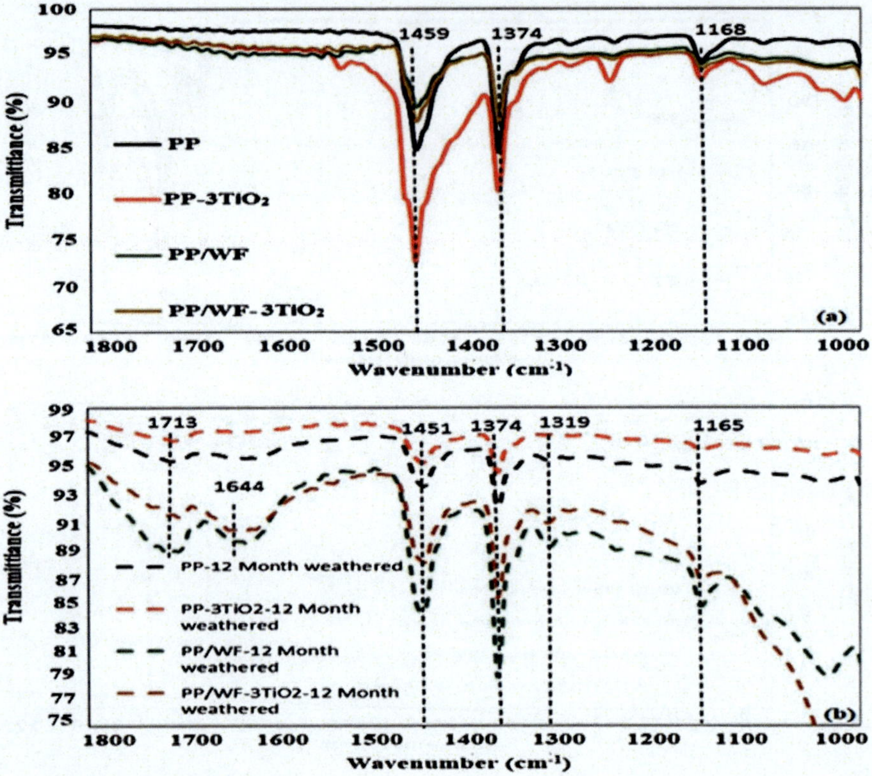

Fig. 10 FTIR (1800–1000) spectra of manufactured composites before (a) and after 12 months (b) of weathering

The changes of FTIR spectra in the region from 1800 to1000 cm^{-1} before and after weathering were illustrated in Fig. 10a and 10b respectively. The PP associated absorption bands at 1451 cm^{-1} (CH_3 asymmetric deformation), 1374 cm^{-1} (CH_3 symmetric deformation), and 1168 cm^{-1} (bending vibration of tertiary carbon) could be clearly seen in all samples (Peng et al. 2014). This suggests that wood flour in composites were well covered with PP layer before exposure.

The intensity of the three groups mentioned above decreased in PP samples. However, intensity of these groups showed some increase in PP based composites. These more intense spectra might be due to the overlapping bands coming from chemical changes of protruded wood particles during weathering ($\sim$1462 cm^{-1}, C-H deformation in cellulose, hemicellulose band $\sim$1375 cm^{-1}, and C-O-C asymmetric vibration in cellulose $\sim$1163 cm^{-1}). After 24 months of weathering, carbonyl band at 1713 cm^{-1} was seen in both PP and PP/WF composites as a result of the photo-degradation of wood and polypropylene. Similar findings were reported by others (Pandey and Pitman 2003; Selden et al. 2004; Butylina et al. 2012). In both PP and PP/WF composites, the presence of TiO_2 slightly reduced the intensity of the all bands associated with weathering process of wood and PP. In another study, after weathering, compared to reference composites, lower carbonyl index values was reported for composites having TiO_2 (Butylina et al. 2016).

4 Conclusions

PP and PP/WF specimens were exposed to an average of 18.7 °C temperature, 57.5% humidity, 35.3 kg/m^2 precipitation and 3.59 m/s wind speed during 24 months of weathering period. The effects of various concentrations of TiO_2 on the selected color, mechanical, morphological and chemical properties of weathered PP and PP/WF specimens were studied.

PP and PP/WF specimens had the highest ΔE^* and ΔL^* values, indicating poor resistance against degredation. Color changes of PP were gradually increased with weathering and reach the color changes of PP/WF at the end of 24 months of weathering. In the case of PP/WF specimens major changes of color occured in the first 6 month period. ΔE^* values of specimens having 3, 6 an 9% TiO_2 were below 10 for both PP and PP/WF samples. This was an indication of higher resistance to degradation during weathering.

Density of test specimens were increased with TiO_2 contents and presence of WF. Although there was an increase in all of the test specimens in the 6 months, the density of the test samples decreased in the following 12–18 and 24 months. In unweathered samples, TS values were reduced noticeably with the addition of WF and various concentration of TiO_2. Large amount of reduction on TS values of PP were observed for weathered samples. No negative changes of TS was observed during weathering of PP and PP/WF specimens having TiO_2. TS values were increased in the first 6 month weathering period and then slight decrease was observed. Regardless of TiO_2 concentration, unweathered and unweathered PP specimens showed lower TM values than PP/WF specimens. Improvement of TM values of PP and PP/WF with TiO_2 addition up to 18 months of weathering were also observed. The EatB values of PP was reduced tremendously with the presence of WF (from PP = 454 ± 0.4% to PP/WF = 4 ± 0.2%). Addition of TiO_2 in PP and PP/WF caused a slight decrease in EatB values. TiO_2 provide some stabilization on EatB values up to 6 month period for PP-6%TiO_2 and PP-6%TiO_2, and up to 12 month period for PP-3%TiO_2 specimens. In the case of IS values, addition of TiO_2 provided higher values for weathered samples. Increased Shore D values with TiO_2 were also observed.

Morhological study was performed on 12 and 24 month weathered PP, PP/WF, PP-3%TiO_2 PP/WF-3%TiO_2 specimens. Study showed that unweathered samples showed regular and smooth surfaces with no significant defects. After 12 months of weathering, clearly distinguishable microcracks were present on all weathered samples. These cracks might be related to polymer chain scission as a result of seasonal changes in moisture and temperature during weathering period. Protruded wood particles on the composite surface were also present on PP/WF specimen. In addition, presence of TiO_2 in samples has changed the shape of the surface cracks.

Number of cracks were increased but their length and depth were less severe. After 24 months of weathering, large cracks and delamination of PP layer were present on the surfaces of all samples. However, this phenomenon was less severe and less pronounced in groups having TiO_2 compared to other groups.

FTIR-ATR spectroscopy analysis of unweathered and 12 month weathered PP, PP-3TiO_2, PP/WF and PP/WF-3TiO_2 samples showed that surface chemical changes on samples were occuered. After 12 months of weathering, intencity of strong band seen on unweathered samples at 2915 cm^{-1} (assigned to alkane CH vibrations of methylene groups of PP) was decreased for both PP and PP/WF samples.

References

ASTM D 256 (2002) Determining the Izod Pendulum Impact Resistance of Plastics. ASTM International, West Conshohocken

ASTM D 638 (2004) Standard Test Method for Tensile Properties of Plastics. ASTM International, West Conshohocken

ASTM D 1641 (2004) Standard Practice for Conducting Outdoor Exposure Test of Varnishes. American Society for Testing and Materials, West Conshohocken, p 1–25

ASTM-D 2244 (2007) Test Method for Calculation of Color Differences Form Instrumentally Measured Color Cordinates. American Society for Testing and Materials, West Conshohocken, p 1–20

Adhikary KB (2008) Wood flour recycled polymer composite panels as building materials. Thesis. Chemical and Process Engineering, University of Canterbury

Beg MDH, Pickering KL (2008) Accelerated weathering of unbleached and bleached kraft wood fiber reinforced polypropylene composites. Polym Degrad Stab 93:1939–1946

Behravesh AH, Shakouri E, Zolfaghari A, Golzar M (2010) Theoretical and experimental study on die pressure prediction in extrusion of wood-plastic composite. J Compos Mater 44(11):1293

Bouza R, Marco C, Naffakh M, Barral L, Ellis G (2011) Effect of particle size and a processing aid on the crystallization and melting behavior of iPP/red pine wood flour composites. Compos Part A 42:935–949

Brebu M (2020) Environmental degradation of plastic composites with natural fillers—a review. Polymers 12(166):2–22

Butylina S, Hyvärinen M, Kärki T (2012) Weathering of wood-polypropylene composites containing pigments . Eur J Wood Prod. 70:719–726

Butylina S, Martikka O, Kärki T (2016) Effect of inorganic pigments on the properties of coextruded polypropylene-based composites. J Thermoplas Compos Mater 31(1):23–33

Catto AL, Montagna LS, Almeida SH, Silveira RMB, Santana RMC (2016) Wood plastic composites weathering: effects of compatibilization on biodegradation in soil and fungal decay. Int Biodeterior Biodegradation 109:11–22

Cesarino I, Dias OAT, Negrão DR, Rocha LLN, Leão AL (2019) Deterioration of Wood Plastics Composites by the White-Rot Fungus Pycnoporus sanguineus. J Compos Sci 3:24 https://doi.org/10.3390/jcs3010024

Chao YY, Hung KC, Xu JW, Wu TL, Wu JH (2020) Effects of acetylated veneer on the natural weathering properties of adhesive-free veneer overlaid wood-plastic composites. Polymers 12 (513):2–10

Chen Y, Stark NM, Tshabalala MA, Gao J, Fan Y (2016) Weathering characteristics of wood plastic composites reinforced with extracted or delignified wood flour. Materials 9:610

Chang ST, Hon DNS, Feist WC (1982) Photodegradation and photoprotection of wood surfaces. Wood Fibre 14(2):104–117

Clemons C (2002) Wood-plastic composites in United States: the interfacing of two industries. For Prod J 52(6):10–18

Çavdar DA, Kalaycioğlu H, Mengeloğlu F (2016) Technological properties of thermoplastic composites filled with fire retardant and tea mill waste fiber. J Compos Mater 50:1627–1634

Çavuş V, Mengeloğlu F (2016) Utilization of synthetic based mineral filler in wood plastics composite. J Achiev Mater Manuf Eng 77(2):57–63

Çavuş V (2020) Selected properties of mahogany wood flour filled polypropylene composites: the effect of maleic anhydride-grafted polypropylene (MAPP). BioResources 15(2):2227–2236

Çavuş V, Mengeloğlu F (2020) Effect of wood particle size on selected properties of neat and recycled wood polypropylene composites. BioResources 15(2):3427–3442

Darie RN, Bodirlau R, Teaca RA, Macyszyn J, Kozlowski M, Spiridon I (2013) Influence of accelerated weathering on the properties of polypropylene/polylactic acid/eucalyptus wood composites. Int. J Polym Anal Charact 2013(18):315–327

Deka BK, Maji TK (2011) Effect of TiO_2 and nanoclay on the properties of wood polymer nanocomposite. Compos A 42:2117–2125

Devi RR, Gogoi K, Konwar BK, Maji TK (2013) Synergistic effect of nano TiO_2 and nanoclay on mechanical, flame retardancy, UV stability, and antibacterial properties of wood polymer composites. Polym. Bull. 70:1397–1413

Ecco LG, Rossi S, Fedel M, Deflorian F (2017) Color variation of electrophoretic styrene-acrylic paints under field and accelerated ultraviolet exposure. Mater. Des. 116:554–564

Eshraghi A, Khademieslam H, Ismaeil G (2016) Effect of weathering on physical and mechanical properties of hybrid nanocomposite based on polyethylene, woodflour and nanoclay maderas. Cienc Tecn 18(4):617–626

Esthappan SK, Kuttappan SK, Joseph R (2012) Effect of titanium dioxide on the thermal ageing of polypropylene. Polym Degrad Stab 97:615–620

Fabiyi JS, McDonald AG, Wolcott MP, Griffiths PR (2008) Wood plastic composites weathering: visual appearance and chemical changes . Polym Degrad Stabil 93:1405–1414

Fabiyi JS, McDonald AG, Morrell JJ, Freitag C (2011) Structural characterization of wood species on biodegradation of wood plastic composites. Compos A 42:501–510

Falk RH, Lundin T, Felton C (2000) The effects of weathering on wood–thermoplastic composites intended for outside applications. In: Proceedings, Durability and Disaster Mitigation in Wood-Frame Housing, Madison, Wisconsin, pp 175–179

Hietala M (2013) Extrusion Processing of Wood-Based Biocomposites. PhD Thesis (121 pages), Department of Engineering Sciences and Mathematics Division of Wood and Bionanocomposites

Kallakas H, Poltimäe T, Süld TM, Kers J, Krumme A (2015) The influence of accelerated weathering on the mechanical and physical properties of wood-plastic composites. Proc Est Acad Sci 64:94–104

Klyosov A (2007) Wood Plastic Composite, vol 13. Wiley Interscience Publication, New Jersey, p 50–416

Krehula LK, Katančić Z, Siročić AP, Hrnjak-Murgić Z (2014) Weathering of high-density polyethylene-wood plastic composites. J Wood Chem Technol 34:39–54

Kumar S, Vedrtnam A, Pawar SJ (2019) Effect of wood dust type on mechanical properties, wear behavior, biodegradability, and resistance to natural weathering of wood-plastic composites. Front Struct Civ Eng 13:1446–1462

Matuana LM, Park CP, Balatinecz JJ (1998) Cell morphology and propert relationships of microcellular foamed PVC/Wood-fiber composites. Polymer Eng Sci 38(1):1862–1872

Matuana LM, Heiden PA (2004) Wood Composites . Encycl Polymer Sci Technol 12:521–546

Matuana LM, Jin S, Stark NM (2011) Ultraviolet weathering of HDPE/wood–flour composites coextruded with a clear HDPE cap layer. Polym Degrad Stabil 96:97–106

Matuana LM (2015) The use of wood fibers as reinforcements in composites Biocomposites reinforced with natural fibers. In: Faruk O, Sain M (eds) Elsevier, Amsterdam ISBN 9781782421276

Mengeloğlu F, Kurt R, Gardner DJ, Oneil S (2007) Mechanical properties of extruded high density polyethylene and polypropylene wood flour decking board. Iran Polym J 16(7):477–487

Mengeloğlu F, Başboğa IH, Aslan T (2015) Selected propeties of furniture plant waste filled thermoplastic composites. Pro Ligno 11(4):199–206

Mengeloğlu F, Çavuş V (2019) Additives used in wood plastic composite manufacturing. Research and Reviews in Agriculture Forestry and Aquaculture Sciences, Gece Publishing, New York, 1, 49–58

Pandey KK, Pitman AJ (2003) FTIR studies of the changes in wood chemistry following decay by brown-rot and white-rot fungi. Int Biodeterior Biodegradation 52(3):151–160

Peng Y, Liu R, Cao J, Chen Y (2014) Effects of UV weathering on surface properties of polypropylene composites reinforced with wood flour, lignin, and cellulose. Appl Surf Sci 317:385–392

Peng Y, Liu R, Cao J (2014) Effects of antioxidants on photodegradation of wood flour/polypropylene composites during artificial weathering. BioResources 9(4):5817–5830

Peng Y, Liu R, Cao J (2015) Effects of UV weathering on surface properties of polypropylene composites reinforced with wood flour, lignin, and cellulose. Appl Surf Sci 332:253–259

Selden R, Nystrom B, Langstrom R (2004) UV aging of poly(propylene)/wood-fiber composites Polym. Compos 25(5):543–553

Shah AA, Hasan F, Hameed A, Ahmed S (2008) Biological degradation of plastics: a comprehensive review. Biotechnol Adv 26(3):246–265

Smith PM, Wolcott MP (2006) Opportunities for wood/natural fiber-plastic composites in residential and industrial applications. For Prod J 56(3):4–11

Soccalingame L, Bourmaud A, Perrin D, Bénézet JC, Bergeret A (2015) Reprocessing of wood flour reinforced polypropylene composites: impact of particle size and coupling agent on composite and particle properties. Polym Degrad Stab 113:72–85

Soccalingame L, Perrin D, Bénézet JC, Bergeret A (2016) Reprocessing of UV-Weathered wood flour reinforced polypropylene composites: study of a natural outdoor exposure. Polym Degrad Stab 133:389–398

Stark NM, Matuana LM, Clemons CM (2004) Effect of processing method on surface and weathering characteristics of wood–flour/HDPE composites. J Appl Polymer Sci 93(3):1021–1030

Stark MN, Rowlands RE (2003) Effects of wood fiber characteristics on mechanical properties of wood/polypropylene composites. Wood Fiber Sci 35:167–174

Stark NM (2007) Considerations in the weathering of wood-plastic composites. In: Proceedings of the 3rd Wood Fibre Polymer Composites International Symposium, Bordeaux, France, 26–27 March

Stark NM, Matuana LM (2007) Characterization of weathered wood-plastic composite surfaces using FTIR spectroscopy, contact angle, and XPS. Polym Degrad Stab 92:1883–1890

Stark NM, Mueller SA (2008) Improving the color stability of wood-plastic composites through fiber pre-treatment. Wood Fiber Sci 40(2):271–278

Teaca C, Rosu D, Bodirlau R, Rosu L (2013) Structural changes in wood under artificial irradiation determined by FTIR spectroscopy and color measurements – a brief review. BioResources 8(1):1478–1507

Tokiwa Y, Calabia BP, Ugwu CU, Aiba S (2009) Biodegradability of plastics. Int J Mol Sci 10:3722–3742

Yang TH, Yang TH, Chao WC, Leu SY (2015) Characterization of the property changes of extruded wood-plastic composites during year round subtropical weathering. Constr Build Mater 88:159–168

Yeh SK, Kim KJ, Gupta RK (2013) Synergistic effect of coupling agents on polypropylene-based wood-plastic composites. J Appl Polym Sci 127:1047–1053

Zaini MJ, Fuad MY, Ismail Z, Mansor MS, Mustafah J (1996) The effect of filler content and size on the mechanical properties of polypropylene/oil palm wood flour composites. Polym Int 40:51–55

Life–Cycle Assessment (LCA) of Plastic–Wood Composites

Rupesh Kumar Malviya, Rajesh Purohit, and Rahul Kumar Singh

Abstract Widespread environmental awareness to achieve product sustainability has triggered great efforts to use more environmentally friendly materials in product design. One of the most promising solutions to meet these needs is the use of wood and plastic composite fibers to reduce reliance on synthetic fibers as reinforcement and fillers for polymer composite construction. Much effort has gone into fully quantifying the benefits of Plastic Wood Composites (PWC) for various applications such as automobiles, building materials, and household appliances. The Life Cycle Assessment (LCA) is one of the efforts of personal watercraft whose main objective is to determine the potential environmental impact of the use of these materials on the environment as a whole. One of the main benefits of conducting life cycle analysis is the ability to provide a more holistic view of environmental impacts, covering the entire life cycle of the product, from the extraction of raw materials to the end. life, allowing justified decisions on the suitability of the use of PWC for specific applications that must be scientifically made. This chapter provides an overview of the LCA method, followed by a description of its uses with PWC in various applications. The advantages and disadvantages of the LCA method compared to personal watercraft are discussed, as well as a final conclusion on the future directions of the LCA application for personal watercraft.

Keywords Environmental impact · Life cycle assessment · Plastic-wood composites

R. K. Malviya · R. K. Singh (✉)
Department of Mechanical Engineering, Rabindranath Tagore University,
Bhopal 462022, MP, India

R. Purohit
Department of Mechanical Engineering, MANIT, Bhopal 462003, MP, India

S. Mavinkere Rangappa et al. (eds.), *Wood Polymer Composites*,
Composites Science and Technology, https://doi.org/10.1007/978-981-16-1606-8_12

1 Introduction

Composites of plastic and wood usually cover a unique amalgamation of filler, thermoplastic (plastic that can be repetitively softened by heating), and essences (e.g., adhesion promoters and lubricants). Additional fillers have also been used, for example, mineral wool, rising from the anticipated engineering properties of wood-plastic composite materials and their manufacturing technology. Plastic and cargo are vital products; for each representative, 30–70% of the total mass of wood-plastic composites. The percentage of filler in a mixture of wood-plastic composites change mechanical properties. For example, increasing the percentage of wood raises the tensile stiffness of wood-plastic composites, while reducing ductility and impact resistance (Abe and Ozaki 1998).

On the track to a sustainable bio-economy, attention has grown in the progress of wood-particle plastic composites, called wood-plastic composites (WPCs). Wood is well thought-out milestones for monetary growth, while the usage of fossil fuel-based thermoplastic materials and additives for wood and plastic composite can have severe environmental influences at most throughout the life cycle. The use of alternate resources redirects the attention of the WPC producer. LCA standards have been applied for synchronized products and have been collected with the physical factors of really manufactured composites to recognize the potential for extra in terms of source quality. Succeeding routes of WPC products to end users reflect the perspective of the recycler. The LCA approach was essential in which systems with the same purposes were generated to confirm an evaluation between end-of-life (EoL) processing systems. The outcomes displayed that WPCs made from secondary materials are the superior ecological and technical alternate. Recycling of composites would be the chosen route for the environment, but the contend of recycling WPCs in the new WPCs is a delicate difficult when comparing the two EoL handling systems. Nevertheless, composite materials are the most essential EoL route because of the present-day recycling guiding principle and the lack of markets for secondary WPC materials (Akhshik et al. 2017).

The rising attentiveness of sustainability everywhere in the world has encouraged numerous creativities to seek and apply extra environment-friendly solutions for everyday accomplishments. One of the clarifications presently extensively studied mentions the usage of wood-plastic composites as an alternate material, in specific synthetic composites and other conventional technical materials. In adding to decreasing raw material costs and decreased properties, wood-plastic composites propose numerous benefits in terms of environmental compatibility, for example, renovation, recyclability, and biodegradability.

For the motives declared above, the existing global market for the use of wood-plastic composites has notable growth and acceptance. The fundamental desires are coming from construction, transportation, and goods Consumption (Aji et al. 2009). The company report states that amongst the motives for the high acceptance of wood-plastic compounds, particularly for the manufacture of internal vehicle elements equated to synthetic compounds, there is good dimensional stability, formability, and strength. High in collisions, in addition to a major accident safety, includes passenger's security (Akhshik et al. 2017).

To encourage better acceptance of polymer wood composites and broader usage, the materials performance is too keenly studied. This technique was developed to measure the possible surrounding effect of products at all stages. The LCA analysis is subject to ISO 14040 and ISO 14044 standards and calculates the potential environmental influence of products from the output of basic materials to the processing, manufacturing, and usage of the products until the end of their valuable life. Cradle-to-grave analysis as well as providing a more comprehensive analysis of product sustainability performance (Alves and Ferrao et al. 2010).

This chapter encloses the comprise life cycle analysis procedure and compilation of investigation consequences on life cycle evaluation of composites of plastic and wood.

2 Review of Life Cycle Evaluation of Composites of Plastic and Wood

The developing apprehension regarding the environmental impact of the product has led to numerous usages of sustainability studies. The product must be suitable while meeting all from desires. All this must be finished with minimal energy and material consumption without causing damage to achieve a sustainable goal (Chiu and Chu 2012). As an outcome, different companies have been used with different approaches, and life cycle analysis is one of the most common techniques of analyzing product durability. Several scantiest use life cycle analysis to measure to design of product for environmental effect including extraction, development, operation and the deactivation of the materials of the product (Corbiere-Nicollier et al. 2001).

The durability of the products is linked with the use of biomaterials, such as wood-plastic composites. Wood-plastic composites are comprised in each type of product. Moreover, wood-plastic composites, called biodegradable materials, are extra environmentally friendly than carbon and fiberglass composites (Curran 2017). Pre-treatment of definite wood-plastics is simple and necessitates minimal handling to prepare fibers for composite materials (Deng et al. 2016). Sugar palm powder is one of the kinds of wood-plastic that does not need secondary treatment, such as flushing of the water or mechanical peeling procedure to extract the dust. Such kind of powder was formerly wrapped around the stem of the plant in a woven form (Dissanayake et al. 2009). In the accumulation of, wood-based plastics are more useful to the environment than synthetic fibers as reinforcements in composite materials. In adding several studies have shown the effective environmental effect of wood-plastic composite products on the substitution of materials. As for the positive environmental impact of wood-plastic composites, most car manufacturers were more attentive in the usage of wood-plastic composites.

2.1 *Framework of Life Cycle Evaluation*

2.1.1 Objective and Scope

Initially, the life cycle evaluation's objective is defined by the purpose of the application and intention to perform the life cycle.

2.1.2 Inventory Analysis

This analysis is required to assemble all valuable data at every stage of the element's life. The facts can affect the life cycle of a product. This includes pre-treatment, in which materials are removed, and post-treatment of the product during the usage phase until the end of its valuable life. At this point, all data is shown critically, as it possess the precision of the outcomes (Hauschild et al. 2018). Data availability is predominantly significant for wood-plastic composites, for which not all the required information is available. Alternatively, materials such as steel data have data available, which can be delivered by reliable information for steel industry (Hawkins et al. 2013).

2.1.3 Impact Assessment

The common elements essential for impact analysis are the selecting of influence groups, group indicators, and characterization models (ISO 2000). The fixed aspects of the impact assessment should be stable with the purposes and possibility of the life cycle assessment. Luz, Caldeira-Pires, and Ferra (Rosa et al. 2013) have inspected a biotic degradation and the potential for generation of photochemical ozone as action categories for polypropylene compounds in sugar cane powder. Likewise, Pegoretti et al. (Duigou et al. 2011) carried out the analyzing of the life cycle of plastic in wood in automotive applications. The impact estimation impact groups are nearly comparable for Luz, Caldeira-Pires, and Ferrao (Rosa et al. 2013).

2.1.4 Interpretation

Interpretation essential accomplish the objective Ultimate within the well-defined framework. The conclusions, references, and limits should be stated in the understanding (Pegoretti et al. 2014). On the basis on this, all main difficulties can be acknowledged, and a consistency and sensitivity analysis can be carried out. Consequently, the expert could fully connect the concluding outcomes and communicate precisely. Appropriate evidence must be well arranged to save time. The last outcomes of the inventory analysis and the impact assessment should be dependable when investigative the sensitivity analysis and the uncertainty analysis.

2.2 *Life Cycle Evaluation of Wood-Plastic Composites*

Also, elements made from wood-plastic composite materials are commonly lighter in the automotive sector. As an outcome, the vehicle, which is equipped with wood-plastic composite elements, has a lower weight and consequently lower fuel consumption.

A study by Joshi et al. (Rosenbaum 2017) has established an essential environmental influence of wood-plastic composites associated with glass powder reinforced composites. The study associated the impact of glass powder reinforced polyamide with polypropylene reinforced with a hybrid of wood/carbon plastic. Polypropylene-carbon reinforced plastic hybrid wood has exposed decreased energy consumption and disposal costs (Sapuan 2014). This would make the hybrid bio-composite extra preferred than the synthetic compound due to its positive environmental impact. Additionally, the manufacturing costs of hybrid bio composites are lesser in energy consumption relation during manufacturing and transport, even though they comprise synthetic materials like carbon dust.

2.2.1 Production Phase

A small number of scholars have analyzed the environmental influence of processing wood-based plastics. The The use of pesticides and other kinds of chemicals has an adverse effect on the environment during plant tissue growth". For this motive, the typical environmental special effects of wood dust, namely eutrophication, followed mostly during the cultivation phase (Summerscales et al. 2010). The usage of fertilizers when rising woody plants leads to an escalation in nitrate and phosphate emissions. And so, the usage of inorganic fertilizers can lead to nitrogen runoff, leading to environmental impacts.

In contrast, in the modern era, kenaf has aroused increasing interest because of its capability to absorb nitrogen and phosphorus contained in sand. However, some scholars have exposed that the adverse effects of pesticides and chemicals during the developing procedure are less than the adverse effects of synthetic dust on the environment throughout the production process. Joshi et al. (2004) established that the effect of wood-plastic composites on eutrophication is less than the eutrophication effect of glass powder composites (Finkbeiner et al. 2006). These remarks are possible apply to several wood-plastics, as their developing procedures are very similar.

2.2.2 Use Phase

The energy intake throughout the procedure stage from thermo mechanical or electromechanical systems. The savings in prime energy and CO2 emissions avoided during the process stage in the usage stage have been estimated. Hansen et al. (Zah et al. 2007) associated the analysis of the life cycle of two internal faces of an Audi A3 (flax reinforced acrylonitrile–butadiene–styrene copolymer, also called ABS) Concering energy intake throughout the use stage. They found that fuel intake throughout the manufacture, usage, and disposal of a car represents 80% of total energy. The environmental influence is assessed, and the wood-plastic composite side panel is desirable to the polymer panel. The special environmental effects for certain materials, energy flow, and emissions are examined (Finkbeiner et al. 2006).

2.2.3 End of Life

In [47] initiate that lower vehicle weight would have fewer influence on the environment with EoL. Likewise, the combustion of wood-plastic composite materials would have a well energy retrieval than the combustion of glass powders (Luz et al. 2010). The glass powder composites are made of inert materials that can inhibit with combustion procedure. This will outcome in the development of slag, that is an ash remains that contains ash and is disposed of in landfills (Liikanen et al. 2019). A study displayed that the wood-plastic composite combustion procedure's benefit lies in the positive sound effects on the environment through energy recovery". Wood-plastic composites could burn entirely and have a relatively high calorific value.

2.3 Summary

In summary, the worldwide generation of waste has improved considerably in recent decades because of suburbanization, and fiscal growth and squander manufacturing. Deceleration is not predictable on the contrary (World Bank 2012). Waste is a challenge in the current biosphere that needs to be met to confirm a sustainable future. Consequently, it is essential to describe processes to find appropriate and other handling approaches for the diverse waste portions to decrease the ecological influence of the waste generated in unlike parts of the world.

3 A Case Study:

Introduction: Equating Wood-Plastic Composites and Glass-Powder Reinforced Polypropylene Composites Car Door Panels
LCA is a tool to find the ecological influence of a specific product during the course of its life cycle. Since diverse products have various influences at diverse phases of their shelf life, a one-step comparison can lead to misleading outcomes considering

the entire life cycle. The automotive industry is the main contributor to environmental contamination, mostly for greenhouse gas emissions.

The assessment of these studies decided that WPCs in automotive applications are more environment friendly than glass powder composites because.

(1) Wood powder manufacture has fewer impacts than glass powder production. They are mostly influenced by the development of solar energy and absorb CO_2 through their growth.
(2) Wood powder occupies more specific space than composites. Subsequently, WPCs need less matrix (plastic) (which has a greater environmental impact) than glass powder composites.
(3) Because of the lighter weight, WPCs reduce the car's fuel consumption and, therefore, decrease emissions during usage (Sivertzen et al. 2003). Energy can be recovered and used by burning wood powder, but not glass powder.

Voluntary policies and adjustments can be promoted in favor of WPCs compared to other conventional composites in this sector. Consequently, preceding studies should be established or refuted by conducting other life cycle assessments based on diverse expectations.

This learning equates the life cycle effect of door panels made of polypropylene reinforced with wood powder and polypropylene (PP) with glass powder.

4 Process of Life-Cycle Assessment

4.1 The Data Collection

The difficulties were resolved by generating community business databases. Though, these databases cannot be precise and sufficient to handle all circumstances. The typical lack of data problem for a part procedure of a system is resolved by cautiously choosing and using similar methods or neglecting less essential procedures.

5 Modeling Life-Cycle Assessment

Life cycle assessments usually start with essential objectives and scope. This is the phase in which the purposes of a life cycle analysis are well-defined, and the limits of the learning are clear. Conclusion the outcome of the life cycle analysis deprived of these limits, the quantity of data a single study can contain is limitless.

When experts effort to link inventory outcomes to environmental effects, these outcomes lose the spatial and temporal associations between sleep and response (changes in body effects). Supreme effect assessment approaches do not and may not reflect these complex authenticities in the far future (UNEP 1996). Besides, most effect valuation approaches use the standard that a better mass of a substance means a better effect. Notwithstanding limits, life cycle assessments deliver info on products' impact on the life cycle that no other tool can deal. The interpretation phase makes a straight forward approximation of uncertainties.

6 Goal and Scope

This life cycle assessment's fundamental purpose is to equate the ecological impact of substituting glass powder reinforced PP car door panels with a wood powder reinforced PP car door panel. The secondary intentions are detecting the effects of replacing glass powder on the main gases of greenhouse and the main reason of contamination in air, recognize the Vital environmental groups in which these two panels have a significant impact and categorize the method and the material that gives the majority to the key categories of the two panels and establish the changes of the life cycle examination due to alteration in the basic hypothesis or the reference situation.

6.1 System Boundaries and Functional Unit

The functional unit is a door panel of a car with a volume of 992 cm^3 for the duration of 200,000 km.

One of the door panels judge against is glass powder and PP (GPP). Another is prepared by wood powder (pulp) and PP (WPP). The powder components characterize 40% and the PP element 60% by weight in the two plates.

WPP and GPP must act upon a similar purpose to be evaluated. This signifies that they must have the similar volume and similar mechanical properties. However, technical data show that, glass powder composites. Nonetheless, the panels are not subject matter to strong stresses during use. Appropriately, dissimilarity in their properties of the resistances has a small effect on their functions. The effects are taken into account in the following three phases: assembly, use and end of life (Fig. 1).

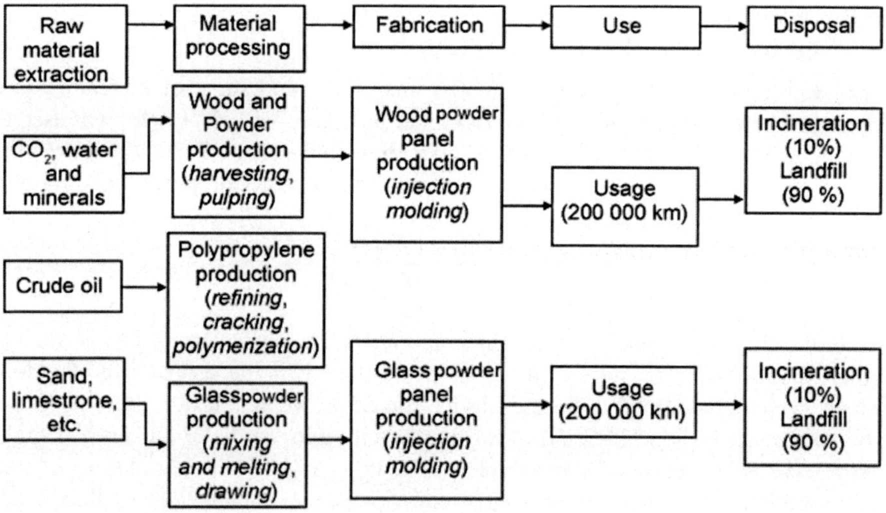

Fig. 1 The system boundaries

6.2 Sources of Data

Primary data was used to estimate the quantity of and variety of unit procedures within a system. The BUWAL 250 database (Renilde 2004) was chosen because it comprises more data on the observed products. Stay away from uncertainties because by the usage of data from diverse resources based on diverse kind of hypothesis.

6.3 Assembly

- Polypropylene
- Glass powder
- Wood Powder (mechanical pulp)
- Transport of raw materials
- Panel production

6.4 Use Phase

A private car has been selected to model this phase, and some of its features are presented below. Equation 1.1 From Heoleian and his coworkers. (1998) was used to obtain the subscription of the weight of door trims to the fuel of an automobile. This equation presumes that the consumption of the fuel is linearly proportional to weight.

$$F = M_t \times L \times \left(\frac{FE}{M_v}\right) \times \left(\frac{\Delta f}{\Delta M}\right) \tag{1.1}$$

In which F represents the fuel consumption for the entire service life of a panel (liter), M_t indicate the mass of the door panel (kg), L is the service life of a door panel, which is assumed to be the service life of the vehicle (200) 000 km), FE is the fuel consumption (11.9 l/100 km for the city and 7.8 l/100 km for the highway), Mv is the mass of the vehicle (1504 kg). Δf/ΔM is a correlation of fuel consumption with mass. A 4.38% (Δf) reduction in fuel is believed to follow every 10% (ΔM) weight reduction (Keoleian et al. 1998). The combined fuel consumption FEcomb, which is 9.62 l/100 km, is calculated using Eq. 1.2 from Sullivan and Hu (1995).

$$FE_{comb} = \frac{1}{\left(\frac{0.55}{FE_{city}}\right) + \left(\frac{0.45}{FE_{hwy}}\right)} \tag{1.2}$$

Where FE_{city} denote city fuel economy and FE_{hwy} signify highway fuel economy.

6.5 End of Life

The Data on recycling, landfill and wood powder combustion come from recycling, landfill and on fire newsprint. Newsprint is formed first and foremost from mechanical powder, comparable to that used in composites. The majority of the waste is send to landfills and a little part can be incinerated. For this reference situation, 10% incineration and 90% sanitary landfill are expected for the two panels.

6.6 Allocation

The incineration procedure assumes that part of the energy is improved from wood powder and PP's combustion. Energy is not recovered when the glass is being on fire. The energy produced by combustion converts the energy that should not be generated in the system (extension of the system's limits).

6.7 *Impact Evaluation Methods*

That lessons use Simapro software to perform analyses. The criteria for selecting the effect assessment technique deliver by the Simapro process selector are also used to choose the best accessible effect valuation technique. Other promising techniques can be established on the relevant effect assessment technique manual.

7 Inventory

This part is the center of attention and identifies the main substances contributing to energy use and air emissions. The quantities of substances are taken into account instead of their ecological impact.

7.1 *Use of Energy*

Figure 2 concluded the energy consumption for the diverse phases of the life cycles. The manufacture of wood powder of 1 kg needs extra energy (15.55 MJ) than 1 kg of glass powder (12.08 MJ). The energy consumption for felling, wood powder processing, and mechanical digestion can be very high (Daz and Patnaik 2000). Though, by installing 1 kg of GPP, it consumes slightly more power than a similar amount of WPP. PP volume has elevated GPP, which need more energy for manufacturing. It also has a heavier weight, which required higher energy while manufacturing. The similar applies to energy for the entire life cycle (assembly, use, and duration). Because of the improved weight, GPP consumes extra energy, particularly during using.

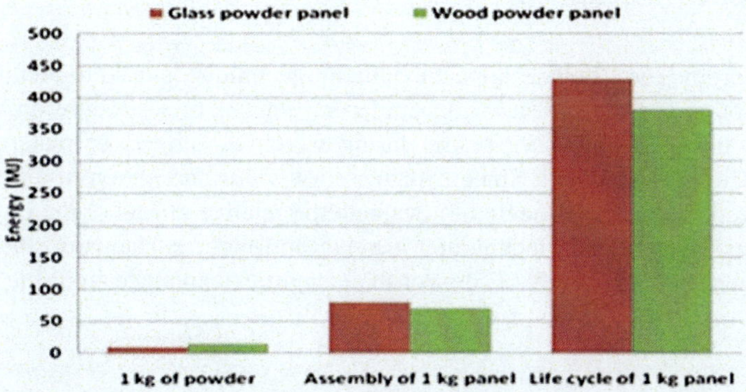

Fig. 2 Cumulative energy uses per panel at different life stages

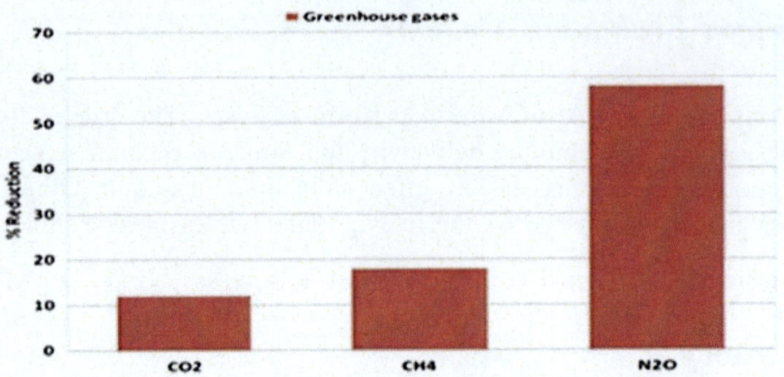

Fig. 3 Reduction in greenhouse gases due to replacement of glass powder by wood powder

7.1.1 Major Greenhouse Gases

Even though CO2 has the lowest global warming possible, it outs perform the critical greenhouse gases evolved during the life span of 2 panels. Figure 3 shows, substituting GPP with WPP has the greatest profit in decreasing N_2O emissions by 57%. This substitution decreases all greenhouse gases.

7.1.2 Major Air Pollutants

The import listed can be well-known as the key air pollutants (Harrizon 2001). These pollutants are decreased by replacing GPP with WPP. This decrease varies from 8.5% for cadmium, from 100.4% for lead to 595.5% for ammonia (NH3). Analysis of the data shows that the manufacture of glass powder leads to large amounts of NH3.

8 Impact Assessment

In this portion, the EI99E/E impact evaluation technique is used to identify the key effect groups and the effects of dissimilar substances or procedure on these categories. Characterization is a phase during which the effects of materials on the atmosphere are quantified. Since categorization values are given in different units, normalized values are presented to evaluate the relative effects of the diverse effect categories. The diverse techniques use various diverse standardization criteria. Consequences with EI99E / E depect the categories explained in Table 1

Table 1 Impact categories for the life cycles of the two panels (Goedkoop and Oele 2004)

Category	Definitions	Unit
Fossil fuels	Surplus energy per extracted MJ fossil fuel as a result of lower quality resources	MJ surplus
Respiratory Inorganics	Respiratory effects resulting from winter smog caused by emissions of dust sulphur and nitrogen oxides and other gases to air	DAILY/ kg
Climate change	Damage resulting from an increase in disease and death caused by climate change	DAILY/ kg
Eco toxicity	Damage to ecosystem quality, as a result of emission of ecotoxic substances to air water and soil	PAF m^2Xyr/ kg

8.1 *Production of Wood Powder*

Wood powder production has less impact on the environment in most impact categories, except carcinogens (Fig. 4). The Wood powder has a great effect on the carcinogen category. Compared to the production of glass powder. The production of glass powder has a double impact compared to the production of wood powder in the fossil fuel category.

In the earlier section, it was shown that wood powder production ingests more energy as than glass manufacture. Hence, we can expect that the production of wood powder will have a greater effect on fossil fuel than the manufacture of glass powder.

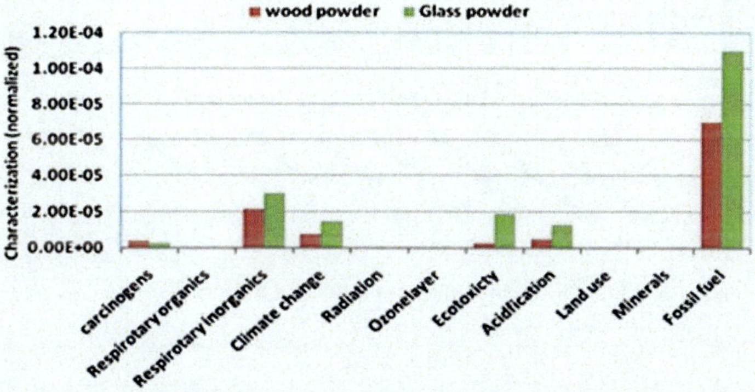

Fig. 4 Comparison among manufacture of 1 kg of glass powder and 1 kg of wood powder

8.2 Assembly of the Panels

The assembly of the 2 panels comprises powder manufacture (Fig. 5), PP production, and plate production. At this end, the two panels show little dissimilarity in their ecological effect, though GPP had a greater effect than WPP in greater cases (Fig. 5). The largest impact is in the fossil fuel type. However, wood powder manufacture had just half the impact of glass powder manufacture in this case (Fig. 5), WPP manufacture had just 15% less effect than GPP manufacture.

8.3 End of Life

On completion of the life, the category of weather change has the maximum impact on the two panels (Fig. 6). This is due to CO2 and NH4 emissions both in combustion and in the discharges of the two panels. On the contrary, the fossil fuel acquires maximum ecological profit. Due to Part to the combustion of the two panels it fixes free heat that can be used to replace the heat generated by the combustion of fossil fuels and the combustion of the WPP offers more advantages in this category than Combustion of GPP.

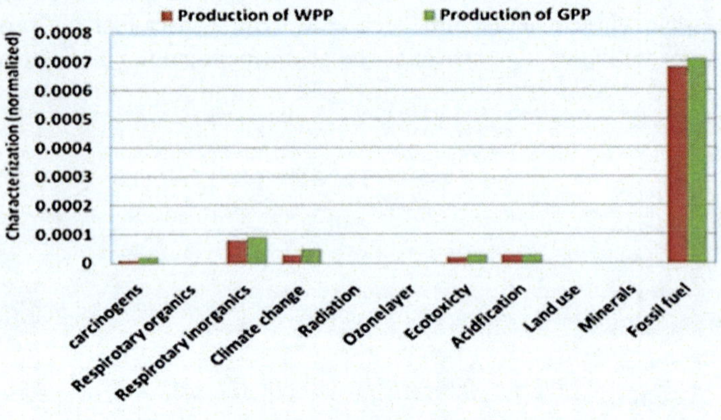

Fig. 5 Comparison among assembly effects of WPP & GPP

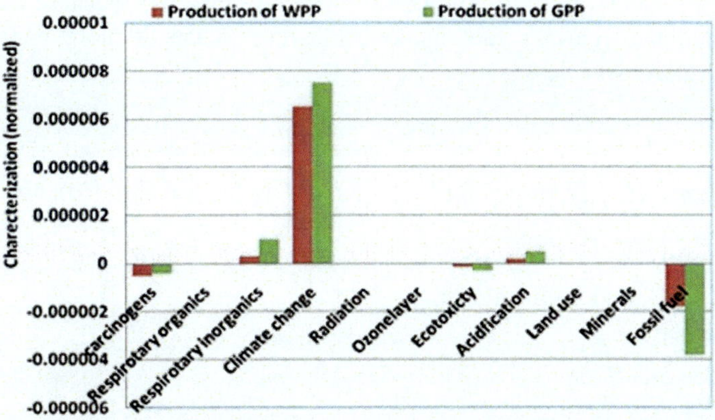

Fig. 6 End-of-life effects of the two plates

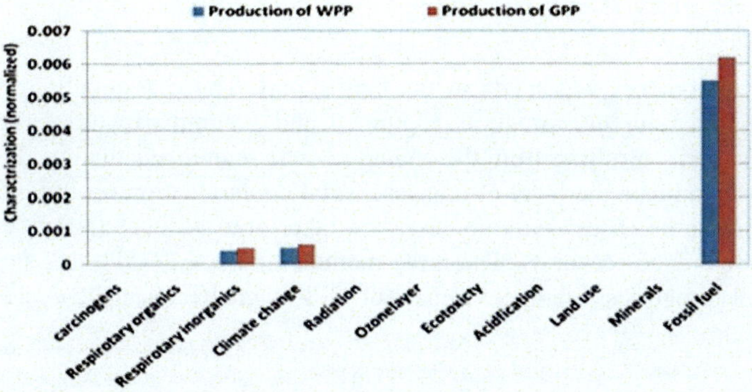

Fig. 7 Comparison among life cycle effect of the two panels

8.4 Life-Cycle Comparisons of the Panels

Here, the sound effects of the complete life cycle of the two panels have been compared. The fossil fuel type has the utmost impact (Fig. 7). The WPP has 12% less effect than GPP. There are a few variations for the other categories on the two panels.

In the use phase that accounts for 86% of life cycle impact of WPP and 85% of GPP, gasoline, a fossil fuel, is used while the car is in motion. The assembly phase underwrites 13.9% for WPP and 14.6% for GPP to the life cycle effect. PP manufacture underwrites 90% to the impact of WPP on the assembly & 86% to the effect of GPP on the assembly. At the end of the valuable life, the end of life adds 0.09% for WPP and 0.15% for GPP to impact the life cycle. Consequently, the ends of life effects are negligible for both panels. The effects of wood powder on declining PP usage per unit volume of composite material (because of its lighter weight) and the vehicle's fuel intake while use makes it more beneficial than powdered glass over here.

8.5 *Elements' Having Role in Life Cycles of the Panels*

- Fossil fuels, and Climate change
- Eco toxicity

9 Interpretation.

9.1 *Sensitivity Analysis*

Essential hypotheses was made in the earlier part. These assumptions have been indicated as a reference situation. Figures 8 and 9 summarize the environmental profit or burdens resulting from the changes in the reference situation. The positive percentages indicate the level of environmental contamination resulting from these changes. In some cases, extreme cases are taken into account (possible deviation from the reference scenario). The three methods in Figs. 10, 11, 12, and 13 show that for all categories, life-cycle impact of GPP is greater than life-cycle impact of WPP.

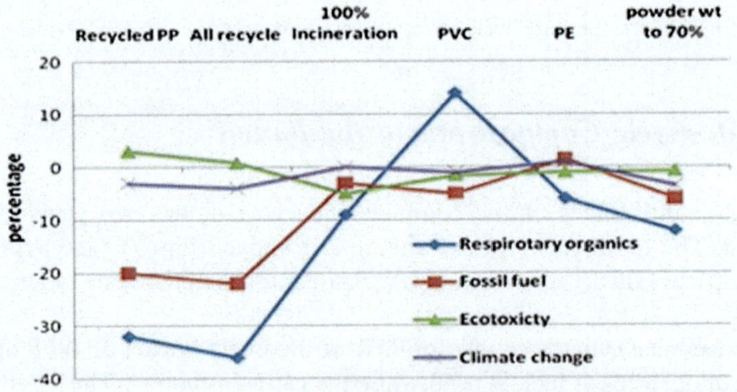

Fig. 8 Environmental profits because of nonconforming to reference scenario in life cycle of WPP

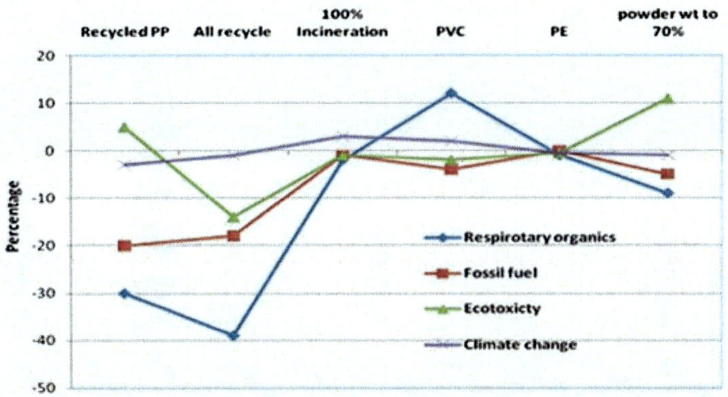

Fig. 9 Environmental profits because of nonconforming to reference scenario in life cycle of GPP

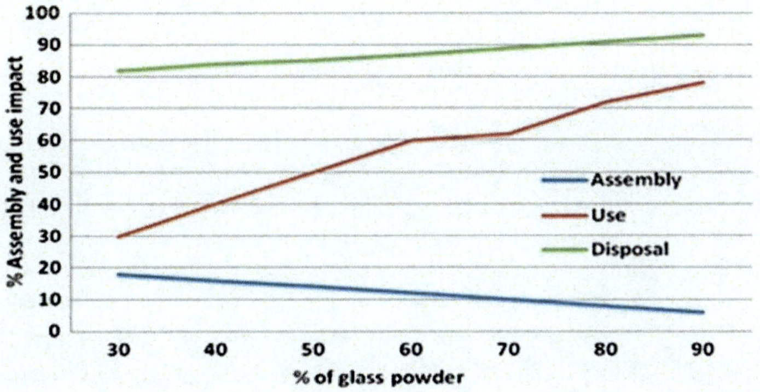

Fig. 10 Effect of weight on the involvement of three life stages to life cycle effects of GPP

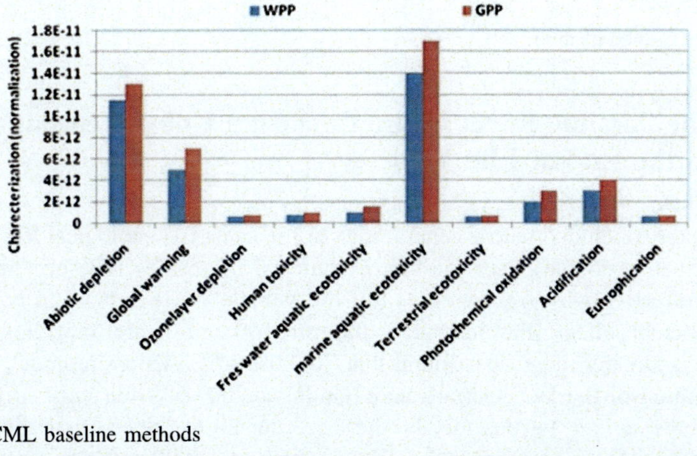

Fig. 11 CML baseline methods

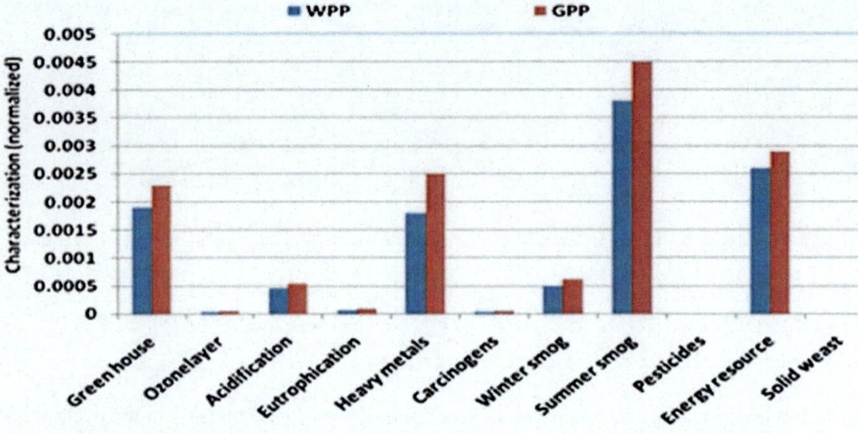

Fig. 12 Eco-indicator 05 method

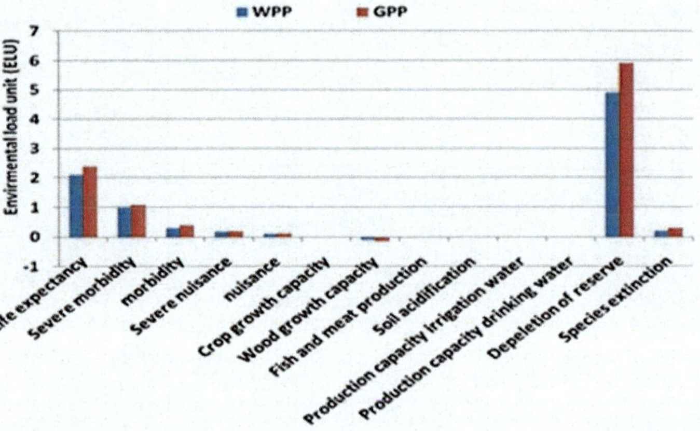

Fig. 13 EPS 2000 method

10 The Possible Effect of the European Union Legislation on the End-of-Life Vehicles

The chapter concluded with a general idea of the latest EU ecological legislation on ELV & their probable effects on motor vehicles' life cycles. The principle of this binding directive was to manage end-of-life vehicle waste better and enhance the environmental effect. Fascinatingly, the present end-of-life recycling rate for vehicles is already more. It is found that 70–80% of ELVs are reprocessed within the dominion of the EU (Gerrard and Handlikar 2007). While most of the reprocessed materials are ferrous metals, the lingering 20 to 25% is first and foremost made up of plastics and their composites, which are challenging to reutilize.

11 Conclusions

The life cycle assessment has presented that substituting glass powder with wood powder in PP compounds utilized for door panels of a car decreases the environmental impact of that. Pollution of air, mainly gases of greenhouse emissions, is a main environmental problem because of the apprehensive effects of these gases on global warming. By substituting glass powder by wood powder, greenhouse gases can be significantly decreased.

The life cycles of two panels mostly influence the groups of fossil fuels, Eco-toxicity, weather alteration & inorganic breathing organs according to the EI99E/E technique. They have the highest influence on fossil fuels. In actual fact the use phase, which controls life cycle effect of 86 and 85% for wood powder polymer composite and glass powder polymer composite, is highly dependent on fossil fuels (gasoline). The production of glass, paste and PP powder is also very energy-intensive and fuel-consuming. PP manufacture governs the sound effects in the assembly phase of the two panels. The replacement of PP with PVC benefits the fossil fuel category, the critical activity category, even though it noticeably increases the inorganic respiratory organs' effects. In this case, at all replacement depends on the group with the highest primacy between the two replacements with PE barely modifies the environmental influence. Though the usage of recycled PP and recycled powder leads to usually important enhancements in the key groups. Specific advantages are achieved by accumulative weight percentage of powder for the two plates in the key groups from 40 to 70%. Nonetheless, this increase tests glass powder composite in the Eco-toxicity category.

References

Abe K, Ozaki Y (1998) Comparison of useful terrestrial and aquatic plant species for removal of nitrogen and phosphorus from domestic wastewater. Soil Sci Plant Nutr 44(4):599–607

Aji IS, Sapuan SM, Zainudin ES, Abdan K (2009) Kenaf fibres as reinforcement for polymeric composites: a review. Int J Mech Mater Eng 4(3):239–248

Akhshik M, Panthapulakkal S, Tjong J, Sain M (2017) Life cycle assessment and cost analysis of hybri11111d fiber-reinforced engine beauty cover in comparison with glass fiber-reinforced counterpart. Environ Impact Assess Rev 65:111–117

Alves C, Ferr ~ ao PMC, Silva AJ, Reis LG, Freitas M, Rodrigues LB, Alves DE (2010) Ecodesign of automotive components making use of natural jute fiber composites. J Clean Prod 18(4):313–27

Broeren MLM, Dellaert SNC, Cok B, Patel MK, Worrell E, Shen L (2017) Life cycle assessment of sisal fibre e exploring how local practices can influence environmental performance. J Clean Prod 149:818–827

Burchart-Korol D (2013) Life cycle assessment of steel production in Poland: a case study. J Clean Prod 54(March):235–243

Business Wire. Biocomposites market by fiber, polymer, product, end-use industry and region - global forecast to 2022-research and markets [Internet]. San Francisco: Research and Markets; October 12, 2017

Chiu MC, Chu CH (2012) Review of sustainable product design from life cycle perspectives. Int J Precis Eng Manuf 13(7):1259–1272

Corbiere-Nicollier T, Laban BG, Lundquist L, Leterrier Y, Manson JAE, Jolliet O (2001) Life Cycle assessment of biofibres replacing glass fibres as reinforcement in plastics. Resour Conserv Recycl 33:267–287

Curran MA (2017) Overview of goal and scope definition in life cycle assessment. In: Curran M, editor. Goal and scope definition in life cycle assessment. LCA Compendium - the complete world of life cycle assessment. Dordrecht: Springer, p 162

Deng Y, Paraskevas D, Tian Y, Van Acker K, Dewulf W, Duflou JR (2016) Life cycle assessment of flax-fibre reinforced epoxidized linseed oil composite with a flame retardant for electronic applications. J Clean Prod 133:427–438

Dissanayake NJ, Summerscales J, Grove SM, Singh MM (2009) Life cycle impact assessment of flax fibre for the reinforcement of composites. J Biobased Mater Bioenergy 3(3):245–248

Finkbeiner M, Inaba A, Tan R, Christiansen K, Kluppel HJ (2006) The new international standards for life cycle assessment: ISO 14040 and ISO 14044. Int J Life Cycle Assess 11(2):80–85

Hansen A, Flake M, Heilmann J, Fleißner T, Fischhaber G (2001) Life cycle studies of renewable raw materials e natural fibre reinforced component and a varnish. In: Kyritsis S, Beenackers AACM, Helm P, Grassi A, Chiaramonti D, editors. Proceedings of 1st World Conference on Biomass for Energy and Industry. 2000 September 13e16; Seville. Lon- don: James & James (Science Publishers) Ltd. p 1036–1039

Hartemink AE, Van Kekem AJ (1994) Nutrient depletion in Ferralsols under hybrid sisal cultivation in Tanzania. Soil Use Manag 10(3):103–107

Hauschild MZ, Bonou A, Olsen SI (2018) Life cycle interpretation. In: Hauschild MZ, Rosenbaum RK, Olsen SI, (eds) Life cycle assessment. Cham: Springer, p. 323e34.

Hawkins TR, Singh B, Majeau-Bettez G, Strømman AH (2013) Comparative environmental life cycle assessment of conventional and electric vehicles. J Ind Ecol 17(1):53–64

ISO 14040:2006eEnvironmental managementd life cycle assessment - principles and framework. (2006). https://www.iso.org/standard/37456.html. Accessed 14 March 2018

ISO 14042:2000eEnvironmental management - life cycle assessment - life cycle impact assessment (2000). https://www.iso.org/standard/23153.html. Accessed 14 March 2018

Joshi SV, Drzal LT, Mohanty AK, Arora S (2004) Are natural fiber composites environmentally superior to glass fiber reinforced composites? Compos Part A Appl Sci Manuf 35(3):371–376

Kumar KP, Sekaran ASJ (2014) Some natural fibers used in polymer composites and their extraction processes: a review. J Reinforced Plast Compos 33(20):1879–1892

La Rosa AD, Cozzo G, Latteri A, Recca A, Bjorklund A, Parrinello E, Cicala G (2013) Life cycle assessment of a novel hybrid glass-hemp/thermoset composite. J Clean Prod 44:69–76

Le Duigou A, Davies P, Baley C (2011) Environmental impact analysis of the production of flax fibres to be used as composite material reinforcement. J Biobased Mater Bioenergy 5(1):153–165

Luglietti R, Rosa P, Terzi S, Taisch M (2016) Life cycle assessment tool in product development: environmental requirements in decision making process. Procedia CIRP 40:202–208

Luz SM, Caldeira-pires A, Ferr ~ ao PMC (2010) Environmental benefits of substituting talc by sugarcane bagasse fibers as reinforcement in polypropylene composites: ecodesign and LCA as strategy for automotive components. Resour Conserv Recycl 54(12):1135–1144

Mani M, Madan J, Lee JH, Lyons KW, Gupta SK (2014) Sustainability characterisation for manufacturing processes. Int J Prod Res 52(20):5895e912

Mansor MR, Salit MS, Zainudin ES, Aziz NA, Ariff H (2015) Life cycle assessment of natural fiber polymer composites. In: Hakeem K, Jawaid MY, Alothman O (eds) Agricultural biomass based potential materials. Springer, Cham, pp 121–141

Liikanen M, Grönman K, Deviatkin I, Havukainen J, Hyvärinen M, Kärki T, Varis J, Soukka R, Horttanainen M (2019) Construction and demolition waste as a raw material for wood polymer composites: assessment of environmental impacts. J Clean Prod 225(10):716–727

Natural fiber composites market by type (wood fiber and non-wood fiber), manufacturing process (compression molding, injection molding, and others), application (building & construction, automotive, and electrical & electronics), and region - global forecasts to 2021 [Internet]. Pune: Markets and Market Research Private Ltd.; April 2016. https://www.marketsandmarkets.com/Market-Reports/natural-fiber-composites-market-90779629. Accessed 14 March 2018

Pegoretti TDS, Mathieux F, Evrard D, Brissaud D, Arruda JRDF (2014) Use of recycled natural fibres in industrial products: a comparative LCA case study on acoustic components in the Brazilian automotive sector. Resour Conserv Recycl 84:1e14

Sommerhuber PF, Wenker JL, Rüter S, Krause A (2016) Life cycle assessment of wood-plastic composites: Analysing alternative materials and identifying an environmental sound end-of-life option. Resour Conserv Recy. https://doi.org/10.1016/j.resconrec.2016.10.012

Rosenbaum RK (2017) Goal and scope definition in life cycle assessment. In: Curran M (ed.) Goal and scope definition in life cycle assessment. LCA Compendium - the complete world of life cycle assessment. Dordrecht: Springer, p 63e120

Sapuan SM (2014) Tropical natural fibre composites: properties, manufacture and applications, 1st edn. Springer, Singapore

Summerscales J, Dissanayake N, Virk A, Hall W (2010) A review of bast fibres and their composites: part 2-Composites. Compos Part A Appl Sci Manuf 41:1336e44.

Tripathi P, Yadav K (2017) Biodegradation of natural fiber & glass fiber polymer composite-a review. Int J Eng Technol 4(4):1224e8

Trusty W (2010) An overview of life cycle assessments: part one of three. Build Saf J https://www.icc-es.org/Education/BSJ_overview_life_cycle_assessment_oct2010.pdf. Accessed 01 Mar 2018

Tudora AC (2014) Life cycle assessment. Methodology and instrument. The Bulletin of the Polytechnic Institute of Jassy-Constr Archit Section. Tomme LX (LXIV), Fascicle 2, p 83e92

Turner JA (1987) Linseed Law: a handbook for growers and advisers. Hadleigh, Suffolk: BASF (UK) Limited

Vidal R, Martínez P, Garraín D (2009) Life cycle assessment of composite materials made of recycled thermoplastics combined with rice husks and cotton linters. Int J Life Cycle Assess 14 (1):73–82

Walden DD, Roedler GJ, Forsberg KJ, Hamelin RD, Shortell TM (2015) INCOSE systems engineering handbook: a guide for system life cycle processes and activities, 4th edn. Wiley, New Jersey

Xu X, Jayaraman K, Morin C, Pecqueux N (2008) Life cycle assessment of wood-fibre-reinforced polypropylene composites. J Mater Process Technol 198(1–3):168–177

Younesi M, Roghanian E (2015) A framework for sustainable product design: a hybrid fuzzy approach based on quality function deployment for environment. J Clean Prod 108:385–394

Zah R, Hischier R, Le AL, Braun I (2007) Curaua fibers in the automobile industry: a sustainability assessment. J Clean Prod 15:1032–1040

Environmental Impact of Wood Based Biocomposite Using Life Cycle Assessment Methodology

P. Ramesh, H. Mohit, and V. Arul Mozhi Selvan

Abstract Awareness against sustainable manufacturing and materials raised globally due to global warming and climate change. Sustainable materials are environmentally benign materials, and which produce lesser impact than conventional materials. Using Life Cycle Assessment (LCA), environmental impact can be computed for product or process. In this chapter, the environmental impact is calculated using LCA methodology for wood-based biocomposite panel. The environmental impact was calculated for Wood chips, and Poly Lactic Acid (PLA) based biocomposite panel. SimaPro software was used to analyze the environmental impact. Two impact assessment methods (EcoIndicator 99 and CML) were used to find the environmental burden of the wood-based composite. And finally, the environmental impact of the wood-based composite is compared with commercially available wood-based panels.

Keywords Life cycle assessment · Environmental impact · Wood composite · SimaPro

1 Introduction

Sustainable materials have a lesser environmental impact than conventional materials. Due to environmental awareness, sustainable products increased in new product development with environmentally friendly materials. Awareness has been

P. Ramesh
Department of Production Engineering, National Institute of Technology, Tiruchirappalli, India

H. Mohit
Center for Nanoscience, Composite Research Center, Chennai, Tamilnadu, India

V. Arul Mozhi Selvan (✉)
Department of Mechanical Engineering, National Institute of Technology, Tiruchirappalli, India
e-mail: arulmozhi@nitt.edu

S. Mavinkere Rangappa et al. (eds.), *Wood Polymer Composites*, Composites Science and Technology, https://doi.org/10.1007/978-981-16-1606-8_13

increased due to global warming and climate change of conventional manufacturing processes and materials. Traditional material may have a higher environmental impact due to processing or material and sometimes both. Bio-based materials had good popularity due to a lesser environmental burden. Composites of polymers reinforced by bio-based materials have got great attention due to the environmental crisis (Xu et al. 2008). Bio-based composites degrade with lesser environmental impact after specific applications. Bio-based composite materials have wide range applications such as construction, furniture making, automobile industry, etc., Generally bio-based materials are obtained from natural resources such as plants (Sisal, jute, wood, etc.) and compound with polymers (Xu et al. 2008). Fabricating polymer composite with lower environmental impact is achievable using bio-based materials and natural polymers. PLA is a biodegradable polymer and derived from starch. The objective of the present study is to analyze the environmental burden of wood-based biocomposite panel. Wood-based composites panels had excellent properties at the same time it has limited mechanical properties, poor water resistance, and limited durability (Segovia et al. 2019). In this chapter, the environmental impact of the wood-based biocomposite panel was analyzed using LCA methodology. Two impact assessment was used to find different environmental impacts (indicators). And finally, a comparative LCA was done with wood-based biocomposite and conventional wood-based panels such as Glued laminated timber (GLT), Laminated timber element (LTE), medium-density fibreboard (MDF), oriented strand board (OSB), particleboard and plywood. The following sections are literature analysis regarding LCA of wood-based composite, LCA methodology, Simplified case study, goal and scope, system boundary, inventory details, impact assessment, results, and conclusions.

2 Literature Survey

Puettmann and Wilson et al. (2007) Compared the energy and environmental impact of different types of wood-based products (wood and composites boards) in two regions (Pacific Northwest (PNW) and Southeast (SE) of united states). LCA was compared for wood-based materials such as Glulam, softwood lumber, laminated veneer lumber (LVL), softwood plywood, and Orientated Strand Board (OSB) in PNW and SE. First, Cradle to gate was compared, and then the system boundary reduced to gate to gate system. From the results, it is found that the SE region had a higher impact in all stages (Harvesting. Product manufacturing, Resin production, and Transportation) in Cradle to gate system boundary. In gate to gate system boundary, 50% of energy demand is increased for most of the wood products in SE region.

Xu et al. (2008) evaluated the environmental impact of wood-based composited and PP. Wood composite made with varying percentage of wood level (10, 30 & 50%). Fabricated wood-based composite compared with PP. Environmental impact and Tensile property of wood composite and PP analyzed. Results showed that 30%

of wood fiber composite had excellent environmental and tensile property than PP. In all impact categories, PP attained 100% impact compared to 30% wood fiber composite.

Silva et al. (2014) Compared the environmental impact of conventional particleboard (%wood composite) and particle board made with sugarcane bagasse (PSB). PSB made with varying percentage level of bagasse (0, 25, 75 & 100). PSB made with urea–formaldehyde resin. Results showed that 40% of the impact got reduced in land use for PSB. PSB made with 75% had good quality and had good environmental performance compared to conventional particleboard.

Rosa et al. (2014) compared the environmental impact of eco sandwich (cork, hemp & bio-resin) and petroleum-based (glass & polyurethane) composite panels. From the results, the eco sandwich composite had a lower impact compared to glass-based petroleum composite. While going for the EoL option for eco sandwiches had a lesser impact in landfill compared to incineration. Petroleum composite attained 100% impact in all categories.

Mahalle et al. (2014) Compared the environmental impact of wood-based biocomposite and petroleum-based polymer (PP). Biocomposite made with wood fire with PLA and TPS (Thermoplastic starch). Two biocomposites (30% wood fibre + 35% PLA + 35% TPS & 30% wood fibre + 70% PLA) are formulation compared with PP to find environmental burdens. From the results, it is found that formulation one (30% wood fibre + 35% PLA + 35% TPS) had a lesser environmental impact than PLA (30% wood fibre + 70% PLA). PP had a higher impact than biocomposite.

Qiang et al. (2014) Evaluated the environmental impact of wood PLA composites. Two different types of wood-based composites were made with a composition of 20 wt% WF + 80 wt% PLA (sample 1) and 20 wt% WF + 55 wt% PLA + 25 wt% PHA (sample 2). From the results, it is found that sample 2 had a lesser impact than sample 1. In GWP alone, sample 2 had a higher impact, and in remaining impact categories, it lesser than sample 1. In energy demand, 1000 kg composite production sample 2 had a lesser impact than sample 1.

Väntsi and Kärki (2015) Performed LCA for four wood flour (WF) based composite. The wood-based composite made with varying percentages of wood, virgin PP (VPP), and recycled PP (RPP). The wood-based composites are 30VPP (70% WF + 30%VPP), 20GF (50% WF + 20% Glass fibre + 30% VPP), 20 MW (50% WF + 20% Mineral wool + 30% VPP) & 30RPP (70% WF + 30% RPP). Incineration and landfill are chosen as End of Life (EoL) option for wood-based composite. The results showed that recycled PP had a lesser impact than other compositions. Incineration of composites had a good advantage than landfill option except in global warming potential.

Feifel et al. (2015) Performed LCA for two different types of wood-plastic composites (WPC) with the differing composition of wood percentage for terrace application. Additionally, bilinga wood and pine wood compared with WPC. Timber Harvesting, debarking, sorting, and drying process are used. Two WPC compositions are polyethylene (PE)with 70% of wood, and the second composition is polyvinylchloride (PVC) with 50% of wood. From the results, it is found that PE based WPC had a lesser impact than PVC based WPC. Due to lower % of wood content in PVC based WPC had a higher impact in all categories. On the other side, pinewood had a lower impact compared to both WPC. bilinga wood had a higher impact in all categories compared to pine wood.

Sommerhuber et al. (2017) Analysed the environmental impact of virgin and recycled HDPE wood-based composite. Composites fabricated with Waste wood, Recycled HDPE, and virgin HDPE. Recycling and energy recovery options were chosen as the EoL option for wood-based composite. From the results, recycled wood-based composite had a lesser impact on all categories. The recycling option had reasonable environmental burdens compared to the energy recovery option.

Yuan and Guo (2017) Compared the environmental impact of two wood-based composites. 1.Hybrid-modified ammonium lignosulfonate (HMAL). 2. Medium-density fibreboard (MDF). The study was done to understand the HMAL process to make sustainable compared to MDF. From the results, GWP of HMAL based wood fiber production had a higher impact than other processes. MDF attained 100% impact in all impact categories compared to HMAL/WF. HMAL process got 24 wt% reduction in H2O2 compared to MDF.

Hossain and Poon (2018) Compared the environmental impact of wood waste with three alternatives. Recycling and reuse of wood waste for particleboard production, wood cement composite, and energy (biofuel production) were compared. Two alternative production processes, i.e., particleboard production and wood cement composites production, are compared with virgin and recycled wood. The study starts with the Waste wood collection, sorting, transporting to recycling site drying, and milling the wood waste to the desired particle size. Production and end of life were observed for three alternatives. From the results, the virgin wood particle board had a higher impact on GHGs emissions than recycled wood particleboard. In the non-renewable energy consumption, reclaimed wood had a lesser impact than virgin wood. For the second scenario, i.e., wood cement composite, recycled wood cement composite had lower GHGs emissions and non-renewable energy consumption than virgin wood cement composite. The third alternative energy production from waste wood had lower GHGs and non-renewable energy consumption compared to other alternatives. From the results, it is understood that waste wood can be used as biofuel and produces a lesser impact than other alternatives.

Qiang et al. (2019) Compared the environmental impact of styrene–butadiene–styrene (SBS) based wood composite with Poly Lactic Acid (PLA) based wood plastic composite. Four scenarios were compared to find environmental impact. i.e., scenario 1. PLA based wood composite (80% PLA & 20% wood). Scenario 2, 3 & 4 had 5% of SBS, 10, 20 & 30% of wood and 85, 75 & 65% of PLA respectively. The results showed that SBS based composite made with an increasing percentage of wood content had good environmental performance than PLA based wood composite.

Beigbeder et al. (2019) evaluated LCA for wood flour (WF) reinforced polypropylene (PP/WF), and flax fibers reinforced polylactic acid (PLA/Fl). The study was done only for the EoL of bio-composites with four scenarios: incineration, landfill, composting, and recycling. From the results, PP/WF composite had excellent environmental performance in incineration, and PLA/FI should be composted in the EoL scenario. The authors suggested that PP/WF had a great environmental advantage in recycling, but if the recycling option is not feasible, then incineration could be a better option.

Segovia et al. (2019) Compared different types of wood-based composite (High-Density Fibreboard (HDF), Medium-Density Fibreboard (MDF), Oriented Strand Board (OSB), Particle Board (PB) and plywood.) panels with aluminum honeycomb panel (AHP). AHP attained 100% impact in all categories. Wood-based composite panel manufacturing showing a better environmental result than the AHP. Especially, OSB manufacturing showed that better environment than any other composite panels.

Table 1 represents the software used, assessment method, and system boundaries of previous studies, and many authors used Cradle to gate boundary.

Table 1 Literature review

S. No	Author	Software	Assessment method	Study type
1.	Puettmann and Wilson et al. (2007)	SimaPro	NA	Cradle to gate
2.	Xu et al. (2008)	SimaPro	EcoIndicator 99	Cradle to gate
3.	Rosa et al. (2014)	SimaPro	CML	Cradle to manufacture
4.	Qiang et al. (2014)	NA	NA	Cradle to gate
5.	Mahalle et al. (2014)	NA	TRACI	Cradle to gate
6.	Silva et al. (2014)	GaBi	CML and USEtox	Cradle to gate
7.	Väntsi and Kärki (2015)	GaBi	CML	Cradle to grave
8.	Feifel et al. (2015)	NA	CML	Incineration
9.	Yuan and Guo (2017)	GaBi	ReCiPe	Cradle to gate
10.	Sommerhuber et al. (2017)	GaBi	CML	Cradle to gate
11.	Hossain and Poon (2018)	SimaPro	IMPACT 2002 +	Cradle to grave
12.	Qiang et al. (2019)	NA	NA	Cradle to gate
13.	Beigbeder et al. (2019)	GaBi	ReCiPe	Cradle to Cradle & Cradle to grave
14.	Segovia et al. (2019)	SimaPro	TRACI & Impact 2002+	Cradle to gate

3 Life Cycle Assessment Methodology

Using LCA methodology, environmental impact for product or process or technology is evaluated. LCA summarizes comprehensive results of environmental impact for a product from raw material extraction to End of Life (EoL). International Organization for Standards (ISO) determined guidelines for conducting LCA with a series of standards 14,040 to 14,044. Society of Environmental Toxicology and Chemistry (SETAC) established LCA in four stages:

3.1 Goal and Scope Definition

In this stage, the aim of LCA is clearly defined with decided system boundaries. System boundary represents, up to what extent the LCA should perform, i.e., Cradle to gate (raw material extraction to factory gate), cradle to grave (raw material extraction to EoL). The functional unit of LCA is also determined in this step. A functional unit represents the quantity of a product through process or technology.

3.2 Inventory

In this stage, essential and required input and output data for system boundaries added. i.e., documents of input data like material, energy, and transport, and output of emissions during each phase.

3.3 Impact Assessment

Characterizing the environmental impacts in different phases (water, air & soil) with units. Impacts are computed based on inventory data for a process or technology. Using the impact assessment method, different environmental indicators are found. (e.g., Carcinogens, Climate change, Ozone layer, Ecotoxicity, Land use, etc.)

3.4 Interpretation

Finding the opportunity to reduce the environmental impact from summarized results of LCA for process or product. Furthermore, finally, necessary actions are provided, alternatives, and recommendations with proper conclusion statements.

4 Simplified Case Study

Wood-based composite panels have wide range applications in the home, automotive industry, etc.. In the market, wood-based composites panels are made with polymers and resins. To improve the strength of wood-based composite panels, petrochemical-based polymers were used to fabricate wood composite. For alternative of petrochemical-based polymers, bio-based polymers are used to reduce environmental impact. In this regard, an environmental impact assessment was done to identify the ecological burdens of wood-based biocomposite panel. PLA and wood chips are to fabricate composite panels.

4.1 Goal and Scope

The objective of this present study is to identify the environmental impact of wood-based biocomposite panel. A cradle to manufacture stage was assumed to find the environmental burdens of wood composite fabrication. Production of 1 kg wood-based biocomposite panel is considered as a functional unit.

Materials for bio-based composite – Wood chips and PLA.

4.2 System Boundary

A cradle to manufacture stage was assumed as a system boundary. Raw material extraction, i.e., wood chips formation, processing, and PLA production to pellet, are considered. Figure 1 depicts the system boundary of the present study (Cradle to manufacture).

4.3 Inventory Data

In this step, according to ISO standards, raw material production, input, and output elements, data is collected. Production of the wood-based composite using an injection molding process is assumed. Wood chips and PLA was used as material for a wood based composite panel. Data were obtained from the EcoInvent database, and Natureworks PLA was chosen as a biopolymer for environmental assessment (Table 2).

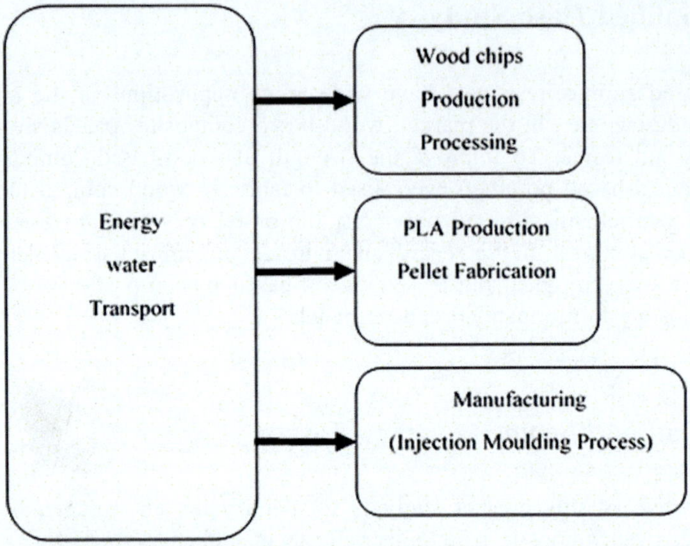

Fig. 1 System boundary for LCA

Table 2 Inventory data

Category	Input	Remarks
Wood-based composite		
Wood chips production	0.5 kg	Ecoinvent data
Nature works PLA production	0.5 kg	Ecoinvent data
Electrical energy	1.5 kWh	Assumed

4.4 Impact Assessment Method

In this stage, assessment methods are defined. SimaPro software was used to find an environmental impact. Two impact assessment methods were used to environmental impact for the wood-based composite. Eco-indicator 99 (I) and CML were used to find environmental impact. Using Eco-indicator, major three damage categories and corresponding environmental indicators are observed (Table 3).

EI 99 (E) - Egalitarian: long time perspective: scientific justification should be added
EI 99 (I) - Individualist: short-time perspective: only proven effects are included
EI 99 (H) - Hierarchist: balanced time perspective: general agreement effects are included

Using CML, major 10 impact environmental indicators are observed (Table 4).

Table 3 Eco-indicator 99 (I) environmental indicators

Damage category	Indicators	Unit	Remarks	Source
Human health	Carcinogenic effects	DALY	Disability adjusted life years	Budavari et al. (2011)
	Respiratory effects (inorganic)	DALY		
	Respiratory effects (organic)	DALY		
	Climate change	DALY		
	Radiation	DALY		
	Ozone depletion	DALY		
Ecosystem quality	Ecotoxicity	PAF*m2yr	Potentially affected fraction of species	
	Acidification/ eutrophication (combined)	PAF*m2yr		
	Land use	PAF*m2yr		
Resources	Minerals	MJ surplus	The difference between the energy needed to extract a resource now and at some point, in the future	

Table 4 CML indicators

Indicators
Abiotic depletion
Acidification
Eutrophication
Global warming
Ozone layer depletion
Human toxicity
Freshwater aquatic ecotoxicity
Marine aquatic ecotoxicity
Terrestrial ecotoxicity
Photochemical oxidation

5 Results

Cradle to manufacture system boundary was used to analyze the environmental impact of the wood-based composite. Wood chips and PLA was used as material, and injection molding process used to manufacture the wood composite panels. Figure 2 represents a network diagram of Cradle to manufacture stage of wood-based composite.

Two impact assessment methods were used to find the environmental burdens of wood composite. From Fig. 2, wood chips contribute a lesser impact than other processes. Because wood chips collected from forest waste residue. Ecoinvent database was used to find the impacts of wood-based composite. From the network diagram (Fig. 2), it is found that PLA contributes 39% of impact during wood-based composite fabrication. The environmental impact assessment was performed with EcoIndicator (EI) and CML method.

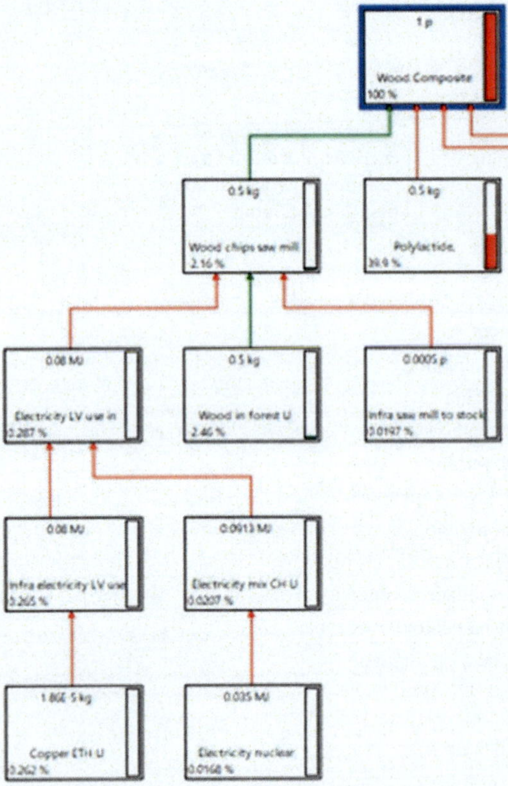

Fig. 2 Network diagram of Cradle to manufacture of wood-based composite

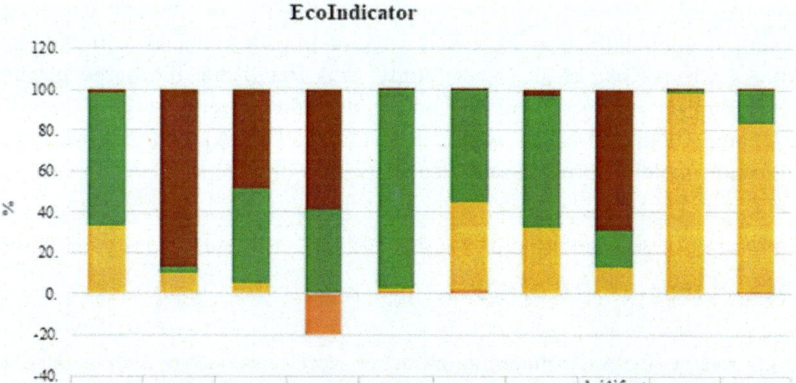

	Carcinogens	Resp. organics	Resp. inorganics	Climate change	Radiation	Ozone layer	Ecotoxicity	Acidification/ Eutrophication	Land use	Minerals
Injection moulding	1.5739	87.0462	48.8516	58.38	0.05	0.05	2.9538	68.7516	2.13E-05	0.1773
Electricity	65.1926	2.7711	45.8808	40.8183	96.8902	54.9618	64.5947	17.9938	1.3246	16.7832
PLA	33.144	10.1687	5.2537	0.8017	1.8778	42.9736	32.2623	13.2322	98.5673	81.8596
Wood_Chips	0.0895	0.0139	0.0139	-20	1.232	2.0646	0.1892	0.0224	0.108	1.1798

Fig. 3 Environmental impact of wood-based composite using EcoIndicator assessment method

5.1 EcoIndicator 99 Assessment Method

EI is a European based impact assessment method. In the first phase, EI was used to find the environmental impact of wood-based composite. Figure 3 represents the environmental impact of wood-based composite using the EcoIndicator assessment method. From Fig. 3, it is understood that PLA contributes more impact than others. Wood chips had a lesser impact on all categories and had negative impacts on climate change. In Respiratory effects (inorganic), injection molding process had a higher impact than others. Polylactic acid had a higher impact in two categories (land use and minerals) because of fertilizers, irrigation, and cultivation of plants, and 98% of land use is observed in the impact category. PLA production occupies more land use for the cultivation of plants (Duigou et al. (2012) and Molins et al. (2018)). Electricity production attained 96% in the radiation indicator.

In climate change, injection molding process had 58%, and electricity production had 40% of impact.

5.2 CML Assessment Method

In the second phase, with the same inventory data, CML assessment method was used to find the environmental burdens of wood composite. From Fig. 4, it is found that wood chips had a lesser impact than others in all categories. In marine aquatic ecotoxicity, electricity production attained 83% in impact category. In the eutrophication indicator, PLA had 56% of impact and had a higher impact than others. From Rosa et al. (2014) study, it is found that bio-based products had a lesser impact on global warming potential. In the present study, in the global warm potential, both bio-based products (wood and PLA) had a lower impact than injection molding process and electricity.

5.3 Comparative Study

In this stage, the environmental impact of wood-based composite panel is compared with six commercially available wood-based panels. EcoIndicator assessment method was used to compare the environmental impacts. Figure 5 represents the comparative environmental impact of a wood based composite. From Fig. 5, it is understood that in all categories, wood-based biocomposite panel had a lesser impact than other panels. Plywood attained 100% impact in eight categories. Plywood manufacturing had a higher environmental impact than any other product (Puettmann and Wilson 2007).

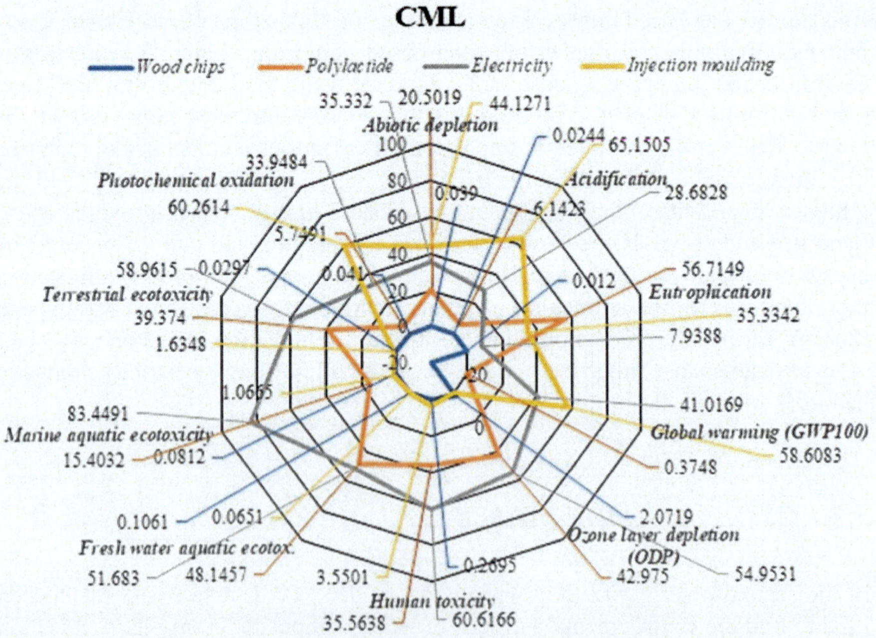

Fig. 4 Environmental impact of wood-based composite using CML assessment method

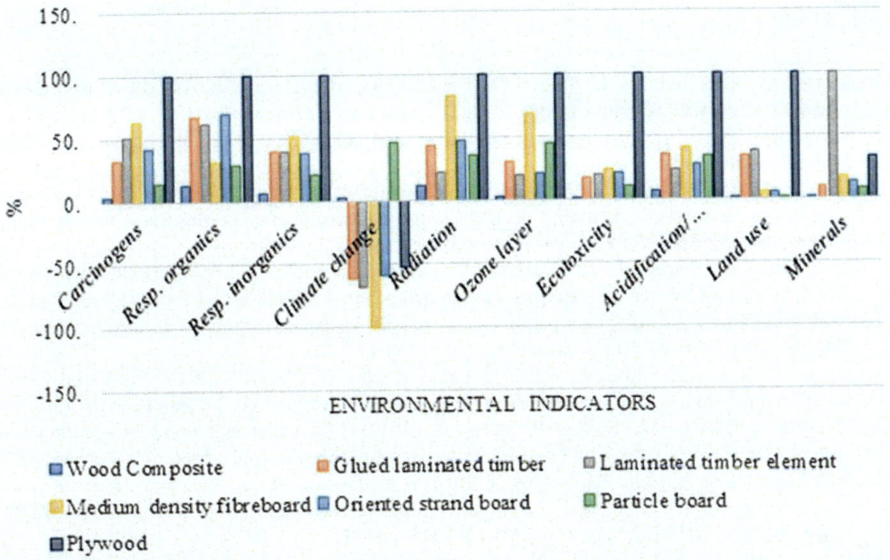

Fig. 5 Environmental impact comparison of wood-based composite using EcoIndicator assessment method

6 Conclusion

Using LCA methodology, the environmental impact is analyzed for wood-based biocomposite panel. Cradle to manufacture stage was assumed, and LCA was performed. Two impact assessment methods were used to find the different environmental burden of wood-based biocomposite panel. The following conclusions are driven by the present study.

- PLA production had a higher impact on land use and minerals.
- Electricity production had a higher impact.
- In climate change, both bio-based materials (wood chips and PLA) had a lesser impact.
- Similarly, in global warming potential, both bio-based materials had a lesser impact.
- In the comparative study, wood-based biocomposite had a lesser impact than conventional wood panels.

References

Beigbeder J, Soccalingame L, Perrin D, Bénézet JC, Bergeret A (2019) How to manage biocomposites wastes end of life? A life cycle assessment approach (LCA) focused on polypropylene (PP)/wood flour and polylactic acid (PLA)/flax fibres biocomposites. Waste Manag 83:184–193

Budavari Z, Szalay Z, Brown N, Malmqvist T, Peuportier B, Zabalza I, Wibke T (2011) Indicators and weighting systems, including normalisation of environmental rofiles. https://www.sintef.no/globalassets/project/lore-lca/deliverables/lore-lca-wp5-d5

Feifel S, Stübs O, Seibert K, Hartl J (2015) Comparing wood–polymer composites with solid wood: the case of sustainability of terrace flooring. Eur J Wood Wood Prod 73(6):829–836

Hossain MU, Poon CS (2018) Comparative LCA of wood waste management strategies generated from building construction activities. J Clean Prod 177:387–397

La Rosa AD, Recca G, Summerscales J, Latteri A, Cozzo G, Cicala G (2014) Bio-based versus traditional polymer composites. A life cycle assessment perspective. J Clean Prod 74:135–144

Le Duigou A, Deux JM, Davies P, Baley C (2012) PLLA/flax mat/balsa bio-sandwich—environmental impact and simplified life cycle analysis. Appl Compos Mater 19(3–4):363–378

Mahalle L, Alemdar A, Mihai M, Legros N (2014) A cradle-to-gate life cycle assessment of wood fibre-reinforced polylactic acid (PLA) and polylactic acid/thermoplastic starch (PLA/TPS) biocomposites. Int J Life Cycle Assess 19(6):1305–1315

Molins G, Álvarez MD, Garrido N, Macanás J, Carrillo F (2018) Environmental impact assessment of polylactide (PLA)/chicken feathers biocomposite materials. J Polym Environ 26 (3):873–884

Puettmann ME, Wilson JB (2007) Life-cycle analysis of wood products: cradle-to-gate LCI of residential wood building materials. Wood Fiber Sci 37:18–29

Qiang T, Chou Y, Gao H (2019) Environmental impacts of styrene-butadiene-styrene toughened wood fiber/polylactide composites: a cradle-to-gate life cycle assessment. Int J Environ Res Public Health 16(18):3402

Qiang T, Yu D, Zhang A, Gao H, Li Z, Liu Z, Chen W, Han Z (2014) Life cycle assessment on polylactide-based wood plastic composites toughened with polyhydroxyalkanoates. J Cleaner Prod. 66:139–145

Segovia F, Blanchet P, Amor B, Barbuta C, Beauregard R (2019) Life cycle assessment contribution in the product development process: case study of wood aluminum-laminated panel. Sustainability 11(8):2258

Silva DAL, Lahr FAR, Pavan ALR, Saavedra YM, Mendes NC, Sousa SR, Sanches R, Ometto AR (2014) Do wood-based panels made with agro-industrial residues provide environmentally benign alternatives? An LCA case study of sugarcane bagasse addition to particle board manufacturing. Int J Life Cycle Assess 19(10):1767–1778

Sommerhuber PF, Wenker JL, Rüter S, Krause A (2017) Life cycle assessment of wood-plastic composites: analysing alternative materials and identifying an environmental sound end-of-life option. Resour Conserv Recycl 117:235–248

Väntsi O, Kärki T (2015) Environmental assessment of recycled mineral wool and polypropylene utilized in wood polymer composites. Resour Conserv Recycl 104:38–48

Xu X, Jayaraman K, Morin C, Pecqueux N (2008) Life cycle assessment of wood-fibre-reinforced polypropylene composites. J Mater Process Technol 198(1–3):168–177

Yuan Y, Guo M (2017) Do green wooden composites using lignin-based binder have environmentally benign alternatives? A preliminary LCA case study in China. Int J Life Cycle Assess 22(8):1318–1326